Medicinal, Aromatic, Spice Plants: Biodiversity, Phytochemistry, Bioactivity and Their Processing Innovation

Medicinal, Aromatic, Spice Plants: Biodiversity, Phytochemistry, Bioactivity and Their Processing Innovation

Guest Editors

Ivan Salamon
Milica Aćimović

MDPI

Basel • Beijing • Wuhan • Barcelona • Belgrade • Novi Sad • Cluj • Manchester

Guest Editors
Ivan Salamon
Department of Ecology
Faculty of Humanities and
Natural Science
University of Presov
Presov
Slovakia

Milica Aćimović
Department of Vegetable and
Alternative Crops
Institute of Field and
Vegetable Crops Novi
Sad—National Institute of the
Republic of Serbia
Novi Sad
Serbia

Editorial Office
MDPI AG
Grosspeteranlage 5
4052 Basel, Switzerland

This is a reprint of the Special Issue, published open access by the journal *Horticulturae* (ISSN 2311-7524), freely accessible at: https://www.mdpi.com/journal/horticulturae/special_issues/aromatic_phytochemistry.

For citation purposes, cite each article independently as indicated on the article page online and as indicated below:

Lastname, A.A.; Lastname, B.B. Article Title. *Journal Name* **Year**, *Volume Number*, Page Range.

ISBN 978-3-7258-3639-0 (Hbk)
ISBN 978-3-7258-3640-6 (PDF)
https://doi.org/10.3390/books978-3-7258-3640-6

Cover image courtesy of Milica Aćimović

Contents

About the Editors

Ivan Salamon

Ivan Salamon is a university teacher and scientist. He is the principal author of the patent for the freeze-drying technology (lyophilization), a unique method of anthocyanin isolation from medicinal plant small fruits, Chamomile variety "LIANKA" with the high content of /-/α-bisabolol and Peppermint variety "KRISTINKA" with the high content of menthol of essential oils. In 2021, he obtained an MBA with an emphasis on management skills in education and R&D.

Milica Aćimović

Principal Research Fellow Milica Aćimović, PhD, is a researcher specializing in medicinal plant biotechnology. Her research primarily focuses on the cultivation technology and postharvest processing of medicinal, aromatic, and spice plants, as well as the investigation of their chemical composition, biological activity, and traditional medicinal knowledge (ethnopharmacology). She is also a specialist in phytotherapy and aromatherapy.

Preface

Dear fellow scientists, professors, students, agricultural producers, and curious readers,

It is with great pleasure that we present this reprint, which offers valuable insights into the fascinating world of medicinal, aromatic, and spice plants. Within these pages, you will explore their rich biodiversity, intricate phytochemistry, bioactivity, and the latest innovations in their processing.

Whether you are interested in learning more about *Juniperus communis, Kopsia fruticosa, Satureja montana, Matricaria recutita, Thymus* sp., *Triticum aestivum, Rosa hybrida, Carissa macrocarpa, Salix alba, Ocimum basilicum, Lavandula intermedia,* or even *Camellia sinensis* and *C. assamica,* this compilation is packed with fascinating details and current advancements related to these plants.

Additionally, we delve into other timely topics in agricultural production, such as the innovative approach of immobilizing lead (Pb) in contaminated soils by combining diammonium phosphate with both organic and inorganic amendments.

We hope this work serves as an informative resource for all those engaged in the study and use of these remarkable plants, as well as those exploring broader environmental challenges in agricultural sciences. Enjoy your journey through this collection, and we invite you to expand your knowledge and curiosity on these captivating subjects.

Ivan Salamon and Milica Aćimović
Guest Editors

horticulturae

Editorial

Medicinal, Aromatic, and Spice Plants: Biodiversity, Phytochemistry, Bioactivity, and Their Processing Innovation

Ivan Salamon

Department of Ecology, Faculty of Humanities and Natural Science, University of Presov, 01, 17th November St., SK-081 16 Presov, Slovakia; ivan.salamon@unipo.sk

1. Introduction

For centuries, drugs were entirely of natural origin, composed of herbs, animal products, and inorganic materials. Early remedies may have combined these ingredients with witchcraft, mysticism, astrology, or religion, but it is certain that those treatments that were effective were subsequently recorded and documented, leading to early herbal remedies. Many widely used pharmaceuticals are derived from plants and other natural sources or are based on traditional knowledge of herbal remedies.

Thanks to numerous scientific and professional meetings, the development of analytical and biological testing methods and related tools, along with the introduction of the large-scale cultivation, processing, and technological innovation of herbal plants, has gained great significance for the economy.

Today, several hundred plant species are used as medicinal, aromatic, and spice herbs. The key commercial species worldwide currently include chamomile, calendula, milk thistle, plantain, balm, mint, and yarrow, but there is potential for the cultivation of a wider range of species across larger areas. Significant opportunities exist for the production and processing of a wide range of herbs in many countries, serving both domestic and export markets.

Medicinal, aromatic, and spice plants are renewable raw materials, offering an alternative to the overproduction of traditional crops in agriculture. The food, pharmaceutical, and cosmetic industries produce important products using these plants. Products based on natural substances are gaining increasing value due to the presence of valuable components, primarily produced through the secondary metabolism of the plants. Many widely used pharmaceuticals are derived from plants and other natural sources or are based on traditional knowledge of herbal remedies. However, these plants can also be used for non-traditional purposes such as phytoremediation, which involves cleaning substrates contaminated by toxic metals using plants.

Citation: Salamon, I. Medicinal, Aromatic, and Spice Plants: Biodiversity, Phytochemistry, Bioactivity, and Their Processing Innovation. *Horticulturae* **2024**, *10*, 280. https://doi.org/10.3390/horticulturae10030280

Received: 28 February 2024
Accepted: 14 March 2024
Published: 15 March 2024

2. Special Issue Overview

The Special Issue of *Horticulturae* entitled *Medicinal, Aromatic, and Spice Plants: Biodiversity, Phytochemistry, Bioactivity, and their Processing Innovation* addresses a somewhat neglected area of biodiversity, including the ethnobotany and breeding of these plants to enhance their quality and productivity. The biosynthesis of natural plant substances and their qualitative–quantitative content plays a vital role in their biological properties, effects, new plant products, and innovative processing, which must be also emphasized in our contributions. Finally, it should be noted that medicinal, aromatic, and spice plants are natural biological resources with the potential to become a new generation of substances for human and animal nutrition and health. The recommended readership of this Special Issue is not only those engaged in research into and the processing of these special crops but also natural product chemists, pharmacogenetics, and users of these plants, given their increasing economic importance.

However, issues related to intellectual and cultural property rights have emerged as significant factors in these research and development fields. A very important aspect of consolidating all original scientific work is the exchange of recent scientific information, results, and knowledge.

2.1. *Diversity of Plants*

Biodiversity is the degree of variation in life forms within a given species, ecosystem, biome, or an entire planet. Biodiversity is a measure of the health of ecosystems. It is estimated that some 10,000 plant species are used medicinally, most of which are used in traditional systems of medicine [1]. However, only a relatively small number of species are used in any significant volume.

Three papers cover wild-growing plant populations, which are heterogeneous both from the perspective of their morpho-phenological properties as well as from the chemical point of view. The original research presents, describes, and analyzes the anatomy of *Kopsia fruticose* Rob. (*Apocynaceae*) and correlates the anatomical features with the habitat and ecology of this plant [2]. Chemotaxonomic and ecological studies of secondary metabolites and their role in ecosystems regarding *Matricaria recutita* L. and *Juniperus communis* L. [3,4] are presented by authors from Albania and Slovakia, respectively. In the first case, chemical type B (/-/-α-bisabololoxide B >/-/-α-bisabololoxide A >/-/-α-bisabolol) was uniquely determined in wild Albanian chamomile populations. A dendrogram was constructed after a hierarchical cluster analysis based on essential oil substances from juniper, which showed four different chemotypes in Slovakia. The quality and chemotypes of juniper berries are very important for the Slovak national beverage "Borovicka" and the distillery industry on a whole in this country.

2.2. *Phytochemistry and Bioactivity*

Medicinal, aromatic, and spice plants continue to attract growing interest from both scientists and the public. Plants that consist of biologically active compounds, such as polyphenols, flavonoids, alkaloids, and polypeptides, have been found to possess many antimicrobial, antioxidant, anti-inflammatory, and antiseptic properties.

Ethnobotany is the study of the relationships that exist between people and plants. Ethnobotanists aim to document, describe, and explain complex relationships between cultures (and uses) of plants, focusing primarily on how plants are used, managed, and perceived across human societies. This includes their use as food, clothing, currency, medicine, and dye and in rituals, construction, cosmetics, and more.

Based on its traditional efficacy, researchers have conducted a series of studies and found that *Carissa macrocarpa* (Eckl.) A. DC. [5] leaf extract showed antioxidant activity with regard to a variety of phytochemicals, such as alkaloids, tannins, phenols, naphthoquinones, flavonoids, saponins, steroids, proteins, carbohydrates, mucilage, gum, and resin. Chemotypes of species of the *Thymus* L. genus [6] in the Carpathians region and their microbial activities were determined. Finally, an important study evaluated the larvicidal properties of the hexane extract of sweet basil (*Ocimum basilicum* L.; family: *Lamiaceae*) [7] leaves against the wild strain of Asian tiger mosquito (*Aedes albopictus*).

2.3. *Herb Cultivation and Processing Innovation*

Medicinal, aromatic, and spice plants are renewable raw materials. Their production is an alternative to the overproduction of the traditional crops in agriculture. Medicinal plants have increasing economic importance, and the food, pharmaceutical, and cosmetic industry produce important goods using these plants [8]. Products based on natural substances are enjoying increasing value. The presently rising consumption of these products is caused by changes in consumer behavior in industrialized countries [9].

Two original R&D papers covered the fields of the introduction of new hydrolats and hydrosols [10] of *Lavandula angustifolia* × *intermedia* and *Rosa hybrida* L., which are products created during the distillation of essential oils. These water-based products contain the fragrant and therapeutic components of the plant material, and they can be used in a variety of ways. These products can be used as natural fragrances, toners, and facial mists, and they can help soothe and hydrate the skin. Hydrosols and hydrolats are also generally gentle and safe for all skin types, making them a good choice for those with sensitive skin.

2.4. Non-Traditional Use of Medicinal Plants in Phytoremediation Technology

In recent years, the practical use of alternative medicine in healing processes has become increasingly popular. Several species of medicinal plants can be used as supplementary nutrition due to their ability to accumulate some essential nutrition elements (e.g., Se, Zn, and Fe) in the edible parts of these plants (so-called phytofortification). On the other hand, however, data related to toxic metal contents (e.g., Cd, Pb, Hg, and Cr) in pharmaceutically utilized parts of the medicinal plants are also considered from the perspective of "food safety " [11]. Considering all of the above-mentioned aspects, the cultivation and use of medicinal plants have to respect the potential hazard connected with environmental contamination, mainly with toxic heavy metals [12].

The influence of the immobilization of Pb in contaminated soils [13] was evaluated by applying organic amendments, in combination with diammonium phosphate (DAP), in a greenhouse experiment. Inorganic bone meal combined with DAP most effectively immobilized Pb in soil. These interesting results indicate that the bark of *Salix alba* contains significant amounts of phenolic compounds of Fe, Ni, and Mn.

3. Conclusions

In recent decades, the usage of medicinal, aromatic, and spice plants has been increasing worldwide, as evidenced by the growing number of products in the pharmaceutical, cosmetic, chemical, food, and related industries. Based on the above-mentioned findings, it can be clearly seen that natural products and their sources, such as medicinal, aromatic, and spice plants, continue to have great importance in the health-care of mankind. It is also clearly seen that natural resources, among them being herbs, are not adequately understood. Their chemical investigation should be intensified. Even plants that have already been studied can provide new biologically active ingredients because of their chemotaxonomic differences, new unknown chemical constituents, newly discovered biological effects, and because of the possibilities of the chemical and/or biological processing of plants and their ingredients to obtain favorable activity (Figure 1).

Due to the current trends of prescription drug costs and consumer desires for natural health-care products worldwide, pharmacies may offer product selection and educate staff in a way that supports conventional and alternative medicine systems, enabling the consumer and the medical practitioner to choose the best medicine, whether conventional or herbal, for the maintenance of health and for the treatment of a medical condition. The sustainable development in these fields lies in the following strategies:

- Developing new and improved varieties and species for soil-climatic conditions, their changes, and international markets;
- Providing agronomic systems for the efficient and sustainable production of high-quality plant raw materials, essential oils, and extracts;
- Devising harvest, post-harvest, extraction, and distillation technology to increase product yields and/or secure specified characteristics required by markets and controlled-registration authorities;
- Standardizing extracts for active natural substances and lyophilizates—their antiviral activity, immunological, and hormonal modulators;
- Standardizing extracts for active natural substances, essential oils, and lyophilizates and their antimicrobial effects on bacteria, yeasts, and microscopic fungi.

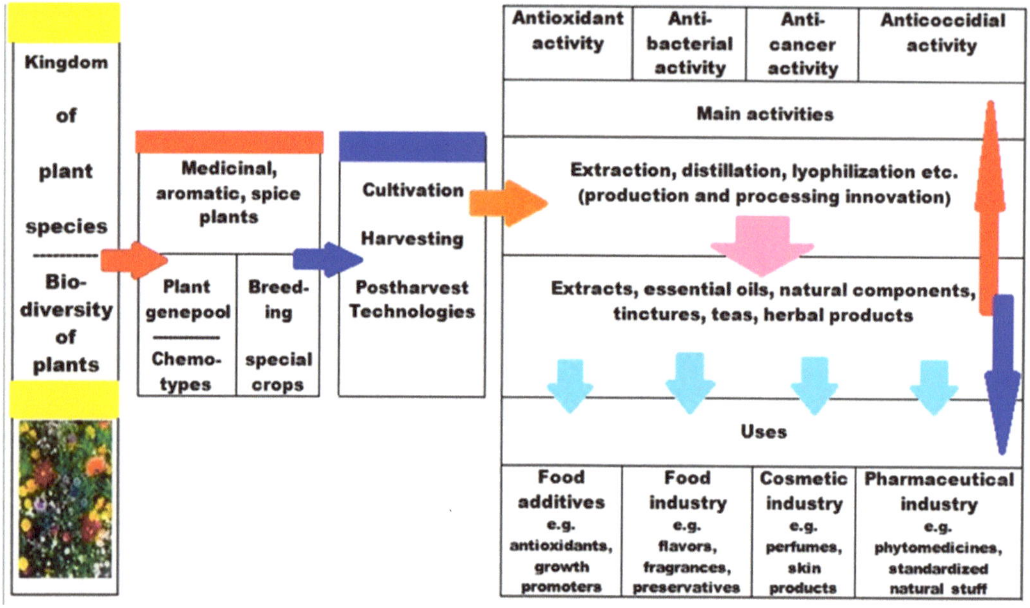

Figure 1. Scheme for introducing wild (autochthonous) plant species from their natural habitat into the cultivation, collection, post-harvest treatment, and technological processing of medicinal, aromatic, and spice plants.

Funding: This research received no external funding.

Institutional Review Board Statement: Not applicable.

Informed Consent Statement: Not applicable.

Data Availability Statement: Not applicable.

Acknowledgments: We gratefully acknowledge all the authors that participated in the Special Issue of Horticulturae entitled *Medicinal, Aromatic, and Spice Plants: Biodiversity, Phytochemistry, Bioactivity, and their Processing Innovation.*

Conflicts of Interest: The authors declare no conflicts of interest.

List of Contributions:

1. Aćimović, M.; Lončar, B.; Stanković Jeremić, J.; Cvetković, M.; Pezo, L.; Pezo, M.; Todosijević, M.; Tešević, V. Weather Conditions Influence on Lavandin Essential Oil and Hydrolate Quality. *Horticulturae* **2022**, *8*, 281. https://doi.org/10.3390/horticulturae8040281.
2. Chan, C.A.; Ho, L.Y.; Sit, N.W. Larvicidal Activity and Phytochemical Profiling of Sweet Basil, (*Ocimum basilicum* L.), Leaf Extract against Asian Tiger Mosquito (*Aedes albopictus*). *Horticulturae* **2022**, *8*, 443. https://doi.org/10.3390/horticulturae8050443.
3. Bajraktari, D.; Bauer, B.; Zeneli, L. Antioxidant Capacity of *Salix alba* (Fam. Salicaceae) and Influence of Heavy Metal Accumulation. *Horticulturae* **2022**, *8*, 642. https://doi.org/10.3390/horticulturae8070642.
4. Ramasar, R.; Naidoo, Y.; Dewir, Y.H.; El-Banna, A.N. Seasonal Change in Phytochemical Composition and Biological Activities of *Carissa macrocarpa* (Eckl.) A. DC. Leaf Extract. *Horticulturae* **2022**, *8*, 780. https://doi.org/10.3390/horticulturae8090780.
5. Tanjga, B.B.; Lončar, B.; Aćimović, M.; Kiprovski, B.; Šovljanski, O.; Tomić, A.; Travičić, V.; Cvetković, M.; Raičević, V.; Zeremski, T. Volatile Profile of Garden Rose (*Rosa hybrida*) Hydrosol and Evaluation of its Biological Activity in vitro. *Horticulturae* **2022**, *8*, 895. https://doi.org/10.3390/horticulturae8100895.

6. Mandici, A.; Cretu, D.E.; Burlica, R.; Astanei, D.; Beniuga, O.; Rosu, C.; Topa, D.C.; Aostacioaei, T.G.; Aprotosoaie, A.C.; Miron, A. Preliminary Study on the Impact of Non-Thermal Plasma Activated Water on the Quality of *Triticum aestivum* L. cv. Glosa Sprouts. *Horticulturae* **2022**, *8*, 1158. https://doi.org/10.3390/horticulturae8121158.
7. Kryvtsova, M.; Hrytsyna, M.; Salamon, I.; Skybitska, M.; Novykevuch, O. Chemotypes of Species of the Genus *Thymus* L. in Carpathians Region of Ukraine; Their Essential Oil Qualitative and Quantitative Characteristics and Antimicrobial Activity. *Horticulturae* **2022**, *8*, 1218. https://doi.org/10.3390/horticulturae8121218.
8. Salamon, I.; Ibraliu, A.; Kryvtsova, M. Essential Oil Content and Composition of the Chamomile Inflorescences (*Matricaria recutita* L.) Belonging to Central Albania. *Horticulturae* **2023**, *9*, 47; https://doi.org/10.3390/horticulturae9010047.
9. Mrđan, S.; Marković, T.; Predić, T.; Dragumilo, A.; Filipović, V.; Prijić, Ž.; Lukić, M.; Radanović, D. *Satureja montana* L. Cultivated under Polypropylene Woven Fabric on Clay-Textured Soil in Dry Farming Conditions. *Horticulturae* **2023**, *9*, 147. https://doi.org/10.3390/horticulturae9020147.
10. Popescu, S.M.; Zheljazkov, V.D.; Astatkie, T.; Burducea, M.; Termeer, W.C. Immobilization of Pb in Contaminated Soils with the Combination Use of Diammonium Phosphate with Organic and Inorganic Amendments. *Horticulturae* **2023**, *9*, 278. https://doi.org/10.3390/horticulturae9020278.
11. Tripathi, S.N.; Sahney, M.; Tripathi, A.; Pandey, P.; Jatav, H.S.; Minkina, T.; Rajput, V.D. Elucidating the Anatomical Features, Adaptive and Ecological Significance of *Kopsia fruticose* Roxb. (Apocynaceae). *Horticulturae* **2023**, *9*, 387. https://doi.org/10.3390/horticulturae9030387.
12. Salamon, I.; Otepka, P.; Kryvtsova, M.; Kolesnyk, O.; Hrytsyna, M. Selected Biotopes of (*Juniperus communis* L.) in Slovakia and Their Chemotype Determination. *Horticulturae* **2023**, *9*, 686. https://doi.org/10.3390/horticulturae9060686.
13. Parveen, A.; Qin, C.-Y.; Zhou, F.; Lai, G.; Long, P.; Zhu, M.; Ke, J.; Zhang, L. The Chemistry, Sensory Properties and Health Benefits of Aroma Compounds of Black Tea Produced by *Camellia sinensis* and *Camellia assamica*. *Horticulturae* **2023**, *9*, 1253. https://doi.org/10.3390/horticulturae9121253.

References

1. Tuhinadri, S.; Samir, K.S. Medicinal Plants, Human Health and Biodiversity: A Broad Review. *Adv. Biochem. Eng. Biotechnol.* **2015**, *10*, 59–110.
2. Middleton, D.J. A Revision of *Kopsia* (Apocynaceae: Rauvolfioideae). *Harv. Pap. Bot.* **2004**, *9*, 89–142.
3. Das, M. *Chamomile—Medicinal, Biochemical, and Agricultural Aspects*; CRC Press: Boca Raton, FL, USA; Taylor & Francis Group: London, UK, 2015; p. 187.
4. Radoukova, T.; Zheljazkov, V.D.; Semerdjieva, I.; Dincheva, I.; Stoyanova, A.; Kacaniova, M.; Marković, T.; Radanovic, D.; Astatkie, V. Essential Oil Yield, Composition, and Bioactivity of Three Juniper Species from Eastern Europe. *Ind. Crops. Prod.* **2018**, *124*, 643–652. [CrossRef]
5. Khan, D. Some Morphological Observation on Leaves of *Carissa macrocarpa* (Eckl.) A. DC. (Family: *Apocynaceae*). *Int. J. Biol. Biotechnol.* **2019**, *16*, 1027–1045.
6. Pirbalouti, A.G.; Bistghani, Z.E.; Malekpoor, F. An Overview on Genus *Thymus*. *J. Herb. Drugs.* **2016**, *6*, 93–100.
7. Khair-ul-Bariyah, S.; Ahmed, D.; Ikram, M. *Ocimum basilicum*: A Review on Phytochemical and Pharmacological Studies. *Pak. J. Chem.* **2012**, *2*, 78–85. [CrossRef]
8. Genovese, F.; Altieri, G.; Admane, N.; Salamon, I.; Di Renzo, G.C. Processing Plants and Technologies for a Sustainable Mediterranean Food Chain. In *The Sustainability of Agro-Food and Natural Resource Systems in the Mediterranean Basin*; Vastola, A., Ed.; Springer: Cham, Switzerland, 2015. [CrossRef]
9. Ibraliu, A.; Dhillon, B.S.; Faslia, A.; Stich, B. Variability of Essential Oil Composition in Albanian Accessions of *Satureja montana* L. *J. Med. Plants Res.* **2010**, *14*, 1359–1364.
10. Tavares, C.S.; Gameiro, J.A.; Roseiro, L.B.; Figueiredo, A.C. Hydrolates: A Review on their Volatiles Composition, Biological Properties and Potential Uses. *Phytochem. Rev.* **2022**, *21*, 1661–1737. [CrossRef]
11. Masarovicova, E.; Kralova, K. Plant-Heavy Metal Interaction: Phytoremediation, Biofortification and Nanoparticles. In *Advances in Selected Plant Physiology Aspects*; Montanaro, G., Bartolomeo Dichio, B., Eds.; IntechOpen Ltd.: London, UK, 2012; pp. 75–102. [CrossRef]
12. Shetty, S.S.; Deepthi, D.; Harshitha, S.; Sankusare, S.; Naik, P.B.; Suchetha, K.N.; Harishkumar, M. Environmental pollutants and their effects on human health. *Heliyon* **2023**, *9*, e19496. [CrossRef] [PubMed]
13. Dagari, M.S.; Badamasi, H. Lead (Pb) Accumulation in Plants Grown on Contaminated Soil. *Int. J. Sci. Res. Sci. Technol.* **2020**, *7*, 401–405. [CrossRef]

horticulturae

MDPI

Article

Selected Biotopes of *Juniperus communis* L. in Slovakia and Their Chemotype Determination

Ivan Salamon [1,*], Pavol Otepka [2], Maryna Kryvtsova [3], Oleh Kolesnyk [4] and Myroslava Hrytsyna [5]

[1] Department of Ecology, Faculty of Humanities and Natural Science, University of Presov, 17th November St. 01, SK-081 16 Presov, Slovakia
[2] European Ecocycles Society, Matrai St. 36, 3200 Gyongyos, Hungary; pavol.otepka@gmail.com
[3] Department of Genetic, Plant Physiology and Microbiology, Faculty of Biology, Uzhhorod National University, Voloshina, 32, 88000 Uzhhorod, Ukraine; maryna.krivcova@uzhnu.edu.ua
[4] Department of Botany, Faculty of Biology, Uzhhorod National University, Voloshina, 32, 88000 Uzhhorod, Ukraine; oleg.kolesnyk@uzhnu.edu.ua
[5] Department Pharmacy and Botany, Faculty of Social Development and Health, Stepan Gzhytskyj National University of Veterinary Medicine and Biotechnologies, Pekarska Str., 50, 79010 Lviv, Ukraine; hrytsynamr@lvet.edu.ua
* Correspondence: ivan.salamon@unipo.sk; Tel.: +421-917984060

Abstract: The objective of this work was to map population of common juniper (*Juniperus communis* L.) in the territory of Slovak Republic. Common juniper is not protected by Slovakian law or the authorities; therefore, there is no law to preserve these plant populations. Biotopes of common juniper consist of light- to heavy-density trees and shrubs. The expansion of its habitat is also connected to human involvement. The loss of juniper naturally in observed plants is caused by a change in land use, loss of feeding pastures for cattle and sheep, and eutrophication of the environment. The current study was focused on monitoring the population of this plant species in the years of 2018–2020, the isolation of essential oils, and the identification of qualitative and quantitative characteristics. It was confirmed that juniper berries usually contain from 0.5 ± 0.05 to 1.8 ± 0.06%, usually 1.2 ± 0.16%, volatile oil depending on geography, altitude, ripeness, and other factors. Volatile oil is made up mostly of monoterpenes, mainly α-pinene (from 37.60 ± 2.23 to 61.00 ± 0.60%), β-myrcene (from 8.03 ± 2.02 to 10.56 ± 0.05%), and sabinene (from 3.50 ± 0.30 to 22.0 ± 0.96%). The dendrogram was constructed after a hierarchical cluster analysis based on the essential oil substances, which showed four different confirmed chemotypes. The essential oil is widely used in medicines, perfumes, insect repellents, insecticides, shoe polish, and in microscopy as a clearing agent of an immersion oil. The quality and chemotypes of juniper berries are very important for the Slovak national beverage "Borovicka" and the distillery industry on a whole in this country.

Keywords: berries; Borovička; juniper; α-pinene; raw material; essential oil; weight

Citation: Salamon, I.; Otepka, P.; Kryvtsova, M.; Kolesnyk, O.; Hrytsyna, M. Selected Biotopes of *Juniperus communis* L. in Slovakia and Their Chemotype Determination. *Horticulturae* **2023**, *9*, 686. https://doi.org/10.3390/horticulturae9060686

Academic Editor: Charalampos Proestos

Received: 14 May 2023
Revised: 2 June 2023
Accepted: 7 June 2023
Published: 9 June 2023

1. Introduction

Juniperus genus has about 60 species of evergreen trees or shrubs, widely dispersed in the Northern Hemisphere, mostly in Eastern parts [1]. Juniper berries of various species have been used for medicinal purposes since antiquity. The berries of several species are consumed as a food by birds. Humans should not consume the berries or oil of different juniper species as some of them may be toxic; the medicinal use of small amounts of oil of some species of juniper has occasionally resulted in death [2,3]. Many of the species are grown as ornamental plants and some are also harvested for timber [4].

Common juniper (*Juniperus communis* L.) is the most widespread of the juniper species; it is indigenous to Eurasia and North America. As a species with a broad ecological amplitude, common juniper is spread across all of Slovakia with an altitude extending from the lowlands up to 1495 m above sea level [5]. It is a shrub or small tree which is 2–20 ft. high and with reddish brown bark which shreds off in papery peels. Leaves (needles) taper to a spiny tip, in whorls of threes with two white bands above (or one white band sometimes divided by a green midrib with a broader green margin). It is a dioecious plant species.

Strictly, a "berry" is a fleshy or pulpy fruit derived from the ovary of a single flower, and the term is inappropriate for the junipers since the fleshy structure containing the seed is derived from three to six fleshy cone scales, each one-ovuled; however, the term berry is universally accepted [6].

Small cones, some male, others female, develop in early summer at the base of a few of the needle-like leaves. Male cones are yellow, while female cones are hard, pungent, and bluish-green with a whitish, waxy coating. The common juniper is an anemophilous species, but a certain degree of entomophily is also quite common. Female cones become fleshy berries, requiring 2 years to ripen, and turning gradually from green to blue-black. Berries are grown on short stalk, round to broadly oval, bluish black, usually with three seeds [7,8].

Juniper berries are one of the most widely used herbal diuretics. Approved in Germany, juniper tea is used for digestive problems and also to stimulate appetite. Their anti-inflammatory and spasm-reducing activity has been confirmed, and it may also contribute to diuretic activity. Fruits eaten fresh or in tea are a folk remedy used as a diuretic and urinary antiseptic for cystitis, carminative for flatulence, antiseptic for intestinal infections, once used for colic, coughs, stomach aches, colds, and bronchitis. Externally, they are used for sores, aches, rheumatism, arthritis, snakebites, and cancer. The content of volatile oil is responsible for diuretic and intestinal antiseptic activity. Diuretic activity results from the irritation of renal tissue [9,10].

Juniper fruits are often fermented for the production of alcoholic beverages; the fruits form the raw material for the production of the Slovak national drink "Borovička". The "brother" of this beverage is the Anglo-Saxon "Gin", which has been popular in the Western World for over 300 years [11]. Berries, fresh or dried, are sometimes used directly as a food, but they are crushed before use. The smell is bittersweet, very reminiscent of gin, pine, and turpentine. The taste is sweet, slightly pungent, with a bitter aftertaste [12,13].

The aim of this study is the monitoring of the habitats of common juniper (*Juniperus communis* L.) in the territory of Slovakia with an emphasis on the determination of natural components. Despite the economic importance of juniper populations, little is currently known about the extent and nature of the essential oil variability and composition. The comprehensive research presents variations in the yield of juniper essential oils and their qualitative–quantitative characteristics in relation to different chemotypes.

2. Material and Methods

2.1. Collection of the Plant Material

The fruits were collected from 25 localities in Slovakia [14] from 2018 to 2020 (Table 1, Figure 1). The biotope consists of open to dense stands of common juniper together with other light-loving species of trees, mainly shrubs, which occur within grass–herb or shrub vegetation communities. Juniper most often spreads on extensively used pastures, and thanks to its sharp needles, it is not threatened by grazing. On the contrary, it has a competitive advantage, because plants typically grow in rocky, infertile soils in fields, meadows, pastures, open woods, and other settings, almost from sea level to Alpine sites.

The preparation of samples after their collection was conducted by cleaning, de-needling, sorting the samples into groups, and drying at a temperature of 38 °C for 6 h in a dryer. The drying process evaporates excess water, and the final moisture content of the plant material for further processing was about 12%. The next step was to determine the weight of 100 juniper berries; this operation was repeated 6 times for each sample. Measurements were made on Sartorius analytical scale of the CPA type.

The fruits were separated and dried in a sheltered, open-air area at a temperature below 32 °C for 14 to 20 days with low humidity of 2 to 5%. The moisture content of the fruit tissue was lowered to 12% to prevent mould infection. The dried materials were cleaned, packed, labelled, and stored in a clean and dry place until further use.

2.2. Isolation of Juniper Essential Oil

Each sample of dried fruits with a weight of 20 g was ground in a blender. The essential oil from this raw material was prepared using hydro-distillation (2 h) in Clevenger-type apparatus according to the European Pharmacopoeia [15]. Hexane was used as an extraction agent. The extracted essential oils were stored under N_2 atmosphere at +4 °C in a dark place before their GC-MS analyses.

2.3. GC/MS Analysis

The main components of the essential oil were determined using a GC-MSD system on a Varian 3090/MS Saturn 2100 T instrument with Split–Splitless injection inlet, injection volume: 2 µL, MSD detector, RX-5MS column, 30 m × 0.25 mm i. d., with internal diameter: 0.25 µm, carrier gas: helium (21 p.s.i.) with a flow rate of 1.50 mL/min BPX-5, 50 m long with an internal diameter of 0.25 mm and with a stationary phase thickness of 0.25 µm. Temperature program: 50 °C = 0 min.; 3 °C/min to 250 °C; 250 °C = 15 min.

Table 1. The basic characters of selected sites in Slovakia with the occurrence of juniper plants.

	Locality	Geographical Latitude	Terrestrial Longitude	Altitude [m]	Aspect	Slope
01	Kišovce—Hôrka	N 49°02′84″	E 20°38′11″	620	northeast	30°
02	Liptovská Lužná 2	N 48°56′30″	E 19°19′12″	700	south	15°
03	Liptovská Lužná 1	N 48°56′31″	E 19°19′15″	730	southwest	25°
04	Selčianske sedlo	N 48°45′53″	E 19°12′24″	380	northeast	10°
05	Kráľová 1	N 48°52′58″	E 20°08′21″	1272	east	35°
06	Liptovská Lužná 3	N 48°56′29″	E 19°19′13″	710	southeast	35°
07	Vihorlat—Užok	N 48°59′05″	E 22°51′53″	380	southwest	12°
08	Horné lazy 2	N 48°48′51″	E 19°36′00″	475	northeast	14°
09	Kráľová 2	N 48°52′51″	E 20°08′25″	1010	south	25°
10	Donovaly	N 48°52′48″	E 19°13′21″	960	north	35°
11	Cerovo	N 48°15′17″	E 19°09′26″	468	northwest	21°
12	Iliaš 1	N 48°41′55″	E 19°18′37″	340	northeast	12°
13	Priechod—West	N 48°46′45″	E 19°13′54″	480	west	20°
14	Priechod—South	N 48°46′41″	E 19°13′49″	420	south	15°
15	Spišský hrad	N 49°00′00″	E 20°46′06″	628	south	30°
16	Priechod—East	N 48°46′40″	E 19°13′50″	390	east	10°
17	Karpaty—Turka	N 49°09′27″	E 23°01′21″	745	southwest	15°
18	Horné lazy	N 48°48′51″	E 19°36′50″	520	northeast	30°
19	Lackov	N 48°19′15″	E 19°11′12″	476	southeast	16°
20	Pravica	N 48°19′01″	E 19°27′26″	356	southwest	12°
21	Chrámec—Teplá dolina 1	N 48°15′35″	E 20°10′56″	248	northeast	22°
22	Iliaš 2	N 48°41′52″	E 19°18′32″	320	east	16°
23	Chrámec—Vlčia dolina 3	N 48°16′24″	E 20°10′54″	250	east	10°
24	Chrámec—Vlčia dolina 4	N 48°16′18″	E 20°10′45″	220	south	10°
25	Chrámec—Teplá dolina 2	N 48°15′34″	E 20°10′60″	196	south	5°

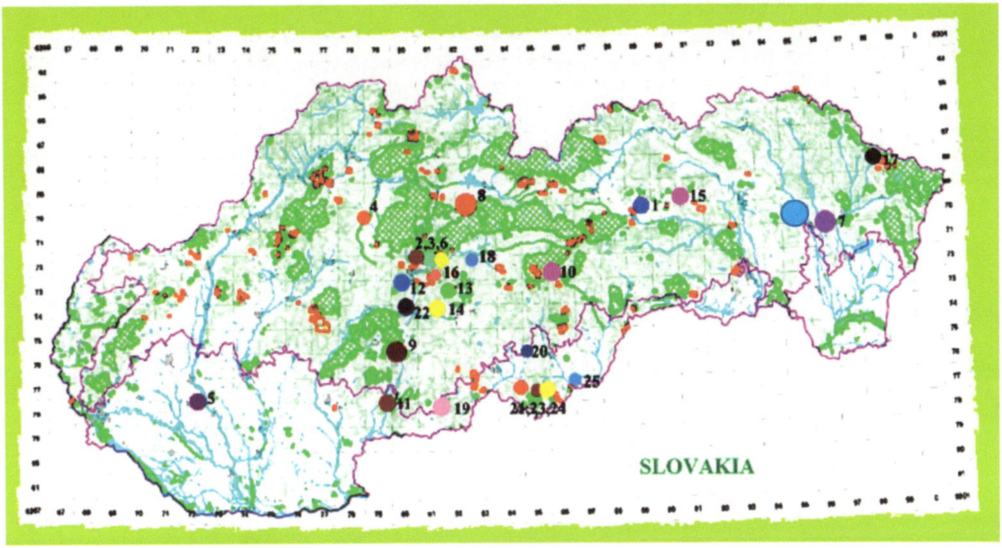

Figure 1. Slovakia—the localities [16] where juniper berries were collected: 1. Kišovce—Hôrka, 2. Liptovská Lužná 2, 3. Liptovská Lužná 1, 4. Selčianske sedlo, 5. Kráľová 1, 6. Liptovská Lužná 3, 7. Vihorlat—Užok, 8. Horné lazy 2, 9. Kráľová 2, 10. Donovaly, 11. Cerovo, 12. Iliaš 1, 13. Priechod—West, 14. Priechod—South, 15. Spišský hrad, 16. Priechod—East, 17. Karpaty—Turka, 18. Horné lazy, 19. Lackov, 20. Pravica, 21. Chrámec—Teplá dolina 1, 22. Iliaš 2, 23. Chrámec—Vlčia dolina 3, 24. Chrámec—Vlčia dolina 4, 25. Chrámec—Teplá dolina 2.

The identification of the individual components of the essential oil was conducted using the retention times of 40 authentic standards of the components supplied by the companies Extrasynthesis, Merck, Fulka, and Sigma-Aldrich and Kovats indexes (used C_5–C_{22} alkanes), and the components were then integrated in NIST 98 library. The spectra of the individual components of the essential oil were compared with the mass spectra using the literature [17,18].

Components were identified using their GC retention times, and the resulting values were comparable to those of the literature. Oil component standards for comparison were supplied by Extrasynthese, s.a., Genay, France.

2.4. Statistical Analysis

Several statistical methods and biometric parameters of the obtained data were used: arithmetic means, standard deviations, Student t-test at the 0.05% level of significance ($n = 6$), and exploratory analysis with a graphic representation of the frequencies of selected quantitative features using variation curves [19]. Statistica Program V12, which was used for processing the measured data, is well-known statistical software developed for mass processing and evaluation of data with a subsequent tabular or graphic presentation. For the biostatistical evaluation of the obtained results, the following statistical tests were used for 130 data ($n = 25$). The Mann–Whitney U-test was used to compare the levels of the substances from two locations (Kisovce 01, Liptovská Lužná 2 02). Due to the low number of samples from the sites, there is a more appropriate choice when detecting significant differences for sites in chemotype D. The Kruskal–Wallis test is a non-parametric analysis of variance that was used when comparing levels from more than 3 sites (for each chemotype separately). The choice of method was made due to the low number of samples from individual locations (5 each). With this method, we can find out if there is a difference in level between any sample pair. The following multiple comparison test is able to specify significant differences between pairs. Statistical analysis was performed using confidence intervals with a significance level $p < 0.05$ using calculation through the mean, standard deviation, and standard error.

The differences between juniper populations for the mean values of the essential oil constituents were detected using an ANOVA. All statistical data were calculated using SAS JMP Statistical Discovery, and after that a dendrogram (ward method) and a relationship diagram (*Juniperus populations* × the essential oil constituents) were made [20].

3. Results

3.1. Fruit Weights and Essential Oil Yield

For a long time, female plants with juniper fruits have been interesting for their harvesting. Common juniper is a highly variable species occurring in several forms (Figure 2), including a small tree up to 0.1 m pyramidal or columnar in shape with an upright bush, even a prostatic bush forming a mat. According to calculated variability in the morphological characteristics of trees and shrubs, different total amounts of harvested fruits were obtained, and, accordingly, significant differences in their size (Figure 3) and weight (Table 2) were proven. These findings prove that the entire spectrum of abiotic and biotic factors affects the ontogenetic development of plants and their populations in individual locations in Slovakia.

Figure 2. Tree forms of juniper bushes (pyramidal, ovoid, and spreading shrub) grown in Slovak montane and sub montane areas.

Figure 3. Shapes of juniper berries—from small to large, from round to elongated, to spindle-shaped.

Table 2. Berry weights (100 pieces/pcs/) and yield of juniper essential oil in % (*v*/*w*—expressed as a dry weight).

	Habitat/Locality	Berry Weights [g]	Essential Oil Yield [%]
01	Kišovce—Hôrka	5.84 ± 0.29	0.5 ± 0.05
02	Liptovská Lužná 2	5.84 ± 0.16	0.6 ± 0.03
03	Liptovská Lužná 1	6.27 ± 0.37	0.6 ± 0.12
04	Selčianske sedlo	6.58 ± 0.07	0.6 ± 0.11
05	Kráľová 1	6.59 ± 0.09	0.6 ± 0.09
06	Liptovská Lužná 3	6.87 ± 0.08	0.7 ± 0.04
07	Vihorlat—Užok	7.05 ± 0.47	0.6 ± 0.15
08	Horné lazy 2	7.22 ± 0.04	0.7 ± 0.06
09	Kráľová 2	7.43 ± 0.19	0.9 ± 0.12
10	Donovaly	7.53 ± 0.21	0.9 ± 0.11
11	Cerovo	7.83 ± 0.26	0.9 ± 0.10
12	Iliaš 1	8.07 ± 0.13	0.9 ± 0.08
13	Priechod—West	8.09 ± 0.17	0.9 ± 0.09
14	Priechod—South	8.33 ± 0.23	1.1 ± 0.04
15	Spišský hrad	8.44 ± 0.23	1.2 ± 0.07
16	Priechod—East	8.69 ± 0.20	1.2 ± 0.10
17	Karpaty—Turka	8.89 ± 0.28	1.2 ± 0.16
18	Horné lazy 1	9.18 ± 0.08	1.2 ± 0.14
19	Lackov	9.24 ± 0.08	1.2 ± 0.12
20	Pravica	9.43 ± 0.19	1.4 ± 0.11
21	Chrámec Teplá dolina 1	10.35 ± 0.16	1.5 ± 0.08
22	Iliaš 2	11.65 ± 0.10	1.5 ± 0.11
23	Chrámec Vlčia dolina 3	14.04 ± 0.27	1.5 ± 0.05
24	Chrámec Vlčia dolina 4	14.64 ± 0.24	1.6 ± 0.07
25	Chrámec Teplá dolina 2	15.08 ± 0.34	1.8 ± 0.06

From the measured values, we found that the samples taken from the Protected Landscape Area—Cerová vrchovina sites had the highest values of weight, where their values ranged from 10.35 ± 0.16 g/100 pcs to 15.08 ± 0.34 g/100 pcs of fruits. The lowest weight values were found in the locations of Poprad—Kišovce Hôrka and Liptovská Lužná 2. In Kišovce, the value reached 5.84 g/100 juniper berries.

The content of juniper essential oil isolated from all dry fruit samples ranging from 0.5 ± 0.05% to 1.8 ± 0.10% (Table 2). The measured values showed that the samples taken from the sites of the PLA Cerová vrchovina, Chrámec Teplá dolina 2 had the highest weights, where the amount of essential oil was 1.8 ± 0.1%. The lowest amount of essential oil, 0.5 ± 0.1%, was found in juniper fruits that were collected at the location Poprad—Kišovce Hôrka.

The box plot of the amounts of juniper essential oil from 25 locations in Slovakia (Figure 4) illustrates the individual amounts of isolated essential oil as a percentage. Individual localities have significant differences in the quantity of essential oil. The mentioned detection documents the importance of site selection for the quantities and quality of the harvest of plant material for the distillery industry. There was up to 1% of essential oil in fruits, the largest number of locations, i.e., 15. There were only five locations with content above 1.5%, which are located up to 200 m above sea level, with the exception of the

Iliaš 2 location, which is located in the valley of the Hron River in Banská Bystrica. The largest quantities of essential oil were clearly isolated from locations in southern Slovakia, where the altitude is up to 150 m a.s.l. and warm to hot weather prevails during the growing season, with maximum average temperatures of 26 °C during the months of July and August.

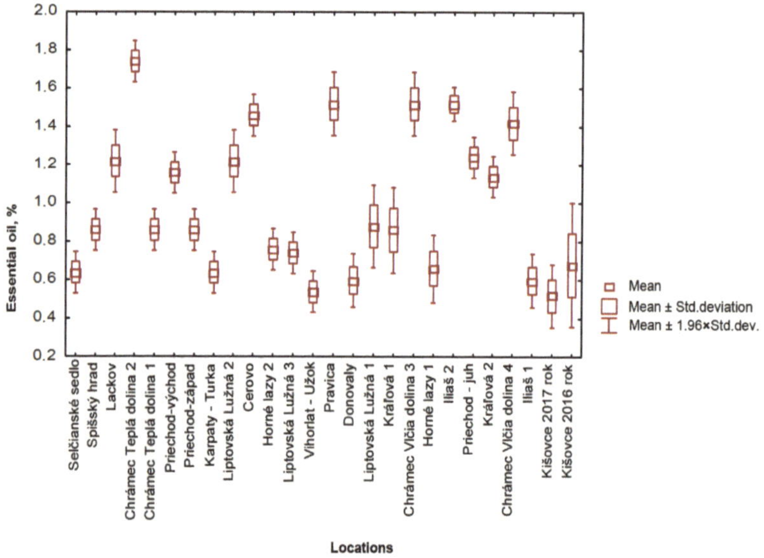

Figure 4. Box graph of quantities of juniper essential oil obtained from locations in Slovakia.

The oil was distilled from dry black, fully ripe berries. It is transparent, fluid, and colourless with a tinge of greenish-yellow. The aroma is similar to that of pine, but more peppery, hot, and balsamic, with a burning, somewhat bitter taste. In accordance with the distillation process, the selected physical parameters of the juniper essential oil were determined (Table 3). The obtained results of the following parameters—bulk density, optical activity, index light refraction, and ash determination—confirmed the findings that the observed parameters follow pharmacopoeia limits.

Table 3. Selected physical and parameters of juniper essential oil.

Physical and Chemical Parameters	Technical Parameters [9]	Determined Parameters
Bulk density	0.860–0.879 g/cm^3	0.876 ± 0.015 g/cm^3
Optical activity	−10°–−30°	−12 ± 1°
Index light refraction	1.457–1.467	1.464 ± 0.001
Ash	max. 0.5%	<0.5%

3.2. The Dependence of Essential Juniper Oil Yield on Fruit Weights

Correlation analysis statistically proved the dependence at the significance level of 0.05 that the size of the fruit affects the amount of essential oil (Figure 5), with higher essential oil contents concentrated in the southern regions of Slovakia. It is interesting, however, that, historically, in this part of the country; an alcoholic drink of the pine type has never been produced, either officially or as a homemade product.

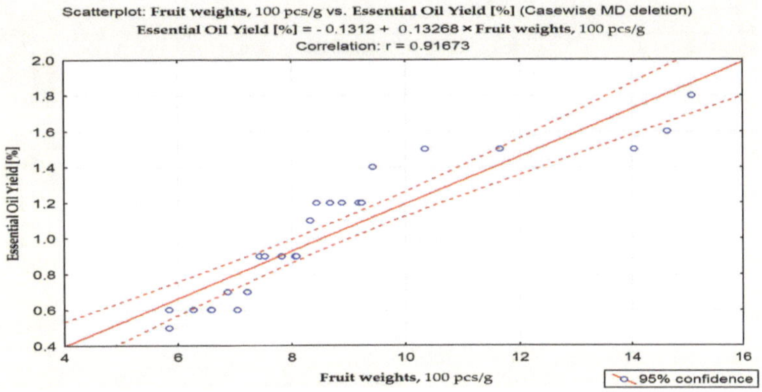

Figure 5. Correlation dependence of the amount of essential oil on fruit weight.

3.3. Essential Oil and Its Qualitative–Quantitative Characteristics

Using GC/MS analyses, 25 different chemical components were identified in the essential oil isolated from juniper berries. For the determination of individual chemical types, only the main siliceous components were selected (Table 4), which are important not only for medicinal utilisation but also for industrial processing in the distillery. Aromatic α-pinene was identified as the first main substance, sabinene, and β-myrcene followed according to the amount in the essential oil.

Table 4. Main components of the essential oil isolated from juniper berries belonging to east and central Slovakia.

Localities		Basic Composition of Essential Oil in %						
		α-Pinene	Sabinene	β-Pinene	β-Myrcene	Limonene	Boranyl acetate	β-Caryophylene
01	Kišovce—Hôrka	61.0 ± 0.60	3.5 ± 0.30	2.5 ± 0.40	10.0 ± 0.30	3.0 ± 0.45	≤0.3	3.5 ± 0.55
02	Liptovská Lužná 2	37.0 ± 0.86	14.5 ± 1.30	1.5 ± 0.41	10.0 ± 0.43	2.0 ± 0.22	≤0.2	9.5 ± 0.45
03	Liptovská Lužná 1	46.0 ± 0.60	10.5 ± 0.68	2.0 ± 0.12	14.0 ± 0.30	3.1 ± 0.35	≤0.3	6.0 ± 0.45
04	Selčianske sedlo	29.0 ± 0.68	12.5 ± 0.68	2.5 ± 0.30	12.0 ± 0.68	2.5 ± 0.21	≤0.3	6.5 ± 0.30
05	Kráľová 1	46.0 ± 1.03	13.0 ± 0.30	2.0 ± 0.17	11.0 ± 0.35	2.2 ± 0.31	≤1.0	12.0 ± 2.05
06	Liptovská Lužná 3	41.0 ± 0.60	14.5 ± 1.30	3.0 ± 0.25	13.0 ± 0.30	3.0 ± 0.21	≤0.3	8.0 ± 0.25
07	Vihorlat—Užok	44.1 ± 0.08	9.50 ± 0.30	2.5 ± 0.12	14.2 ± 0.16	4.50 ± 1.31	≤0.7	7.9 ± 0.30
08	Horné lazy 2	37.0 ± 0.86	9.5 ± 0.30	2.7 ± 0.35	10.0 ± 0.30	2.0 ± 0.45	≤0.5	9.0 ± 1.00
09	Kráľová 2	46.0 ± 0.68	9.5 ± 0.30	2.9 ± 0.25	16.0 ± 0.66	2.1 ± 0.15	≤0.5	8.7 ± 0.75
10	Donovaly	44.0 ± 0.86	8.0 ± 0.30	2.0 ± 0.09	9.0 ± 0.30	6.0 ± 0.15	≤0.5	8.5 ± 0.45
11	Cerovo	36.0 ± 0.61	9.5 ± 0.30	2.5 ± 0.20	16.0 ± 0.43	4.0 ± 0.55	≤0.3	5.5 ± 0.50
12	Iliaš 1	49.0 ± 1.91	8.0 ± 0.42	2.2 ± 0.06	11.0 ± 0.30	3.2 ± 0.21	≤1.0	6.1 ± 0.35
13	Priechod—West	38.0 ± 0.68	15.0 ± 0.43	3.0 ± 0.27	16.0 ± 0.43	2.5 ± 0.10	≤0.5	10.0 ± 1.50
14	Priechod—South	47.0 ± 0.86	12.5 ± 0.30	1.9 ± 0.35	11.0 ± 0.42	1.7 ± 0.15	≤0.3	9.1 ± 0.45
15	Spišský hrad	32.0 ± 0.86	20.5 ± 0.68	2.5 ± 0.10	8.0 ± 0.43	7.5 ± 0.15	≤0.5	5.5 ± 0.25
16	Priechod—East	37.0 ± 1.35	10.5 ± 0.68	3.0 ± 0.15	9.0 ± 0.43	3.5 ± 0.55	≤0.5	17.0 ± 1.05
17	Karpaty—Turka	47.0 ± 1.52	9.21 ± 0.42	1.42 ± 0.19	13.50 ± 0.95	2.40 ± 1.50	≤0.2	11.5 ± 1.30
18	Horné lazy 1	48.0 ± 0.60	10.0 ± 0.30	2.0 ± 0.15	15.0 ± 0.30	2.7 ± 0.31	≤0.3	4.3 ± 0.15
19	Lackov	32.0 ± 1.09	13.5 ± 0.68	2.4 ± 0.42	20.0 ± 0.96	3.5 ± 0.25	≤0.5	8.5 ± 0.50
20	Pravica	43.0 ± 1.35	12.0 ± 0.96	1.0 ± 0.06	14.0 ± 0.30	2.0 ± 0.15	≤0.5	8.2 ± 0.49

Table 4. *Cont.*

	Localities	Basic Composition of Essential Oil in %						
		α-Pinene	Sabinene	β-Pinene	β-Myrcene	Limonene	Boranyl acetate	β-Caryophylene
21	Chrámec, Teplá dolina 1	37.0 ± 0.68	19.5 ± 0.68	2.0 ± 0.15	13.0 ± 0.30	2.0 ± 0.31	≤0.3	4.5 ± 0.22
22	Iliaš 2	47.0 ± 0.60	14.5 ± 0.68	1.5 ± 0.12	16.0 ± 0.30	0.7 ± 0.09	≤0.3	3.3 ± 0.18
23	Chrámec, Vlčia dolina 3	45.0 ± 0.86	16.0 ± 0.42	2.6 ± 0.22	9.0 ± 0.30	4.7 ± 0.25	≤0.5	3.1 ± 0.15
24	Chrámec, Vlčia dolina 4	49.0 ± 1.35	22.0 ± 0.96	2.0 ± 0.15	2.0 ± 0,30	5.1 ± 0.15	≤0.5	3.0 ± 0.25
25	Chrámec, Teplá dolina 2	37.0 ± 0.68	20.0 ± 0.43	3.0 ± 0.25	7.0 ± 0.30	1.5 ± 0.11	≤0.3	9.0 ± 0.72

3.3.1. α-Pinene Content

α-Pinene is an organic compound of the terpene class, one of the two isomers of pinene. It is an alkene that contains an active four-membered ring. It is found in the oils of many conifer species, including juniper (*Juniperus communis* L.).

The content of aromatic α-pinene in the essential oil varies from 29.0 ± 0.68% at the Selčianske sedlo location to 41.0 ± 0.60% at the Liptovská lužná location, and from 50.0 ± 0.60% at the Iliaš location up to 61.0 ± 0.60% at the Kišovce-Hôrka location. These are locations located in both eastern and central Slovakia (Figure 6).

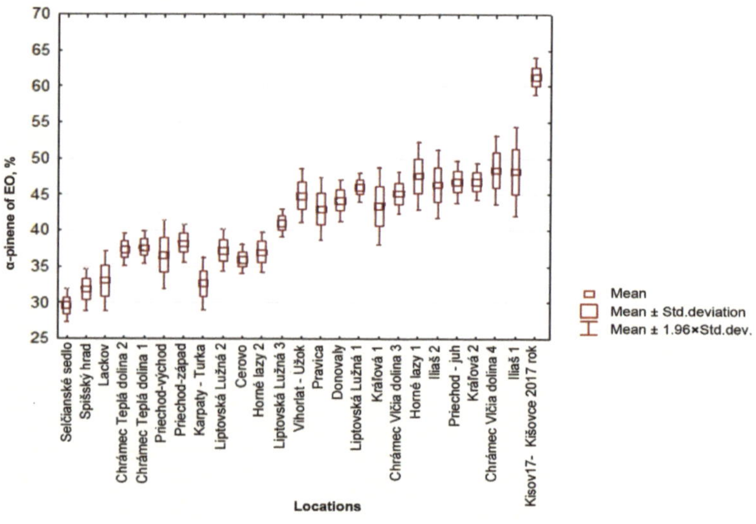

Figure 6. Box graph of quantities of α-pinene essential oil from localities in Slovakia.

These findings proved that the tolerance of Slovak juniper populations in connection with the content of α-pinene in essential oil is significant. It means that the ecological amplitude and thus the adaptation of plants to the specific properties of the biotope, such as temperature, salinity, and nitrogen content in the soil, are wide with considerable ecotolerance. The adaptation of plants to climate conditions is usually in close connection to the entire average of the ecological amplitude of the species in correlation towards the existing conditions.

3.3.2. Sabinene Content

From the point of view of the quantitative amount, the second natural substance in essential oil is the monotropic hydrocarbon sabinene. Chromatographic records confirmed that its quantity interval ranged from 3.5 ± 0.30% at the Kišovce—Hôrka location through to 11.0 ± 0.42% at the Liptovská Lužná location, and up to 22.0 ± 0.96% at locality Chrámec-Vlčia dolina (Figure 7).

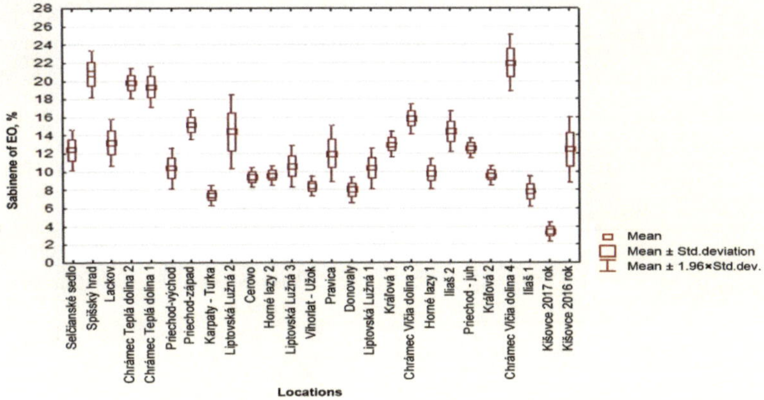

Figure 7. Box graph of quantities of sabinene essential oil according to localities in Slovakia.

3.3.3. β-Myrcene Content

As the third main monoterpene component, β-myrcene was determined. Its content ranged from 2 ± 0.30% at the Chrámec Vlčia dolina location through to 12.0 ± 0.68% at the Selčianske sedlo location, and up to 20 ± 0.96% at the Lackov location in the essential oil isolated from Slovakian juniper berries (Figure 8).

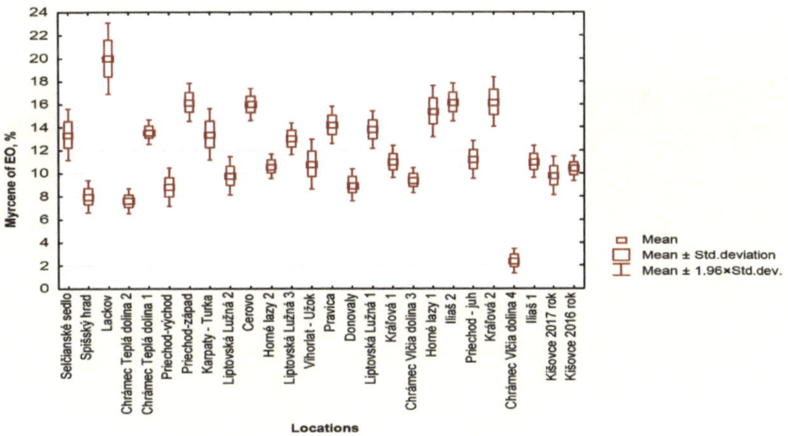

Figure 8. Box graphs of β-myrcene content in the essential oil of juniper berries from localities in Slovakia.

Table 5 presents the qualitative–quantitative characteristics of complete GC-MS analyses of juniper fruit essential oils from three different locations in Slovakia. The interpretation of these results is interesting from several aspects. The interval of α-pinene content ranges from the highest content of 61.00 ± 0.60% (locality: Kišovce-Hôrka) to the lowest 37.60 ± 2.23% (locality: Chrámec—Vlčia dolina). There is a very large difference in the content of sabinene as its content in the essential oil from the locality Kišovce-Hôrka was

3.50 ± 0.30% and for the other localities (Chrámec—Teplá dolina and Chrámec—Vlčia dolina) it was over 19.00%. The myrcene content showed no statistical differences at all three mentioned locations. A significant difference in the content of essential oil was also noted for α-caryophyllene. In the connection with the overall results of the essential oil content, it is confirmed that the natural plant species of common juniper occurs in a wide range of environmental factors and their influence. This species has wide ecological amplitude, and we refer to it as indifferent in certain conditions. The biotopes of the juniper population are of great importance in forestry because they reliably indicate the properties of the locality, the potential of the plant community, and its production potential.

Table 5. The marked differences in chemical composition of juniper essential oil from tree localities in Slovakia [%].

Component as % of Essential Oil *	GC-MS t_R (min.) **	Localities of Juniper Fruit Collection		
		Kišovce—Hôrka	Chrámec—Teplá Dolina	Chrámec—Vlčia Dolina
α-thujone	*8.04*	trace	trace	trace
α-pinene	*8.16*	61.00 ± 0.60	46.50 ± 3.35	37.60 ± 2.23
β-pinene	*13.27*	2.50 ± 0.30	2.51 ± 0.92	2.57 ± 0.92
sabinene	*14.31*	3.50 ± 0.30	19.67 ± 2.08	19.77 ± 1.67
β-phellandrene	*16.12*	0.34 ± 0.03	0.07 ± 0.02	0.08 ± 0.02
camphene	*16.49*	0.22 ± 0.05	0.24 ± 0.06	0.26 ± 0.05
β-myrcene	*17.13*	10.01 ± 0.30	8.03 ± 2.02	10.56 ± 0.05
limonene	*19.57*	2.50 ± 0.30	5.06 ± 0.92	2.27 ± 1.14
γ-terpinene	*19.62*	0.85 ± 0.06	0.31 ± 0.06	0.35 ± 0.07
cis-sabinene hydrate	*20.43*	trace	trace	trace
terpinolene	*23.76*	0.33 ± 0.08	0.37 ± 0.05	0.40 ± 0.06
trans-sabinene hydrate	*24.09*	trace	trace	trace
1-terpinene-4-ol	*24.44*	0.12 ± 0.02	0.95 ± 0.09	1.56 ± 0.92
β-elemene	*38.03*	0.05 ± 0.01	0.23 ± 0.01	0.43 ± 0.02
methyl citronelate	*38.75*	trace	trace	trace
boranyl acetate	*38.97*	0.10 ± 0.01	0.43 ± 0.09	0.40 ± 0.09
γ-elemene	*41.25*	1.09 ± 0.06	1.25 ± 0.07	1.74 ± 0.47
(Z)-β-farnesene	*43.33*	1.33 ± 0.27	1.44 ± 0.29	1.52 ± 0.29
α-caryophyllene	*45.19*	3.50 ± 0.30	6.51 ± 2.01	5.01 ± 1.66
germacrene D	*45.98*	0.99 ± 0.17	2.42 ± 0.95	1.81 ± 0.96
γ-elemene	*46.21*	trace	trace	trace
α-gurjunene	*47.59*	1.41 ± 0.08	1.22 ± 0.97	2.05 ± 0.96
γ-cadinene	*49.41*	0.27 ± 0.07	1.04 ± 0.14	0.25 ± 0.05
β-cadinene	*50.11*	1.12 ± 0.13	0.27 ± 0.07	0.10 ± 0.03
δ-cadinene	*50.22*	0.44 ± 0.03	0.16 ± 0.04	0.21 ± 0.08
elemole	*51.09*	trace	trace	trace
germacrene D-4-ol	*51.42*	trace	trace	trace
spathulenole	*52.35*	trace	trace	trace
shyobunole	*53.76*	trace	trace	trace

Table 5. *Cont.*

Component as % of Essential Oil *	GC-MS t_R (min.) **	Localities of Juniper Fruit Collection		
		Kišovce—Hôrka	Chrámec—Teplá Dolina	Chrámec—Vlčia Dolina
abietadiene	*54.06*	trace	trace	trace
4-epi-abietale	*55.87*	trace	trace	trace
abieta-7,13-diene-3-on	*56.22*	trace	trace	trace
Total [%]		100.00	100.00	100.00

Note: *—data are expressed as area in % of the 100.00% of all identified peaks; **—retention times.

3.4. Hierarchical Cluster Analysis

In order to provide additional insights into the chemotypes of juniper essential oils, a hierarchical cluster analysis based on the constituents was carried out. The dendrogram of the analysis is shown in Figure 9. Based on this analysis, there are four different confirmed chemotypes:

- Chemotype A, dominated by α-pinene (29–33%) > sabinene (14–22%) > myrcene (8–22%);
- Chemotype B, dominated by α-pinene (35–47%) > sabinene (9–21%) > myrcene (9–17%);
- Chemotype C, dominated by α-pinene (49–54%) > sabinene (8–24%) > myrcene (11–18%);
- Chemotype D, dominated by α-pinene (60–62%) > sabinene (3–4%) > myrcene (10–11%).

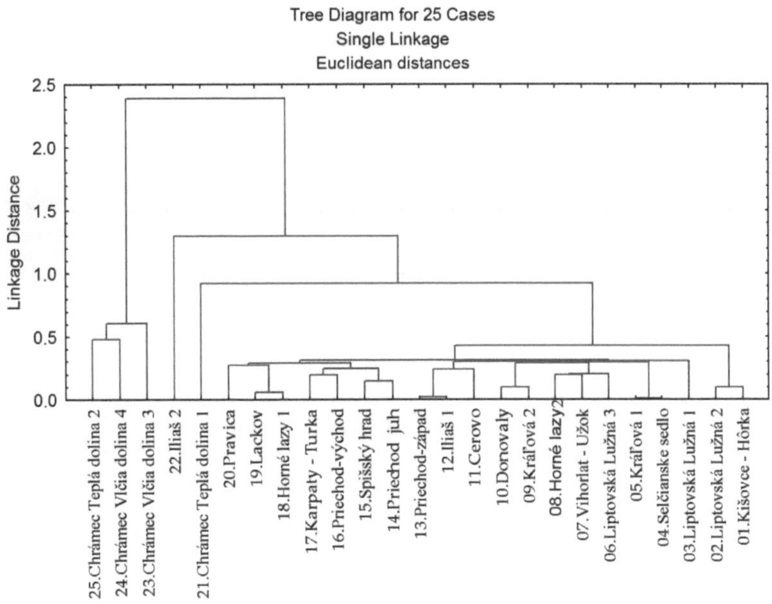

Figure 9. Dendrogram of relationships among juniper populations in Slovakia and their qualitative–quantitative characteristics of essential oils.

The results obtained from the cluster analysis show the existence of a high variability within the essential oils and differences among the following juniper populations collected in different areas in Slovakia (Figure 9). From the 25 populations submitted to multivariate analysis, 4 well-defined groups of essential oils were differentiated using cluster analysis.

Four subclusters can be observed: the first subset contains tree populations collected in Chrámec—Teplá dolina 2, Chrámec—Vlčia dolina 4, and Chrámec—Vlčia dolina 3; the second subset includes the sample collected: Iliaš 2; the third subset includes the sample collected: Chrámec—Teplá dolina 1; the forth subset has twenty populations: Pravica, Lackov, Horné lazy 1, Karpaty—Turka, Priechod—East, Spišský hrad, Priechod South, Priechod—West, Iliaš 1, Cerovo, Donovaly, Kráľová 2, Horné lazy 2, Vihorlat—Užok, Liptovská Lužná 3, Kráľová 1, Selčianske sedlo, Liptovská Lužná 1, Liptovská Lužná 2, and Kišovce—Hôrka.

4. Discussion

Common juniper, *Juniperus communis* L., is a chamaephyte or phanerophyte depending upon growth form and habitat types, respectively [21,22]. In connection with that, the diameter of juniper berries ranged from 3 to 8 mm. From this detection, it is possible to conclude that the most fruits with the largest average size (6–7 mm) were collected from the locations Cerovo, Horné Lazy 2, Priechod—East [23]. In the case of the harvesting of junipers from plant populations in the Chrámec—Teplá dolina location, the diameter of berries was confirmed to be up to 6.5–9 mm, and only 5–7 mm at the Spiš castle on Ostrá hora. Significant differences in the average were also determined in their weight, when 100 juniper fruits in the first case weighed 15.08 ± 0.34 g versus 8.44 ± 0.23 g. Fruits harvested under the forest and near the village of Kišovce-Hôrka in the Tatra valley have a weight of only 5.84 ± 0.19 g/100 juniper berries.

Natural factors (such as soil structure, plant density, relief, sum of precipitation, and groundwater level) play an important role for the different structure and functions of juniper populations [24], which was confirmed by the significant differences in the weight of berries harvested in Slovakia. Similarly, as part of a harvesting expedition carried out in the region of the Stredna Gora mountain range, near the town of Kazalnak, in Bulgaria, fruits from two species were obtained: common juniper (*Juniperus communis* L.) and red juniper (*Juniperus oxycedrus* L.). The weight of 100 fruits of the first species was 10.19 ± 0.14 g, while the weight of 100 larger red juniper berries was 29.37 ± 0.12 g [25].

In cooperation with the laboratory Dr. Panghyova, interesting results were detected regarding the juniper berry carbohydrate contents and the amount of relevant essential oil. A higher content of essential oils was determined in fruits containing fewer carbohydrates. The contents of juniper essential oil isolated from all the dry fruit samples of all localities in Slovakia ranged within the interval from 0.50 ± 0.10% (416.7 g/kg carbohydrates) to 1.80 ± 0.10% (300.2 g/kg carbohydrates) [22].

Matovic et al. [26], in a research article from Serbia, published that the content of essential oil in juniper fruits varies from 2.13 to 3.25%. Another publication from 2011 [26] reported that the essential oil content of juniper berries harvested in a forest region ranged from 2.30 to 2.66%. Our results show that the essential oil yield ranged from 0.5 ± 0.05% to 1.8 ± 0.10%. It has been confirmed that in the Balkan Peninsula, juniper berries contain higher amounts of essential oil. Comparing the yield of essential oil in different parts of Europe, it is important for the future development of the cultivation of this plant species to obtain enough raw material of a suitable quality for the distillery industry in Slovakia.

The number of published works focuses primarily on the qualitative–quantitative composition of isolated essential oil from fruits harvested from autochthonous populations of common juniper (*Juniperus communis* L.) in various areas of their occurrence. However, in most cases this analysis of essential oils content is linked to the biological properties of the medicinal plant, and the chemotype is not investigated [27,28].

The influence of the length of distillation on the amount of α-pinene was monitored by Chatzopolou et al. [29] in Greece. It was concluded that the most suitable time period for juniper distillation is 1 h, when 27.8% of α-pinene was isolated in the essential oil, for example, compared to 3 h distillation 22.4% and 6 h for only 9.7% of α-pinene.

A study from Bulgaria reports on the essential oil composition of common juniper populations, while the predominant component was α-pinene with an interval of its content ranging from 21.4 to 38.4%, sabinene from 10.5 to 19.6%, limonene from 1.8 to 5.5%, and terpinen-4-ol from 3.2 to 7.5% [30]. As can be seen, the main component α-pinene predominates and presents a spectrum of other components (sabinene, limonene, terpinen-4-ol), while the important myrcene as well as β-pinene is almost missing [31].

Extensive research on the composition of essential oil isolated from needles (narrow, linear, and hard leaves) and unripe green berries was carried out in Lithuania [32,33]. Analysis using capillary gas chromatography as a higher content of β-pinene in the essential oil compared to α-pinene was shown. The analyses of the essential oil obtained from green juniper fruits did not confirm these conclusions, while the α-pinene content was up to 44.0 ± 0.60% and only 2.5 ± 0.30% for β-pinene, sabinene: 17.5 ± 0.68%, myrcene: 5.0 ± 0.31%, α-caryophyllene: 5.5 ± 0.11%, and 1-terpinen-4-ol: 1.50 ± 0.32%.

In relation to biological tests, Matovic et al. [34] confirmed the microbial effects of juniper from Serbia in protection against various pathogenic microorganisms (*Agrobacterium tumefaciens*, *Bacillus subtilis*, *Esherichia coli*, and *Pseudomonas flurescens*). They reported that the composition of the essential oil was 30.76% α-pinene, 19.37% sabinene, and 16.42% myrcene. In one of the Slovak samples from the Kišovce-Hôrka locality, there was a determined content of 61% α-pinene, 21% sabinene, and 18% myrcene (group D). At low exposure levels, α-pinene is a human bronchodilator and is highly bioavailable with 60% of absorption by human lungs with rapid metabolism and redistribution. In addition, this monoterpene has anti-inflammatory and antimicrobial effects.

In the last decade, chemotyping of common juniper (*Juniperus communis* L.) populations was done in Lithuania [35]. The essential oil content in the harvested juniper berry samples from 34 biotopes ranged from 0.4 to 2.9%. The most dominant chemotype was α-pinene; sabinene chemotype was proved only at three locations. The value of sabinene, as the main silica component, in these cases ranged from 34.1 to 40.8%, α-pinene from 11.7 to 27.8%, and myrcene from 4.3 to 12.8%. The Slovak α-pinene chemotypes contain 3.5 to 22.0% of sabinene in the essential oil. Juniper berries collected in the surroundings of the capital city of Lithuania, Vilnius, have high values of α-pinene contents (38.5–59.9%) in the essential oil, which are very similar to those in Slovakia (group D, location Kišovce-Hôrka, content of α-pinene: 29–61%). High values of α-pinene (53.6–62.3%) in juniper oil have been reported by Orav et al. [34] in berries harvested from stands at several locations in Estonia.

Juniper berry samples were harvested in Slovakia during 2012–2014 at five locations: Hôrka, Mil'poš, Lačnov, Zbojné, and Kamienka (parts of northeastern Slovakia) [36]. The percentage of essential oil varied from 0.6 to 1.9%; its α-pinene content ranged from 31.0 to 49.0%. The results confirmed the presence of a α-pinene chemotype. In the investigation, when studying 25 locations in Slovakia, the range of the content of essential oil was at a similar interval, but the values of α-pinene were in a much wider range. It was confirmed that the genetic diversity at the level of the common juniper population accurately characterised the chemotype diversity and its structure in Slovakia.

Fejer et al. [36] reported that the values for the β-myrcene content ranged from 7.00 ± 0.5% (Mil'poš) to 16.67 ± 3.88% (Hôrka), and for the essential oil component sabinene from 2.33 ± 1.89% (Hôrka) to 7.17 ± 8.61% (Mil'poš). In the investigation, for essential oil isolated from 25 locations, the amounts of myrcene were confirmed from 2.0 ± 0.30% (Chrámec Vlčia dolina 4) to 20.0 ± 0.96% (Lackov), and in the case of sabinene from 3.5 ± 0.30% (Kišovce-Hôrka) to 22.0 ± 0.96% (Chrámec—Vlčia dolina 4). This proved that the tolerance level of Slovak populations of the common juniper species in connection with the contents of the mentioned two components of the essential oil is extensive. Their ecological amplitude of biosynthesis and thus the adaptation of plants to the specific conditions of the given biotope is found with considerable eco-tolerance.

5. Conclusions

Juniper populations have undoubtedly declined many times in Slovakian local areas due to changes in climate, competition, and human activities, including the clearing and abandonment of land and changes in grazing. In recent decades, the domestic production of black juniper berries was almost completely replaced by imports from Balkan countries.

The presented investigation shows the results of the production capacity of juniper plant populations depending on the size and weight of the fruits. Sufficient attention was carried out to the isolation and content of essential oil and its qualitative–quantitative characteristics. Based on the statistical processing of the data, it can be concluded that there is an α-pinene chemotype of juniper plant populations in Slovakia. From the monitoring of the occurrence of populations during the field research and experimental works, it was confirmed that the genetic diversity at the level of juniper populations in the territory of Slovakia characterises the chemotype species diversity and its structure as an α-pinene chemotype.

In future, great attention must be paid towards the analyses of juniper distillates, which form the basis of the alcoholic beverage, in connection with all already mentioned detections. From the production aspect of juniper berries, the most plentiful is to establish "new" plantations by picking up (digging) young juniper plants from their autochthonous populations and transplanting them to the maintained rural landscape.

Author Contributions: I.S. processed the experimental data, performed the biochemical–biological analysis, and drafted the manuscript. P.O. carried out almost all the technical details and performed numerical calculations for the suggested experiment with plant population's collection in Slovakia. M.H. devised the project of Juniper biodiversity, the main conceptual ideas, and proof outline. M.K. and O.K. worked on the hierarchical cluster analysis of results in regard to qualitative–quantitative analyses of juniper essential oil composition. All authors discussed the results and commented on the manuscript. All authors have read and agreed to the published version of the manuscript.

Funding: This research was supported by the Slovak Research and Development Agency (SRDA), the project: APVV-14-0843: "R&D of possibilities of growing juniper (*Juniperus communis* L.) for the production of fruits".

Data Availability Statement: The data are openly available in a public repository that issues datasets with DOIs.

Conflicts of Interest: The authors declare that the research was conducted in the absence of any commercial or financial relationships that could be construed as a potential conflict of interest.

References

1. Adams, R.P. *The Juniperus of the World: The Genus Juniperus*, 3rd ed.; Trafford Publishing: Victoria, BC, Canada, 2011; 422p; ISBN 1-49072-325-0.
2. Vitalini, S.; Iriti, M.; Vaglia, V.; Garzoli, S. Chemical Investigation and Dose-Response Phytotoxic Effect of Essential Oils from Two Gymnosperm Species (*Juniperus communis* var *saxatilis* Pall. and *Larix decidua* Mill.). *Plants* **2022**, *11*, 1510. [CrossRef]
3. Steiner, L.; Jenny, U.; Hirsbrunner, G.; Welkenhorst, M. Phytotherapeutic Treatments of Gynecological Diseases and Fertility Disorders in Cattle—A Veterinary Historical Analysis. *Schweiz. Arch. Tierheilkd.* **2022**, *164*, 645–659. [CrossRef]
4. Semeyutina, A.V.; Tsoi, M.V.; Sapronova, D.V.; Sapronov, V.V. Research of the Decorative Advantages of Coniferous Trees and Shrubs for Landscaping. *Ecol. Eng. Environ. Technol.* **2022**, *23*, 125–134. [CrossRef]
5. Kormutak, A.; Bolecek, P.; Galgoci, M.; Gomory, D. Longevity and Germination of *Juniperus communis* L. Pollen after Storage. *Nat. Portf.-Sci. Rep.* **2021**, *11*, 12755. [CrossRef]
6. Thomas, P.A.; El-Barghathi, M.; Polwart, A. Biological Flora of the British Isles: *Juniperus communis* L. *J. Ecol.* **2007**, *63*, 335–344. [CrossRef]
7. Small, E. *Culinary Herbs*, 1st ed.; National Research Council (NRC) Research Press: Ottawa, ON, Canada, 1997; 710p; ISBN 0-660-16668-2.
8. Ward, L.K. Lifetime Sexual Dimorphism in *Juniperus communis* var. communis. *Plant Species Biol.* **2007**, *22*, 11–21. [CrossRef]
9. Albrecht, U.; Madisch, A. Therapeutic Potentials Associated with Biological Properties of Juniper Berry Oil (*Juniperus communis* L.) and its Therapeutic Use in Several Diseases—A Review. *Bioact. Compd. Health Dis.* **2022**, *5*, 174–185. [CrossRef]
10. Frydrysiak, E.; Kunicka-Styczynska, A.; Smigielsky, K.; Frydrysiak, M. The Impact of Several Essential Oil Applied to Non-woven Viscose on Bacteria that Cause Lower Urinary Tract Infections-Preliminary Studies. *Molecules* **2021**, *26*, 6854. [CrossRef] [PubMed]

11. Salamon, I.; Otepka, P. Juniper Berries, *Fructus juniperi*, as raw-material for the production of the "Borovička"—Alcoholic Beverage. *Acta Hortic.* **2018**, *1205*, 45–57. [CrossRef]
12. Salamon, I.; Kryvtsova, M.; Hrytsyna, M. Chemical and Phyto-therapeutically properties of essential oils from three *Juniperus* species. *Int. J. Phytomed. Relat. Ind.* **2020**, *12*, 220–226. [CrossRef]
13. Salamon, I.; Petruska, P. Quality of *Juniper* Essential Oil (*Oleum juniperi*) in the South Slovakia and its Curative and Industrial Utilization. *Indian J. Pharm. Educ. Res.* **2017**, *51*, 136–140. [CrossRef]
14. Polák, P.; Saxa, A. *Priaznivý stav Biotopov a Druhov Európskeho Významu*; ŠOP SR: Banská Bystrica, Slovakia, 2005; 736p.
15. *European Pharmacopoeia (Ph. Eur.)*, 10th ed.; EDQM Council of Europe: Strasbourg, France, 2005; 3856p.
16. Markovic, S.M.; Vragolovic, N.M.; Ristic, M.J. Fractionation of the Essential Oil from Juniper (*Juniperus communis* L.) Berries by Hydrodistillation and Rectification. *Hem. Ind.* **2017**, *71*, 9–15. [CrossRef]
17. Adams, R.P. *Identification of Essential Oil Components by Gas Chromatography/Mass Spectrometry*; Allured Publishing Corporation: Carol Steam, IL, USA, 1995; ISBN 978-1-932633-21-4.
18. Falasca, A.; Caprari, C.; Felice, V. GC-MS Analysis of the Essential Oils of *Juniperus communis* L. Berries Growing Wild in the Molise Region: Seasonal Variability and in vitro Antifungal Activity. *Biochem. Syst. Ecol.* **2016**, *69*, 166–175. [CrossRef]
19. Palaniswamy, U.R.; Palaniswany, K.M. *Handbook of Statistics for Teaching and Research in Plant and Crop Science*, 1st ed.; Food Products Press, The Haworth Reference Press: New York, NY, USA; London, UK; Oxford, UK, 2006; 624p; ISBN 978-1-56022-293-4.
20. Kezik, J.; Hill, M. *Data Management Essentials Using SAS and JMP*; Cambridge University Press: Cambridge, UK, 2016; pp. 122–132.
21. Jociene, L.; Krokaite, E.; Rekasius, T.; Vilcinskas, R.; Judzentiene, A.; Marozas, V.; Kupcinskiene, E. Ionomic Parameters of Populations of Common Juniper (*Juniperus communis* L.) Depending on the Habitat Type. *Plants* **2023**, *12*, 961. [CrossRef] [PubMed]
22. Swierkosz, S.; Reczynska, K. Differentiation of Natural Scrub Communities of the *Cotoneastro-Amelanchieretum* Group in Central Europe. *PLoS ONE* **2022**, *17*, e0266868. [CrossRef]
23. Panghyová, E.; Kiss, E.; Šalamon, I.; Jankovič, J. Borievky zo Slovenska. *Trendy Potravinárstve* **2017**, *XXII*, 92–95.
24. Šalamon, I.; Jeliazkov, V.; Petruška, P. Borievka obyčajná (*Juniperus communis* L.) v CHKO Cerová vrchovina a jej súčasný výskum. In *Venkovská Krajina. Sborník z 15. Ročníku Mezinárodní Mezioborové Konference Konané*; Černušáková, L., Ed.; Ekologický Institut Veronica: Hostětíně, Czech Republic, 2017; pp. 68–74; ISBN 978-80-7458-095-6.
25. Zheljazkov, V.D.; Cantrell, C.C.; Semerdjieva, I.; Radoukova, T.; Stoyanova, A.; Maneva, V.; Kacaniova, M.; Astatkie, T.; Borisova, D.; Dincheva, I.; et al. Essential Oil Composition and Bioactivity of Two Juniper Species from Bulgaria and Slovakia. *Molecules* **2021**, *26*, 3659. [CrossRef]
26. Vaičiulytė, V.; Ložienė, K. Variation of Chemical and Orphological Characters of Leaves and Unripe Cones in *Juniperus communis*. *Bot. Lith.* **2013**, *9*, 37–47. [CrossRef]
27. Matovic, M.; Bojovic, B.; Juskovic, M. Composition and Antimicrobial Activity of *Juniperus communis* L. Physiological Metabolites from Serbia. *Int. J. Life Sci. Med. Sci.* **2011**, *1*, 5–8.
28. Matovic, M.; Bojovic, B.; Juškovic, M. Composition of Essential Oils from Three Classes of Juniper Fruit from Serbia. *J. Med. Plants Res.* **2011**, *5*, 6160–6163. [CrossRef]
29. Garcia, D.; Zamora, D.; Gomez, J.M.; Jordano, P. Geographical Variation in Seed Production, Predation and Abortion in *Juniperus communis* throughout its Range in Europe. *J. Ecol.* **2000**, *88*, 435–446. [CrossRef]
30. Chatzopoulou, P.; Katsiotis, S. Procedures Influencing the Yield and the Quality of the Essential Oil from *Juniperus communis* L. Berries. *Pharm. Acta Helv.* **1995**, *70*, 247–253. [CrossRef]
31. Adams, R.P.; Tashev, A.N. The Volatile Leaf Oils of Three *Juniperus communis* Varieties from Bulgaria. *Phytologia* **2018**, *95*, 302–309. [CrossRef]
32. Vilčinskas, R.; Vilkaitytė, A.; Zybartaitė, L.; Paulauskas, A.; Kupčinskienė, E. Genetic Characteristics of *Juniperus communis* L. from Diferent Ecotypes. *Biologija* **2013**, *59*, 129–134.
33. Butkiene, R.; Nivinskiene, O.; Mockute, D. α-Pinene Chemotype of Leaf (Needle) Essential Oils of *Juniperus communis* L. Growing Wild in Vilnius District. *Chemija* **2005**, *16*, 53–60.
34. Butkiene, R.; Nivinskiene, O.; Mockute, D. Two Chemotypes of Essential Oils Produced by the Same *Juniperus communis* L. Growing Wild in Lithuania. *Chemija* **2009**, *20*, 195–201.
35. Orav, A.; Kailas, T.; Muurisepp, M. Chemical Investigation of the Essential Oil from Berries and Needles of Common Juniper (*Juniperus communis* L.) Growing Wild in Estonia. *Nat. Prod. Res.* **2010**, *24*, 1789–1799. [CrossRef]
36. Fejer, J.; Grulova, D.; Eliasova, A.; Kron, I.; De Feo, V. Influence of Environmental Factors on Content and Composition of Essential Oil from Common Juniper Ripe Berry Cones (*Juniperus communis* L.). *Plant Biosyst. Int. J. Deal. All Asp. Plant Biol.* **2018**, *152*, 304–312. [CrossRef]

horticulturae

MDPI

Article

Elucidating the Anatomical Features, Adaptive and Ecological Significance of *Kopsia fruticosa* Roxb. (Apocynaceae)

Shakti Nath Tripathi [1,2], Manju Sahney [1], Arpita Tripathi [3], Praveen Pandey [4,*], Hanuman Singh Jatav [5], Tatiana Minkina [6] and Vishnu D. Rajput [6,*]

1 Department of Botany, University of Allahabad, Prayagraj 211002, India
2 Department of Botany, Nehru Gram Bharati (Deemed to Be University), Prayagraj 221505, India
3 Faculty of Education, Teerthanker Mahaveer University, Moradabad 244001, India
4 Division of Plant Breeding and Genetic Resource Conservation, CSIR-Central Institute of Medicinal and Aromatic Plants, Lucknow 226015, India
5 Department of Soil Science and Agricultural Chemistry, Sri Karan Narendra Agriculture University, Jobner 303329, India
6 Academy of Biology and Biotechnology, Southern Federal University, 344090 Rostov-on-Don, Russia
* Correspondence: pandeypraveen1986@yahoo.com (P.P.); rvishnu@sfedu.ru (V.D.R.)

Abstract: Anatomical characteristics play a pivotal role in understanding the evolutionary relationship among various plant taxa and identifying species of pharmacological importance. Though the anatomical features of the family Apocynaceae have been widely recognized, there is limited research on the stem wood anatomy of *Kopsia fruticosa*, whereas nothing was previously known about its root wood anatomy. The present work describes and analyses its anatomy and correlates the anatomical features with the habitat and ecology of this plant. The oval shape of the young stem and the presence of unicellular trichomes, stone cells in the pith region, laticiferous canals, calcium oxalate crystals, and vascular bundles of two different sizes, viz., smaller in the broad, flattened region and more prominent on the two narrow sides, are remarkable features of the plant, which collectively may often be helpful in distinguishing *K. fruticosa* from other species of this genus. Apart from the previously known qualitative characteristics of the family Apocynaceae, the coalescence of pit aperture and storied pattern of vessels of *K. fruticosa* are newly observed features of the subfamily Rauvofiòideae. On the other hand, in the root wood, vessels are wider (33–64 μm), less frequent (about 53% more in the stem), and shorter, and the rays are larger (21–46 cells in height) and more frequent than those of stem wood; these are the valuable findings which strongly support the non-climbing nature of the studied plant. The vulnerability and mesomorphy indices for stem wood are 0.914 and 349, respectively, indicating plants' adaptation toward a mesic habitat. The correlation of the anatomical traits of plants with the habitat and ecology represents their survivability in different situations. Consequently, anatomical features such as intraxylary phloem, vessel grouping, the storied pattern of vessels, the simple perforation plate, and intervascular vestured pits suggest that plants can tolerate drought. We firmly believe that the present study's outcome can fulfil the research gaps of this hardy plant.

Keywords: *Kopsia fruticosa*; wood anatomy; ecological significance; vulnerability index; mesomorphy index

Citation: Tripathi, S.N.; Sahney, M.; Tripathi, A.; Pandey, P.; Jatav, H.S.; Minkina, T.; Rajput, V.D. Elucidating the Anatomical Features, Adaptive and Ecological Significance of *Kopsia fruticosa* Roxb. (Apocynaceae). *Horticulturae* **2023**, *9*, 387. https://doi.org/10.3390/horticulturae9030387

Academic Editor: Luigi De Bellis

Received: 18 February 2023
Revised: 14 March 2023
Accepted: 15 March 2023
Published: 16 March 2023

1. Introduction

Biodiversity refers to the range and diversity of life, including all the plants, animals, microorganisms, habitats, and ecosystem functions on planet Earth. Angiosperms (flowering plants) are the most prominent clades of existing seed plants [1]. They have tremendous evolutionary plasticity and evolved to inhabit a wide range of ecological niches and have a variety of growth forms, from minute aquatic plants, herbs, shrubs, and climbers to gigantic trees. This clade is a highly diversified and abundant group of embryophytes, with around

260,000 extant species classified into 416 families [2]. They occupy every habitat on earth except the highest mountain tops.

Plants develop anatomical features and adaptive mechanisms, enabling them to function and thrive in various environmental circumstances. The anatomical characteristics play a significant role in identifying species of pharmacological importance since the commercialized material is scratched and challenging to locate. Moreover, these characteristics can provide essential clues to understanding the various mechanisms and adaptations in plants of different growth forms. In addition, the plant roots' structure, function, and interactions with soil are fascinating breakthroughs that will reveal a great deal about our natural ecosystems and global water, mineral, and carbon cycles and contribute to the long-term security of food supplies [3]. Furthermore, the roots of several perennial plants can pose additional hurdles to biologists owing to their size, depth, and complexity in manipulations. Consequently, it is crucial to study the roots' anatomical structure to understand plants' physiological processes. However, most of our current knowledge of roots is based on research from annual, short-lived plants, primarily focusing on the roots of young plants. Therefore, scientific work on the perennial root is unquestionably at the cutting edge of root biology and function. It will surely be a promising area of research in the future.

Kopsia fruticosa Roxb. is an important species of the Apocynaceae (frequently referred to Dogbane or Oleander family) with an enormous evergreen shrub and spreading growth habit. It is classified as a member of the subtribe Kopsiinae (Vinceae tribe) of the Rauvofiideae subfamily in the most recent Apocynaceae taxonomy [4]. This species is widely distributed in tropical and sub-tropical regions of the world, native to Myanmar, and has been domesticated in India, Malaysia, Indonesia, Thailand, and the Philippines [5]. Some genera of this family are essential timber, while many are grown as ornamentals; non-medicinal uses include food, fodder, wood, poisons, dye, and perfume. The production of milky sap by nearly all species is one of the Apocynaceae family's most notable distinguishing traits. In traditional medicine, they are frequently used to cure malaria, fever, diabetes, gastrointestinal illness, and pain in the Asia-Pacific region [6] and were used as a source of arrow poison, while in cases of tertiary syphilis, ulcerated nostrils are poulticed with the root and other *Kopsia* species [7]. Most of the indigenous species of this plant have indole alkaloids that exhibit a variety of pharmacological actions. Several alkaloids, viz., kopsine, fruticosine, fruticosamine, venacarpine A and B, kopsorinine, and kopsifolines A–F, have been identified from the leaves of *K. fruticosa* [8,9]; among them, kopsine was the first isolated alkaloid. However, kopsorinine was extracted from the stem bark, and eleven additional identified alkaloids were found. Leaf and stem bark extracts were subjected to further isolation work, resulting in novel alkaloids of kopsofinone, kopsimalines A–E, kopsinicine, and kopsiloscines H–J [10].

Several researchers have provided the general anatomical features of the family Apocynaceae [11–13] and the wood anatomy of the subfamily Rauvofiòideae [14,15]; however, the literature lacks detailed information on the wood anatomy of *K. fruticosa*. Moreover, nothing has been reported pertaining to its root wood anatomy. Consequently, the present study aimed to investigate the following: (i) the novel anatomical features of the subfamily Rauvofiòideae and (ii) the clear stem and root anatomy of *K. fruticosa* to understand its ecological adaptation in the xeric habitat.

2. Materials and Methods

Plant materials were collected for the present investigation from the district Prayagraj, Uttar Pradesh, India (25.28° N latitude, 81.54° E longitude, and an altitude of 103 m), and Almora, Uttarakhand, India (29.30° N latitude and 79.2° E longitude and an altitude of 1668 m). In order to eliminate air bubbles from the xylem of the samples for better preservation, they were taken to the lab and evacuated using a vacuum pump. Following a 24 h fixing period, materials were placed in formaldehyde alcohol acetic acid for additional processing and preservation. The sections were cut from the sub-apical toward the lower portion at distinct levels. Three types of sections were cut to study the woods, viz., transverse sections (TS), radial longitudinal sections (RLS), and tangential longitudinal sections (TLS). Nevertheless, the TS of the stem were cut at different levels, from the sub-apical to the lower portion. Hand sections as well as microtome sections were prepared. Subsequently, the samples were dehydrated for microtomy using the series of tertiary butyl alcohol (TBA). The dry materials were steadily unfiltered through paraffin wax (58 °C) for three days and immersed in paraffin wax. The sections were stained in safranin (2.0% in 50% ethanol) for 5 min, then rinsed with distilled water, dehydrated twice in a series of alcohols for 3 min each, in 50%, 70%, 95%, and 100%, and were stained with fast green (1.0% in absolute ethanol) for 2 min then entered into xylene and absolute alcohol solution series (25:75, 50:50, and 75:25) and finally in xylene. Then, sections were mounted in Canada balsam. Individual elements of the xylem were isolated by macerating suitable pieces of root and stem wood in Jeffrey's fluid containing equal quantities of 10% nitric acid and 10% chromic acid [16]. All the chemicals and dyes used in the present work were from Sigma-Aldrich®, Saint Louis, MO, USA, Chemicals Private Limited. Macerated slices were thoroughly washed with water and stained with 1.0% aqueous safranin to quantify the dimensional features, viz., vessel element length and xylem fibres. However, tangential dimensions of the vessel lumen were measured on transverse sections. An Olympus binocular compound microscope (Model: CH2i) and a Leica binocular compound microscope (Model: DM2500) were used for examining the anatomical sections and for photography, respectively.

The numerical values given in the descriptions are the mean of thirty measurements. Maximum and minimum values are the single highest or lowest measurements recorded within a species. Eventually, ocular micrometre measurements were converted into μm (Table 1).

Table 1. Conversion of ocular lens measurements.

Lens Objective	
(1 Ocular Unit)	**(μm)**
4×	25
10×	10
40×	2.5
100×	1

The cell counts, measures, and anatomical descriptions follow the recommendations of the IAWA [17]. According to IAWA Committee, fibre length and vessel size (diameter) have been classified into three distinct groups (Table 2).

Table 2. Grouping of fibre lengths and vessel size.

Fibre Lengths			Vessel Size		
Class	**Subclass**	**Lengths (μm)**	**Class**	**Subclass**	**Tangential Diameter (μm)**
	Extremely short	Less than 500		Extremely small	Less than 25
Short	Very short	500–700	Small	Very small	25–50
	Moderately short	700–900		Moderately small	50–100
Medium	Medium	900–1600	Medium	Medium	100–200
	Moderately long	1600–2200		Moderately large	200–300
Long	Very long	2200–3000	Large	Very large	300–400
	Extremely long	Over 3000		Extremely large	More than 400

Following Carlquist [18], vessel diameter was divided by vessel frequency to determine wood vulnerability (v), while mesomorphy (m) was computed by multiplying the vulnerability with the average length of the vessel element. Low v and m values (<1 and 50, respectively) imply xeric adaptation, while their high values (>1 and 800, respectively) point to the transformation towards mesic conditions. The average libriform fibre length is divided by the average vessel element length to determine the F/V ratio. The sequences of three numeric values given in the text represent the minimum, mean, and maximum values, respectively.

3. Results

3.1. Stem

As observed in a transverse section, the *K. fruticosa* young stem appears to be approximately oval. The single-layered epidermis is composed of compactly arranged oval cells and covered by a thin layer of cuticle. Furthermore, trichomes are unicellular and vary in length from 92 to 412 μm. The cortex region is formed from a 2–5-celled collenchymatous hypodermis, which surrounds the 4–7-celled parenchymatous cortex that bears starch grains. The collenchymatous portion of the cortex contains solitary stone cells, while the pericycle region has isolated large patches of fibres. Numerous vascular bundles form a continuous oval ring traversed by narrow rays. In the broad, flattened region, the vascular bundles are smaller, while the two narrow sides' vascular bundles are larger in size. Vascular bundles are conjoint collateral, endarch, and open. A short distance away from the vascular bundles on the inner side adjacent to the pith are strands of intraxylary phloem. Larger vascular bundles of the two narrow sides are opposed by larger patches of the intraxylary phloem, while smaller patches of the intraxylary phloem oppose smaller bundles. The presence of calcium oxalate crystals (in the form of prismatic crystals), and laticiferous canals can be observed in the cortex, pericycle, and pith regions. The pith region is made up of parenchymatous cells without intercellular space and has enormous clusters of stone cells. In the stone cells, prismatic crystals are occasionally observed (Figure 1B–H).

In the nodal as well as inter-nodal portions of the young stem, the shape of the vascular cylinder transforms; a section through the foliar apex is oval and exhibits two unilacunar trace nodes. As the leaves are opposite one another, two leaf gaps are apparent. In the region just below the node, there is a continuous vascular cylinder having two raised notches (at the position of the leaf gaps), which steadily depresses and develops two new notches at a right angle to the first notches at the narrower portion of the stem (Figure 1A–C).

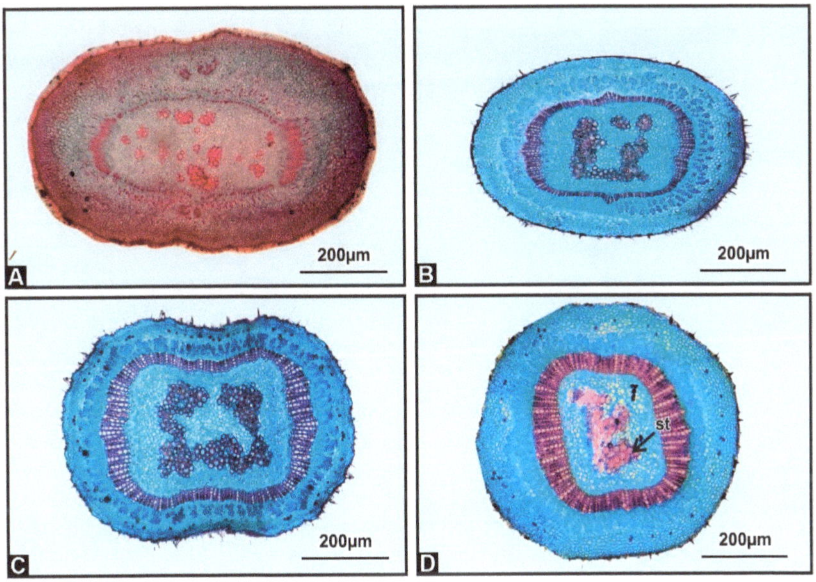

Figure 1. *Cont.*

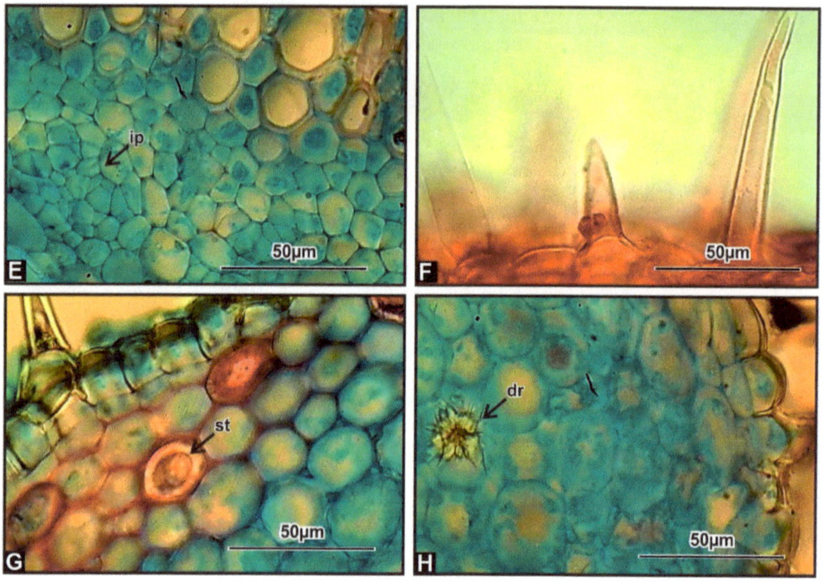

Figure 1. Anatomy of *Kopsia fruticosa* Roxb. stem: (**A**) TS of the stem through the node (foliar apex); (**B–D**) TS of the young stem from sub-apical toward the lower portion at distinct levels showing the development of secondary growth; (**E**) magnified portion exhibits intraxylary phloem; (**F**) magnified image exhibits unicellular hairs; (**G**) magnified image exhibits the stone cells in collenchymatous region; (**H**) magnified image exhibits the druses crystal (a type of calcium oxalate crystal) in pith region; ip—intraxylary phloem; dr—druses; st—stone cell.

3.2. Stem Wood

The vessel arrangement is a diffuse porous pattern with a frequency of 44–47–49/mm^2. Individual vessel elements varied from very small to moderately small (31–43–56 µm) in diameter and 202–382–552 µm in length (Table 3). They are angular in the cross-section and are often radial multiples of 2–11. Vessel perforation plates are always simple and present either terminal or sub-terminal in position. Intervascular pits are vestured, alternate, and small to minute, but the arrangement pattern is occasionally coalescent. Vessels are sometimes arranged in a storied pattern (Figure 2A,C–E).

Table 3. Dimensions of wood elements (stem and root) of *Kopsia fructicosa;* values are the mean ± standard error (n = 30).

Wood Elements	Stem Wood	Root Wood
Vessel length (µm)	382 ± 3.95	302 ± 1.48
Vessel diameter (µm)	43 ± 0.30	39 ± 0.27
Vessel Frequency (Number/mm^2)	47 ± 0.06	31 ± 0.05
Fibre tracheid length (µm)	196 ± 2.19	195 ± 2.15
Libriform fibre length (µm)	752 ± 6.44	611 ± 5.81

The values of the vulnerability (v) and mesomorphy (m) indices are 0.914 and 349, respectively. Both septate as well as non-septate libriform fibres are present, which range from extremely short to medium (402–752–1105 µm) in length, and fibre tracheids are 152–196–500 µm in size, while tracheids are totally absent (Table 3, Figure 2G). The ratio of F/V is 1.97. The parenchyma is axial, mostly apotracheal, diffuse, and occasionally scanty paratracheal. Rays are typically uniseriate (5–9 cells tall), although they can sometimes also be compound (5–9 cells in height), and are heterocellular with a higher frequency of upright cells than square cells. The axial parenchyma and ray cells exhibit rhombic crystals (Figure 2B,D–F).

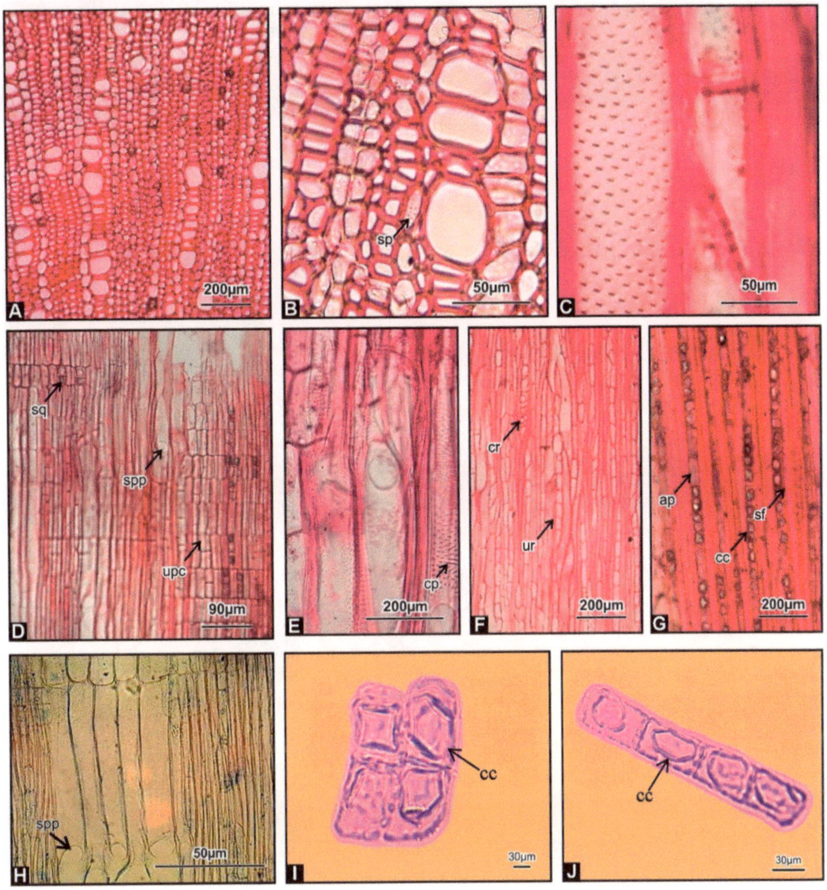

Figure 2. Wood anatomy of *Kopsia fruticosa* Roxb. stem: (**A**) TS revealing a radial arrangement of vessels; (**B**) magnified image revealing scanty paratracheal parenchyma; (**C**) TLS revealing vestured pits (alternate); (**D**) RLS revealing simple perforation plate, upright and square cells; (**E**) TLS image of simple perforation plate and coalescent pit; (**F**) image of uniseriate and compound rays; (**G**) image of rhomboidal shape calcium oxalate crystals in axial parenchyma; (**H**) RLS revealing storied arrangement of vessels; (**I**) square parenchyma cells with chambered calcium oxalate crystals; (**J**) upright parenchyma cells with chambered calcium oxalate crystals; sp—scanty paratracheal parenchyma; spp—simple perforation plate; cc—calcium oxalate crystal; ap—axial parenchyma; upc—upright cell; sq—square cell; sf—septate fibre; cp—coalescent pit; cr—compound rays; ur—uniseriate rays.

3.3. Root Wood

The vessel arrangement is a diffuse porous pattern with a frequency of 29–31–33/mm^2. Individual vessel elements ranged from very small to moderately small (33–39–64 µm) in diameter and 262–302–452 µm in length (Table 3). They are angular in the cross-section and are often radial multiples of 2–5 while occasionally solitary or in the cluster. Vessel perforation plates are always simple and present either terminal or sub-terminal in position. Intervascular pits are vestured, alternate, and small to minute but occasionally the arrangement pattern is coalescent (Figure 3A,C–E,K).

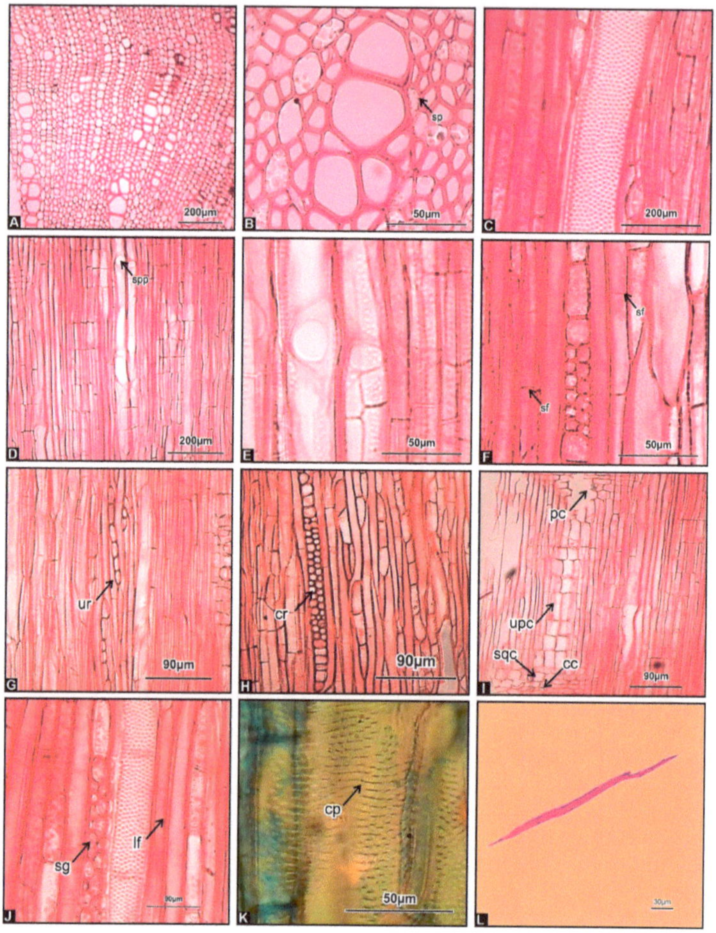

Figure 3. Wood anatomy of *Kopsia fruticosa* Roxb. root: (**A**) TS revealing a radial arrangement of vessels; (**B**) magnified portion showing scanty paratracheal parenchyma; (**C**) image showing alternate vestured pits; (**D**) RLS revealing simple perforation plate; (**E**) magnified portion of simple perforation plate; (**F**) image showing septate fibres; (**G**) TLS image of uniseriate rays; (**H**) section showing compound rays; (**I**) section showing upright, square, and procumbent cells with rhomboidal calcium oxalate crystals; (**J**) section showing rays with starch grains and libriform fibres; (**K**) magnified portion showing coalescent pits in vessels; (**L**) a libriform fibre; sp—scanty paratracheal parenchyma; spp—simple perforation plate; sf—septate fibre; ur—uniseriate rays; cr—compound rays; upc—upright cell; sqc—square cell; pc—procumbent cell; sg—starch grain; lf—libriform fibre; cc—chambered crystal; cp—coalescent pits.

The vulnerability and mesomorphy indices are 1.26 and 380, respectively. Both septate and non-septate libriform fibres are present. They are extremely short to medium in length (302–611–1000 µm). Fibre tracheids are 105–195–410 µm in size. Bifurcate fibres are also present, but tracheids are totally absent (Table 3, Figure 3J,L). The ratio of F/V is 2.02. The parenchyma is axial, predominantly apotracheal, and diffuse; however, sometimes scanty paratracheal parenchyma is also present. Rays are frequently compound (21–46 cells in height) and heterocellular with square, upright, and procumbent cells. Rhomboidal-shaped calcium oxalate crystals and starch grains are visible in the parenchymatous cells (Figure 3B,G–J).

4. Discussion

Plant anatomy is a critical component for species identification as well as being essential for understanding growth processes and nutrient and water mobilization in trees [19]. Therefore, it is crucial to clarify the wood anatomy, including the different cell types that form it and their function, organization, and structural characteristics. The present study evaluates, compares, and explains the anatomy of the wood (stem and root) of *K. fruticosa* from two different altitudes in India. Previously reported qualitative features of the family Apocynaceae, such as stone cells, internal phloem, laticiferous canals, and calcium oxalates (prismatic and druses), have also been noted in the current study [11]. Intraxylary or internal phloem is a significant feature that describes the sieve components that distinguish between the pith and the proto-xylem [20–22]. Internal phloem was first discovered in the family Cucurbitaceae and later in Solanaceae and Apocynaceae. Subsequently, its existence is confirmed in several other species. The secondary xylem has a functional and evolutionary significance closely linked to plant architecture and changing environmental circumstances. Ecological factors can impact all dimensions of xylem structure, either qualitatively or quantitatively. Further, because of the complexity of the dicotyledonous wood system, several characteristics may be used in their identification, i.e., the presence or absence of vessels, distribution of vessels in tissues, types of rays, distribution of axial parenchyma, and types of perforation plate and pits and their arrangements. The presence of distinct intraxylary phloem strands that occur from the starting of primary growth in the young stem of *K. fruticosa* may be associated with an increased translocation of photosynthates and also be an adaptive functionality to protect the sieve elements, which become nonfunctional during an unconducive climate or in hot summers [23]. Furthermore, crystals of calcium oxalate and stone cells are seen in the stem. Stone cells serve as special repositories of plant secretory products and metabolic waste [24], whereas calcium oxalate crystals are recognized for their participation in various essential functions, viz., tissue calcium regulation, metal detoxification, and herbivore defence [25]. In addition, the presence of collenchymatous hypodermis and sclerenchymatous pericycles supply tensile strength to the young stem.

Among the anatomical features seen in this study, radial multiple of vessels, the diffuse porous structure of wood, the simple perforation plate, apotracheal and scanty paratracheal axial parenchyma, and heterocellular rays are the previously identified traits of the stem wood of the family Apocynaceae [11]. However, intervascular pitting of Apocynaceae is reported to be vestured, alternate, and small to minute, while in vessels of *K. fruticosa*, including the above characteristics, coalescence of pit aperture and the storied pattern of vessels are newly observed features of the subfamily Rauvofiòideae. In intervascular pitting, the coalescence of apertures gives an impression of scalariform pitting. Sidiyasa and Baas [26] also reported such pits in the *Alstonia* of the subfamily Apocynoideae.

Like stem wood, root wood exhibits these qualitative features except for the storied pattern of vessels, which occurs only in stem wood. It is important to mention that while the plants were collected from two different altitudes, no significant morphological variations were observed. However, an assessment of the quantitative comparison of stem and root wood confirms significant differences in the length, frequency, and diameter of vessels as well as rays' distribution. The rays are broader and more frequent in the root wood than stem wood, while the vessels are shorter, more comprehensive, and less frequent. The stem wood has a mean vessel frequency of approximately 53% higher than root wood. In stem woods' axial and ray parenchyma, rhomboidal-shaped chambered calcium oxalate crystals are consistently seen but are quite meagre and infrequent in root wood.

As we know, plants' internal structure (young stem and wood) is a key feature of their successful survival in different habitats. Yang et al. [27] correlated the relationship between wood structure and environmental conditions as a survival strategy. Variations in the hydraulic architecture of the xylem, particularly quantitative features such as vessel frequency, vessel grouping, type of perforation plates, vessel element length, and width, are correlated with ecological parameters such as water availability, temperature, and geographical variables (e.g., latitude and altitude). A vessel, also known as the trachea, is built of a series of axially superimposed cells of which intervening end walls have perforations. Vessel diameter is an important factor influencing the conduction of water. It is well established that maximal conductivity and safety from embolism are strongly related to vessel frequency and diameter. A higher frequency of vessels in the stem might be helpful in balancing water transport ability [28,29]. One of the fundamental organizing principles of hydraulic architecture has been noted as the tendency of the roots of trees and shrubs to have wider vessels than the stem [30]. However, the reduced lumen of vessels in the stem aids the mechanical strength of the stem. In the root wood of *K. fruticosa,* vessels are wider than stem wood. It has been reported that due to the presence of wide vessels, the roots of diffuse porous plants are potentially more vulnerable to drought-induced embolism than the stem [31]. In that condition, the presence of starch grains in the parenchyma cells of root wood of *K. fruticosa* is helpful in the removal of embolism, thus increasing the conduction of water in vessels. Simultaneously, it is observed that a storied pattern of vessels is present in the stem wood of *K. fruticosa*. Carlquist [32] mentioned that the arrangement of a storied pattern of vessels indicates the tendency of the plant to survive in desert areas.

In addition, the arrangement of vessels is a characteristic feature of the secondary xylem, which helps in the identification of angiospermous wood. In the stem and root wood of *K. fruticosa,* vessels are arranged radial, multiple, and in a diffuse porous pattern. It was reported that the species in which growth begins early during the growing season are ring porous, whereas species in which growth begins late are diffuse porous [33]. Similar relationships between evergreen and deciduous habits have been discovered. Diffuse porous dicotyledons are generally evergreen, but ring porous dicotyledons are typically deciduous [34].

Perforation plates and inter-vessel pits are particularly crucial for water transport, making up the link across vessel elements inside individual vessels and adjoining vessel walls. Pits in the walls of vessels vary in their shape, size, and arrangement, especially in relation to the type of adjoining cell. Intervascular pits between contiguous vessels are the most conspicuous and taxonomically significant. Vessels in the stem and root wood of *K. fruticosa* possess simple perforation plates and vestured pits. It is supposed that vessel elements with simple perforation plates are more potent in water conduction due to low friction. Vestured pits show protuberances from the secondary cell wall [35]. These are the bordered pits with pit cavities wholly or partially lined with projections from the secondary wall [17]. Vestured pits are hypothesized to serve an essential role in the prevention of cavitations and in improving vessel performance by helping to repair embolisms during drought conditions [36–38]; they are also a sound taxonomic characteristic.

5. Conclusions

This work, for the first time, reported insights into featuring the young stem and wood anatomy of *K. fruticosa.* Apart from the previously known characteristics of the family Apocynaceae, the coalescence of pit aperture in vessels and the storied pattern of *K. fruticosa* are newly observed features of the subfamily Rauvofiòideae. It has a dual therapeutic and poisonous characteristic. It is common to observe that misidentified plants lead to mortality; therefore, specific anatomical features may be helpful in identifying plants. Several anatomical features of the young stem and wood of the *K. fruticosa*, such as the oval shape of the young stem, the presence of unicellular trichomes, the shape of the vascular cylinder in the nodal as well as inter-nodal portions, vascular bundles of two different sizes (smaller in the broad, flattened region and more prominent on the two narrow sides), and the cluster of stone cells in the pith region, may be collectively helpful in distinguishing it from other species of this genus. The correlation of the anatomical features with the habitat and ecology represents the plant's survivability in different conditions. Moreover, it has been documented that low vulnerability index and mesomorphy index values (<1 and 50, respectively) represent an adaptation to xeric conditions, while their high values (>1 and 800, respectively) show transformation to mesic conditions. Accordingly, the presence of v and m values of 0.914 and 349 in the present investigation for stem wood indicate an adaptation of the plant towards mesic conditions. Meanwhile, anatomical features such as vessel grouping, simple perforation plate, intervascular vestured pits, and intraxylary phloem suggest that plants can tolerate drought. Hence, the observed anatomical characteristics provide the framework for research on interspecific and intraspecific variations in woody species, which help us to understand the structural adaptations of the plants with respect to adverse environmental conditions. This framework may also be employed for insight into the plants' physiological processes and for solving taxonomic problems and may be crucial for pharmacology as well as for elucidating phylogenetic relationships.

Author Contributions: Conceptualization, S.N.T. and M.S.; Methodology, S.N.T. and M.S.; Formal analysis, A.T.; Investigation, S.N.T.; Writing—Original Draft Preparation, S.N.T., P.P. and A.T.; Writing—Review and Editing, P.P., H.S.J., T.M. and V.D.R.; Supervision, M.S. All authors have read and agreed to the published version of the manuscript.

Funding: This research received no external funding.

Institutional Review Board Statement: Not applicable.

Informed Consent Statement: Not applicable.

Data Availability Statement: Not applicable.

Acknowledgments: We are thankful to D.K. Chauhan, Department of Botany, University of Allahabad, for their help in the identification of the plant. The authors would like to acknowledge the Ministry of Science and Higher Education of the Russian Federation (no. 075-15-2022-1122) by the Strategic Academic Leadership Program of the Southern Federal University ("Priority 2030").

Conflicts of Interest: The authors declare that they have no conflict of interest.

References

1. Hernández-Hernández, T.; Wiens, J.J. Why are there so many flowering plants? A multiscale analysis of plant diversification. *Am. Nat.* **2020**, *195*, 948–963. [CrossRef] [PubMed]
2. Angiosperm Phylogeny Group IV. An update of the Angiosperm Phylogeny Group classification for the orders and families of flowering plants: APG IV. *Bot. J. Linn. Soc.* **2016**, *181*, 1–20. [CrossRef]
3. Ryan, P.R.; Delhaize, E.; Watt, M.; Richardson, A.E. Plant roots: Understanding structure and function in an ocean of complexity. *Ann. Bot.* **2016**, *118*, 555–559. [CrossRef]
4. Endress, M.E.; Liede-Schumann, S.; Meve, U. An updated classification for Apocynaceae. *Phytotaxa* **2014**, *159*, 175–194. [CrossRef]
5. Wong, S.K.; Lim, Y.Y.; Chan, E.W. Botany, uses, phytochemistry and pharmacology of selected Apocynaceae species: A review. *Phcog. Commn.* **2013**, *3*, 2–11. [CrossRef]
6. Wiart, C. *Medicinal Plants of Asia and the Pacific*, 1st ed.; CRC Press/Taylor and Francis: Boca Raton, FL, USA, 2006; p. 336.
7. Mauger, A.; Jarret, M.; Kouklovsky, C.; Poupon, E.; Evanno, L.; Vincent, G. The chemistry of mavacurane alkaloids: A rich source of bis-indole alkaloids. *Nat. Prod. Rep.* **2021**, *38*, 1852–1886. [CrossRef]
8. Kam, T.S.; Choo, Y.M. Venalstonine and dioxokopsan derivatives from Kopsiafruticosa. *Phytochemistry* **2004**, *65*, 2119–2122. [CrossRef]
9. Kam, T.S.; Lim, K.H. Alkaloids of *Kopsia*. In *The Alkaloids: Chemistry and Biology*; Elsevier: Amsterdam, The Netherlands, 2008; Volume 66, pp. 1–111.
10. Subramaniam, G.; Hiraku, O.; Hayashi, M.; Koyano, T.; Komiyama, K.; Kam, T.S. Biologically active aspidofractinine alkaloids from Kopsiasingapurensis. *J. Nat. Prod.* **2008**, *71*, 53–57. [CrossRef]
11. Metcalfe, C.R.; Chalk, L. *Anatomy of the Dicotyledons: Leaves, Stem, and Wood in Relation to Taxonomy with Notes on Economic Uses*; Clarendon: Oxford, UK, 1950; p. 1500.
12. El-Taher, A.M.; Gendy, A.E.; Alkahtani, J.; Elshamy, A.I.; Abd-El Gawad, A.M. Taxonomic implication of integrated chemical, morphological, and anatomical attributes of leaves of eight Apocynaceae taxa. *Diversity* **2020**, *12*, 334. [CrossRef]
13. Beckers, V.; Rapini, A.; Smets, E.; Lens, F. Comparative wood anatomy and origin of woodiness in subfamilies Secamonoideae and Asclepiadoideae (Apocynaceae). *Taxon* **2022**, *71*, 1230–1250. [CrossRef]
14. Baas, P.; Lens, F.; Wheeler, E.A. Wood anatomy. In *Flora Malesiana. Apocynaceae (Subfamilies Rauvolfioideae and Apocynoideae)*; Nooteboom, H.P., Ed.; Series 1, Seed plants; National Herbarium of the Netherlands: Leiden, The Netherlands, 2007; Volume 18, pp. 16–18.
15. Lens, F.; Endress, M.E.; Baas, P.; Jansen, S.; Smets, E. Wood anatomy of Rauvolfioideae (Apocynaceae): A search for meaningful non-DNA characters at the tribal level. *Am. J. Bot.* **2008**, *95*, 1199–1215. [CrossRef]
16. Johansen, D.A. *Plant Microtechnique*, 1st ed.; McGraw-Hill Book Company, Inc.: London, UK, 1940; p. 530.

17. IAWA. List of microscopic features of hardwood identification. *IAWA Bull.* **1989**, *10*, 219–332.
18. Carlquist, S. Ecological factors in wood evolution: A floristic approach. *Am. J. Bot.* **1977**, *64*, 887–896. [CrossRef]
19. Kramer, J.P.; Koslowski, T. *Fisiologia das Árvores*; Fundação Calouste Gulbenkian: Lisboa, Portugal, 1960; p. 745.
20. Patil, V.S.; Rao, K.S.; Rajput, K.S. Development of intraxylary phloem and internal cambium in *Ipomoea hederifolia* (Convolvulaceae). *J. Torrey Bot. Soc.* **2009**, *136*, 423–432. [CrossRef]
21. Patil, V.S.; Koyani, R.D.; Sanghvi, G.V.; Rajput, K.S. Structure and development of internal phloem in *Solanum pseudocapsicum* (Solanaceae). *IAWA J.* **2014**, *35*, 1–11. [CrossRef]
22. Zózimo, E.; Tamaio, N.; Vieira, R.C. Development of intraxylary phloem in the stem of *Combretum rotundifolium* (Combretaceae). *IAWA J.* **2011**, *32*, 14–24. [CrossRef]
23. Rajput, K.S.; Kapadane, K.K.; Ramoliya, D.G.; Thacker, K.D.; Gondaliya, A.D. Inter- and Intraxylary Phloem in Vascular Plants: A Review of Subtypes, Occurrences, and Development. *Forests* **2022**, *13*, 2174. [CrossRef]
24. Rao, A.R.; Malaviya, M. The distribution, structure and ontogeny of sclereids in *Dendropthœ falcata* (Lf) ettings. *Proc. Indian Acad. Sci.* **1962**, *55*, 239–243. [CrossRef]
25. Nakata, P.A. Advances in our understanding of calcium oxalate crystal formation and function in plants. *Plant Sci.* **2003**, *164*, 901–909. [CrossRef]
26. Sidiyasa, K.; Baas, P. Ecological and systematic wood anatomy of *Alstonia* (Apocynaceae). *IAWA J.* **1998**, *19*, 207–229. [CrossRef]
27. Yong, S.; Furukawa, I.; Jiang, Z. Anatomical variations in the woody plants of and areas. In *Desert Plants Biology and Biotechnology*; Ramavat, K.G., Ed.; Springer: New York, NY, USA, 2010; pp. 136–155.
28. Carlquist, S. *Comparative Wood Anatomy: Systematic, Ecological, and Evolutionary Aspects of Dicotyledon Wood*, 2nd ed.; Springer: Berlin/Heidelberg, Germany, 2001; p. 448.
29. Metcalfe, C.R.; Chalk, L. *Anatomy of the Dicotyledons*, 2nd ed.; Clarendon Press: Oxford, UK, 1983; p. 1500.
30. Jaquish, L.L.; Ewers, F.W. Seasonal conductivity and embolism in the roots and stems of two clonal ring-porous trees, *Sassafras albidum* (Lauraceae) and *Rhus typhina* (Anacardiaceae). *Am. J. Bot.* **2001**, *88*, 206–212. [CrossRef] [PubMed]
31. Longui, E.L.; Galão, A.T.D.; Rajput, K.S.; Melo, A.C.G. Anatomical investigation of root, stem and branch wood in 10-year-old Inga laurina in the context of anatomical adaptation to hydraulic and mechanical stresses. *An. Biol.* **2018**, *40*, 31–39. [CrossRef]
32. Carlquist, S. Wood anatomy of Compositae: A summary, with comments on factors controlling wood evolution. *Aliso* **1966**, *6*, 25–44. [CrossRef]
33. Mowbray, T.B.; Oosting, H.J. Vegetation gradients in relation to environment and phenology in a southern Blue Ridge gorge. *Ecol. Monogr.* **1968**, *38*, 309–344. [CrossRef]
34. Tomlinson, P.B.; Craighead, F.C. Growth-ring studies on the native trees of sub-tropical Florida. In *Research Trends in Plant Anatomy*; Ghouse, A.K.M., Yunus, M., Eds.; Tata McGraw-Hill Publishing Company: New Delhi, India, 1972; pp. 39–51.
35. Jansen, S.; Smets, E.; Baas, P. Vestures in woody plants: A review. *IAWA J.* **1998**, *19*, 347–382. [CrossRef]
36. Dickison, W.C. *Integrative Plant Anatomy*; Harcourt Academic Press: San Diego, CA, USA, 2000.
37. Jansen, S.; Baas, P.; Gasson, P.; Smets, E. Vestured pits: Do they promote safer water transport? *Int. J. Plant Sci.* **2003**, *164*, 405–413. [CrossRef]
38. Jansen, S.; Baas, P.; Gasson, P.; Lens, F.; Smets, E. Variation in xylem structure from tropics to tundra: Evidence from vestured pits. *Proc. Natl. Acad. Sci. USA* **2004**, *101*, 8833–8837. [CrossRef] [PubMed]

horticulturae

Article

Immobilization of Pb in Contaminated Soils with the Combination Use of Diammonium Phosphate with Organic and Inorganic Amendments

Simona M. Popescu [1], Valtcho D. Zheljazkov [2,*], Tess Astatkie [3], Marian Burducea [4] and Walter C. Termeer [5]

1 Faculty of Horticulture, University of Craiova, Str. A.I. Cuza nr. 13, RO-200585 Craiova, Romania
2 Crop and Soil Science Department, Oregon State University, 3050 SW Campus Way, 109 Crop Science Building, Corvallis, OR 97331, USA
3 Faculty of Agriculture, Dalhousie University, 50 Pictou Road, Truro, NS B2N 5E3, Canada
4 Research and Development Station for Aquaculture and Aquatic Ecology, Alexandru Ioan Cuza University of Iasi, Carol I Bd. 11, RO-700506 Iasi, Romania
5 Fundy Compost, Ltd., 1332 Pleasant Valley Rd, Brookfield, NS B0N 1C0, Canada
* Correspondence: valtcho.Jeliazkov@oregonstate.edu; Tel.: +1-541-737-5877

Abstract: The intensive use of lead (Pb)-based insecticides (lead arsenate, $PbHAsO_4$) has led to Pb accumulation in agricultural soil, endangering human health through the possibility of transferring it to the food chain. The aim of this study was to evaluate the potential for the immobilization of Pb in the soil by applying organic (sludge, biocompost, yard compost, and peat) and inorganic (bonemeal, zeolite, lime, and wood ash) amendments, in combination with diammonium phosphate (DAP) in a greenhouse experiment. Two amendment rates were used: low and high, and three rates of DAP: 0 (zero), low (0.25 g of DAP/kg soil), and high (1.25 g DAP/kg soil). The results showed that the dry yield of carrot (*Daucus carota* susp. *sativus*) was the highest for the organic amendments in combination with the low rate of DAP. The high rate of inorganic amendments also increased the yield. Applications of inorganic bonemeal, inorganic lime, and inorganic wood ash yielded the lowest Pb tissue concentration (TC), and organic peat had the highest Pb TC. Inorganic bonemeal combined with DAP most effectively immobilized Pb in soil.

Keywords: heavy metals; metal toxicity; soil amendments; lead contamination; plant uptake; liming

Citation: Popescu, S.M.; Zheljazkov, V.D.; Astatkie, T.; Burducea, M.; Termeer, W.C. Immobilization of Pb in Contaminated Soils with the Combination Use of Diammonium Phosphate with Organic and Inorganic Amendments. *Horticulturae* **2023**, *9*, 278. https://doi.org/10.3390/horticulturae9020278

Academic Editor: Domenico Ronga

Received: 7 February 2023
Revised: 14 February 2023
Accepted: 16 February 2023
Published: 18 February 2023

1. Introduction

Contamination of the environment with heavy metals has become one of the biggest health concerns all over the world due to their persistence in the environment and accumulation in the food chain posing significant threats to human health [1,2]. The major pathway of human exposure to heavy metals is soil to plant transfer where vegetables take up heavy metals by absorbing them from contaminated soils [3]. Heavy metal contamination of soil may occur due to anthropogenic activities through irrigation with contaminated water, the addition of fertilizers and metal-based pesticides, emissions from the metallurgical industry, transportation, and harvesting process [4,5]. Among the most common heavy metal contaminants, lead (Pb) has been reported as a major concern due to its high stability in soil, and accumulation in plants and animals, and it is considered the second most dangerous hazardous substance on the priority list of the U.S. Environmental Protection Agency [6]. It is estimated that approximately half of the human Pb intake is through food, with around half originating from plants [7]. The toxic level of Pb in plants inhibits germination, suppresses growth parameters, reduces the rate of photosynthesis, and alters the levels of photosynthetic pigments, transpiration, gaseous exchange in leaves, and total chlorophyll production [8,9]. It is well known that Pb is highly toxic for humans due to its interference with several biochemical processes, contributing to oxidative stress [10–12]. Because of its high toxicity, the concentrations of Pb in soil and vegetables are restricted by legislation.

The European Union has set standards for Pb at 0.1 mg kg^{-1} (f.w.) for fruits and roots and 0.3 mg kg^{-1} (f.w.) for leafy greens [13]. So far, there are no national health-based standards for Pb in vegetables, fruits, or other staple food crops in the United States [14], although FDA monitors and regulates Pb concentrations in foods and in consumer products.

Previous research has shown that the amount of phytoavailable heavy metal forms in soil and the level of their accumulation in plants depend on soil properties such as: pH, organic matter content, redox potential, cation exchange capacity, and soil texture [15], as well as plant species and root system, growth stage, type of metal, environmental conditions, and agricultural practices [16].

Among the remediation solutions for Pb-polluted soils to reduce the mobility and phytoavailability of this metal, the application of amendments has gained much attention in recent years as an environmentally friendly and low-cost agricultural management practice [17]. Codling [18] reported that the application of phosphorus (P) and iron (Fe) plus phosphorus as amendments on soils contaminated with Pb arsenate increased water extractable Pb concentration making it less accessible to plants. In another study, Ngole [19] found that Pb bioavailability for carrot plants in a sludge-amended soil decreased slightly with an increase in the sludge amendment rate. Recently, Guo et al. [20] reported that dolomite, slaked lime, and limestone applied as amendments to Pb-contaminated soil, significantly reduced Pb content in rice plants. In addition, amendments such as biochar, slag, and ferrous manganese can also successfully reduce the toxicity, leachability, and mobility of Pb in the environment [21].

Some agricultural soils in Nova Scotia, Canada, have high levels of Pb, due to the historic application of $PbHAsO_4$ insecticide in apple orchards several decades ago [22]. Pb arsenate was the most extensively used insecticide as a foliar spray to control codling moth in tree fruit orchards in countries throughout the world, including the USA, Canada, Australia, New Zealand, England, and France [23]. It remained the preferred insecticide for codling moth control because of its high efficacy and lower phytotoxicity and it was being applied in Nova Scotia as late as 1981 [24] until it was officially banned in 1988 [25].

Carrot is a major specialty cash crop in Nova Scotia. Previous studies have shown that carrots, just like other vegetables, take up Pb from contaminated soils by accumulating this metal in their tissues [26–28]. Chisolm [22] reported that Pb concentration in carrots grown in a lead arsenate-contaminated soil exceeded the Canadian tolerance of 2.0 ppm in fresh vegetables. The use of phosphate in Pb immobilization from water or soil is an accepted technique. Furthermore, amendments that contain P can transform the Pb fractions in soil, from highly available to bounded forms, such as pyromorphite Pb5(PO4)3X where X=F, Cl, Br, OH. These Pb compounds are stable under a wide range of pH and Eh. Numerous phosphate materials of natural or synthetic origin have been used to immobilize Pb: apatite and hydroxyapatite, rock phosphate, monoammonium phosphate, diammonium phosphate, biosolids, etc. [29].

The hypothesis in this study was that the application of different organic and inorganic amendments would bind Pb to soil and make it unavailable to plants cultivated under a lead arsenate-polluted soil from Canning, Nova Scotia.

Therefore, the objective of this study was to evaluate the potential for the immobilization of Pb in the soil by applying organic (sludge, biocompost, yard compost, and peat) and inorganic (bonemeal, zeolite, lime, and wood ash) amendments, in combination with diammonium phosphate (DAP) under greenhouse conditions. The Pb availability was assessed using carrot plants.

2. Materials and Methods

2.1. Greenhouse Experiment

The experiment was conducted in the Cox greenhouse of the Faculty of Agriculture of Dalhousie University (formerly Nova Scotia Agricultural College) under natural daylight with day temperatures of 22 to 25 °C and night temperatures of 18–19 °C. The soil used in this study was a sandy loam with 47% sand, 48.2% silt, and 4.8% clay, pH 6.2, and a cation exchange capacity of 17.9. The total Pb concentration of the soil was 109 mg/kg. The soil was collected from the surface layer (0–20 cm) in Canning, NS, Canada, that has a lead arsenate, $PbHAsO_4$ application history.

Plastic pots (20-cm diameter and 15-cm high, Classic 600; Nursery Supplies, Inc., Fairless Hills, PA, USA) were filled with 2 kg of air-dried soil each.

Certified seeds of carrots (*Daucus carota* L. Red Core Chantenay) were direct-seeded into the pots. After 10 days, seedlings were thinned to 6 per pot.

2.2. Experimental Design

A factorial experimental design with 3 replications was used in this experiment. Plants were fertilized with potassium (K) as potassium chloride (2 g/kg) and nitrogen (N) as ammonium nitrate (1.25 g/kg) through incorporation in the soil.

Treatments were represented by 3 application rates of diammonium phosphate (DAP): 0 (zero); low (0.25 g/kg), and high (1.25 g/kg), calculated to represent zero, low and high fertilizer application rates under field conditions; and soil amendments (organic and inorganic), also added at zero, low and high application rates (Table 1). The zero rates of application represents untreated pots in which no treatments (neither amendments nor fertilizers) were added.

Table 1. The application rates of soil amendments to immobilize Pb.

Inorganic	Low *	High *	Organic	Low *	High *
Bonemeal	10	100	Sludge	26	130
Zeolite	10	100	Biocompost	29	116
Lime	5	10	Yard Compost	29	116
Wood Ash	10	20	Peat	3	6

* dry weight basis in t ha^{-1} except the peat was added in units of cubic feet per 4 m^2.

The sludge and biocompost were obtained from Fundy Compost (http://www.fundycompost.com/index.php accessed on 10 June 2021) in Nova Scotia. The sludge originated from the city of Halifax, NS, Canada. The low rate of sludge was selected based on crop N requirements (130 kg ha^{-1}), and assuming a 25% availability of N in composts. The high rate was set at four times the amount of N required.

The yard compost was supplied by Peter Peill of Minas Seed Ltd. (Canning, NS, Canada). The low rate for each of the composts was decided based on crop N requirements (130 kg ha^{-1}), and assuming a 15% availability of N in composts. The high rate was set at four times the amount of N required.

Sphagnum peat moss (ASB Greenworld) was applied to the soil on a 50% (low) and 100% (high) v/v basis. The calculated volume of soil assumes an incorporating depth of 0.3 m over the plot area.

The bonemeal and zeolite were purchased from Digby O & E farms Ltd. (Drayton, ON, Canada). The rates of bonemeal and zeolite were applied to add 0.5% w/w and 5% w/w, according to the recommendations.

The lime (Easy Spread dolomitic limestone) was supplied by Mosher Limestone Co., Ltd. (Upper Musquodoboit, NS, Canada). The rates of lime were applied to change the soil pH from 6.2 to 6.5 (low) and to 7.0 (high) based on the calculation in [30].

The wood ash was supplied by the Faculty of Agriculture power plant facilities. Since the effectiveness of wood ash in changing the soil pH was assumed to be half of that of lime, the low and high rates of wood ash were doubled that of the lime rates.

The elemental compositions of the immobilization amendments were determined as described previously [31,32] (Tables 2 and 3).

Table 2. Mean elemental (mg kg^{-1}) composition of applied immobilization amendments.

Organic	P	Ca	Mg	Na	K	Mn	Fe	S
Sludge	7533	8823	1601	407	978	270	4310	6355
Sludge Compost	3336	3560	893	180	1102	434	3356	2857
Yard Compost	3624	7168	2514	440	3344	239	2784	4222
Peat	115	2048	1039	451	51	10	1201	2898
Inorganic								
Zeolite	424	16,680	6042	3553	1991	178	16,957	174
Bonemeal	42,740	88,444	1597	2217	185	nd *	41	1756
Wood Ash	3918	73,544	8592	744	6983	10,577	2032	453

* nd represents no concentrations detected.

Table 3. Mean elemental (mg kg^{-1}) composition of applied immobilization amendments.

Organic	Pb	Cu	Zn	Cd	Cr	Ni	B
Sludge	18.25	431	328	0.45	8.77	5.42	4.09
Sludge Compost	7.78	103	90	nd *	3.67	4.51	4.89
Yard Compost	14.55	26	90	nd	0.61	3.03	9.67
Peat	1.66	2	7	nd	nd	1.26	5.86
Inorganic							
Zeolite	8.10	48	34	0.18	5.90	10.12	150
Bonemeal	1.42	2	49	nd	nd	0.61	0.61
Wood Ash	7.33	35	166	1.36	3.36	6.20	57.5

* nd represents no concentrations detected.

2.3. Determination of Pb in Plant Samples

The plants were harvested 78 days after establishment, when carrots reached a marketable stage. The carrots were washed thoroughly to remove all soil particles, dried in a drying oven at 70 °C for 72 h, until a constant weight, and the dry weight was recorded. The concentration of Pb in tissue and soil samples was determined as described previously [31,32]. Briefly, heavy metal concentrations in tissue, soil amendments, and soil samples were determined by an inductively coupled argon plasma spectrometer (ICAP) model 61 (Thermo Jarrell Ash, Franklin, MA, USA) following nitric acid digestion as described previously [31,32]. Because of the expected relatively low concentrations of heavy metals in tissue, soil, and soil amendment samples, larger samples of 4 g were digested for 8 h in 250-mL digestion tubes. The available Pb concentration in soil was determined using extraction with 1 M $Mg(NO_3)_2$ to extract the exchangeable (bioavailable) Pb fraction [31,32].

2.4. Determination of Plant Nutrients in Soil Samples

The concentration of plant-available nutrients in soil samples was determined on ICAP by the Nova Scotia Soil Testing Laboratory following the Mehlich 3 extraction, which is a standard procedure in Atlantic Canada.

2.5. Statistical Analysis

The main and interaction effects of Amendment (Bio-compost, Bone Meal, Lime, Peat, Sludge, Wood ash, Yard-compost and Zeolite), Application Rate (Zero, Low and High) and diammonium phosphate (DAP: No DAP [Zero DAP], Low DAP and High DAP) on carrot dry yield, and the concentrations of Pb and P in the tissue (Pb TC, P TC), Pb and P tissue uptake (Pb TU, P TU), and soil available Pb and P (Pb SA, P SA) were determined by conducting Analysis of Variance (ANOVA) of an 8 × 3 × 3 factorial design with 3 replications.

The validity of normal distribution and constant variance assumptions on the error terms were verified by examining the residuals as described in Montgomery [33,34]. Some of the response variables required square root transformation; however, the means shown in the tables are back-transformed to the original scale. Independence assumption on the error terms was ensured by the proper randomization performed during the experiment.

All analyses were completed using the Mixed Procedure of SAS 9.4 (SAS Institute Inc., Cary, NC, USA). For significant (p-value < 0.05) effects, multiple means comparisons were completed by comparing the least squares means of the corresponding treatment combinations. Letter groupings were generated using the Tukey-Kramer method at a 5% level of significance for the main and two-way interaction effects, but for the three-way interaction effect, a 1% level of significance was used to reduce the potential overinflation of Type II experimentwise error rate due to the large number (72) of treatment combinations being compared. Figures 1–3 were produced using Minitab 21 software (State College, PA, USA).

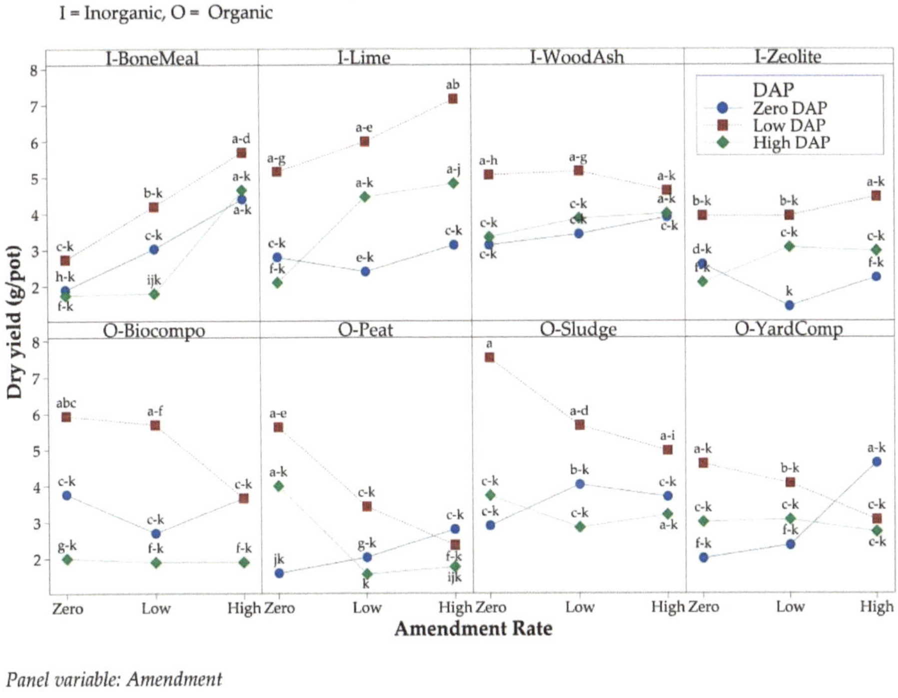

Figure 1. Amendment*Rate*DAP Interaction effect on carrot dry yield (g pot^{-1}). The amendments are grouped as Inorganic (I) and organic (O). Means sharing the same letter are not significantly different.

Panel variable: Amendment

Figure 2. Amendment*Rate*DAP Interaction effect on P tissue uptake plot. The amendments are grouped as Inorganic (I) and organic (O). Means sharing the same letter are not significantly different.

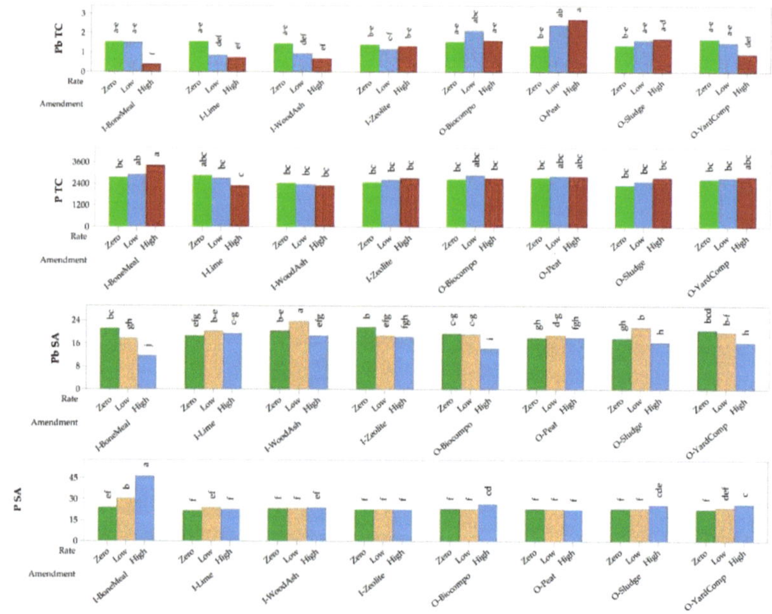

Figure 3. Amendment*Rate Interaction effect on Pb tissue concentration (TC), P TC, Pb soil available (Pb SA), and P SA. Within each response variable, means sharing the same letter are not significantly different.

3. Results

The ANOVA results indicated that there was a significant three-way (Amendment*Rate*DAP) interaction effect on carrot dry matter yield and P tissue uptake (P TU) (Table 4). The two-way interaction effect of Amendment and Rate was significant on Pb tissue concentration (Pb TC), P TC, Pb soil available (Pb SA), and P SA (Table 4). The Amendment by DAP interaction effect was significant only on Pb tissue uptake (Pb TU), but the Rate by DAP interaction effect was significant on Pb TC and Pb TU (Table 4). DAP had no interaction effect on P TC, Pb SA, or P SA, suggesting the differences among the three DAPs were consistent on these three response variables regardless of the amendment and rate.

Table 4. ANOVA *p*-values that show the significance of the main and interaction effects of Amendment (A), Application Rate (R), and Diammonium phosphate (DAP) on Dry yield (Yield), and tissue concentration (TC), tissue uptake (TU), and soil available (SA) Pb and P.

Source of Variation	Yield	Pb TC	P TC	Pb TU	P TU	Pb SA	P SA
A	0.001	0.001	0.001	0.001	0.001	0.001	0.001
R	0.043	0.003	0.131	0.068	0.002	0.001	0.001
A*R	0.001	**0.001**	**0.001**	0.141	0.001	**0.001**	**0.001**
DAP	0.001	0.471	**0.009**	0.001	0.001	**0.001**	**0.001**
A*DAP	0.001	0.066	0.081	**0.044**	0.009	0.550	0.537
R*DAP	0.001	**0.006**	0.541	**0.001**	0.011	0.851	0.908
A*R*DAP	**0.001** *	0.799	0.956	0.120	0.005	0.534	0.498

* Significant effects that require multiple means of comparison are shown in bold.

3.1. *Dry Yield of Carrot Plants*

Low DAP amendment rates gave the highest dry yield of carrot plants in all organic amendments (Figure 1). However, within the low DAP, the amendment rate played different roles among the organic and inorganic amendments. Carrot dry matter yield increased as the amendment rate of inorganic fertilizer increased, but decreased as the amendment rate of organic amendments increased. Among the inorganic fertilizers, increasing the amendment rate of bonemeal and lime showed a high increase in yield, and stayed about the same for wood ash and zeolite.

Although yield was lower when high DAP amendment rates were applied, the pattern in which the inorganic and organic amendments responded was similar to that of low DAP. With zero DAP, a high amendment rate of both inorganic and organic amendments boosted yield to some extent.

The release of nutrients from inorganic amendments is faster than that from organic amendments, leading to a faster growth rate. However, the highest carrot dry matter yield was obtained from the low DAP rate and no sludge, suggesting that all essential nutrients were available in this particular soil. The results from the three-way interaction effect on P tissue uptake (Figure 2) are consistent with those on carrot dry yield (Figure 1) except that the P tissue uptake from inorganic bonemeal was extreme.

3.2. *Amendments' Effect on Pb Immobilization*

A comparison of the treatment combinations of Amendment and Rate (interaction effect of Amendment and Rate) in terms of tissue concentration (TC) of Pb and P, and Pb and P soil available (SA) is shown in Figure 3.

When the amendment rate was zero, there was no significant difference in tissue concentration of Pb among the eight inorganic and organic amendments. However, when the rate was low the inorganic and organic amendments showed some differences, and when the rate was high, they showed significant differences, with inorganic bonemeal, inorganic lime, and inorganic wood ash having the lowest Pb TC, and organic peat having the highest Pb TC (Figure 3).

The amendment rate by DAP interaction effect on Pb TC showed that DAP had a more pronounced effect when there is no amendment, with high DAP and no amendment rate giving the highest Pb TC (Table 5).

Table 5. Mean Pb tissue concentration (Pb TC in mg kg^{-1}), and Pb tissue uptake (Pb TU mg kg^{-1}) obtained from the 9 combinations of Amendment Rate and DAP (Rate*DAP interaction effect) and Pb TU obtained from 24 combinations of Amendment and DAP (Amendment*DAP interaction effect).

Rate	DAP	Pb TC	Pb TU	Amendment	Pb TU from the 3 DAP Levels		
					No	Low	High
No	No	1.16 b *	2.66 d				
No	Low	1.43 ab	6.92 a	I-BoneMeal	3.13 cd	4.07 a–d	2.06 cd
No	High	1.85 a	4.80 abc	I-Lime	3.20 cd	3.73 bcd	2.82 cd
Low	No	1.66 ab	4.32 bcd	I-WoodAsh	3.62 bcd	4.06 a–d	3.99 a–d
Low	Low	1.35 ab	6.08 ab	I-Zeolite	1.79 d	5.70 abc	4.25 a–d
Low	High	1.41 ab	3.68 cd	O-Biocompo	5.52 abc	8.19 ab	3.81 a–d
High	No	1.27 ab	4.34 bcd	O-Peat	4.05 a–d	8.35 a	4.70 a–d
High	Low	1.12 b	4.54 bc	O-Sludge	5.83 abc	8.39 a	4.39 a–d
High	High	1.08 b	2.85 cd	O-YardComp	3.49 bcd	5.17 abc	4.30 a–d

* within each of Pb TC and Pb TU, means sharing the same lower case letter are not significantly different.

Having both amendment and DAP at high rates led to the lowest Pb TC. On the other hand, for Pb tissue uptake, low DAP was more favorable than high DAP when there was no amendment. The interaction between amendment and DAP revealed that organic amendment combined with low DAP maximized Pb TU, and the inorganic amendments gave consistently low Pb TU regardless of the DAP level used (Table 5).

In this study, a significant increase in Pb tissue uptake was found when the low rate of DAP was applied in combination with the organic amendments sludge, peat, and biocompost (8.39, 8.35, and 8.19 mg kg^{-1}, respectively) (Table 5). The lowest Pb uptake was obtained from zeolite amendment and without DAP application (1.79 mg kg^{-1}), while the high rate of DAP in combination with bonemeal and lime also significantly reduced Pb TU (2.06 mg kg^{-1} and 2.82 mg kg^{-1}).

Regardless of the amendment and its rate, zero DAP gave the highest P tissue concentration (2765 mg kg^{-1}) and soil available Pb, but high DAP gave the highest (26.61 mg kg^{-1}) soil available P (Table 6).

Table 6. Mean P tissue concentration (P TC in ug g^{-1}), soil available Pb (Pb SA), and soil available P (P SA) obtained from the three levels of DAP showing the main effect of DAP.

DAP	P TC	Pb SA	P SA
No	2765 a *	18.86 a	23.32 b
Low	2585 b	18.92 a	23.86 b
High	2659 ab	18.28 b	26.61 a

* Within each column, means sharing the same lower case letter are not significantly different.

4. Discussion

4.1. Dry Yield of Carrot Plants

Organic amendments, in addition to their essential nutrient content, and liming can improve soil structure, porosity, water retention capacity, pH, buffering capacity [34–36]. These characteristics improve the physiological state of the plant by increasing the chlorophyll content and photosynthesis and ultimately increasing biomass [37]. For instance, sludge was reported to increase the dry mass of basil due to a high content of OM and a whole array of micro and macronutrients [38,39].

The decrease of biomass in plants exposed to high Pb concentrations in soil could be due to the production of oxidative stress at the photosynthetic apparatus [40,41]. According to Zhou et al. [42], the photosynthetic-related parameters (net photosynthesis, stomatal conductance, transpiration rate) of *Ligustrum lucidum* decreased as a result of the deformation of the chloroplasts and separation of the thylakoid membrane from the cell wall.

4.2. Amendments' Effect on Pb Immobilization

Lead availability is highly dependent on soil physico-chemical characteristics (particle size, pH, cation exchange capacity) [40]. Although both organic and inorganic amendments led to a reduction in Pb mobility in the soil, except for peat (Figure 3), the mode of action is different.

Organic amendments increase soil OM, which determines the immobilization of metals in the soil by the formation of organo-mineral complexes that restrict Pb translocation in the plant [43]. The application of inorganic amendments causes an increase in soil pH that leads to the decrease of Pb mobility by the formation of hydroxides and carbonates precipitates, as was shown for lime, that increased the absorption of HM by varying the variable charge on soil and reducing the extractability [44]. Previously, it was proposed that Pb absorption in soil follows the Langmuir relation, increasing when pH rises from 3.0 to 8.5 [45]. Furthermore, in the case of soil with a pH of 5.5–7.5, Pb availability to plants is controlled by phosphate or carbonate precipitates [46].

On the other hand, the mean P TC obtained from all three rates and the eight amendments was pretty consistent, except for the high rate of inorganic bonemeal amendment, which gave the highest P TC (Figure 3). Inorganic bonemeal also showed the largest variation among the rates in terms of soil available Pb and soil available P, with the zero rate giving the highest and the high rate giving the lowest soil available Pb. Conversely, the zero rate of bonemeal gave the lowest and the high rate gave the highest soil available P (Figure 3). Bonemeal was most efficient at immobilizing Pb in soil and in reducing its transfer to the carrot plants, probably by forming P precipitates due to an increased P concentration (42,740 mg kg^{-1}) in its composition. Similarly, Cao et al. [47] found that phosphate application decreased Pb water solubility and its phytoavailability. Application of a high rate of DAP increased the P concentration in soil which further led to a reduction in the Pb availability in soil, probably by the formation of insoluble Pb phosphate precipitates [47].

The results of this study demonstrated the usefulness of inorganic and organic amendments in reducing the Pb availability for carrot plants through the mechanisms of changing pH and formation of insoluble precipitates. Similarly, it has been shown that the use of organic and inorganic amendments resulted in an immobilization of Pb and Cd in soil, reducing their bioavailability for wheat and maize crops [43]. Since these amendments contain essential nutrients for plants and have the capacity to improve the physical and chemical properties of the soil, a synergistic effect may be responsible for the reduced Pb tissue uptake. Therefore, future research is needed to elucidate the mechanism of action at the physiological level.

5. Conclusions

The use of organic and inorganic amendments may be a solution for the treatment of Pb-contaminated soils in Nova Scotia, and elsewhere. In this study, the use of organic and inorganic amendments in combination with phosphorus induced the immobilization of Pb in the soil and reduced its bioavailability for carrot plants. Amendment application was made based on the nutrient availability for plants as well as to increase the soil OM content and pH. The results revealed that bonemeal in combination with diammonium phosphate was most effective at immobilizing Pb. Furthermore, inorganic bonemeal, inorganic lime, and inorganic wood ash resulted in the lowest Pb tissue concentration, and organic peat resulted in the highest Pb tissue concentration. This study demonstrated the usefulness of applying organic and inorganic amendments to help tackle the Pb soil pollution in soils with a history of lead arsenate application. Future research is needed to verify the results under field conditions, and to elucidate the physiological mechanisms in carrot plants responsible for reducing the phytotoxic effects of Pb when organic and inorganic amendments are used.

Author Contributions: Conceptualization, V.D.Z.; methodology, V.D.Z., W.C.T. and T.A.; software, T.A.; validation, T.A. and V.D.Z.; formal analysis, T.A. and V.D.Z.; investigation, V.D.Z.; resources, V.D.Z. and T.A.; data curation, V.D.Z. and T.A.; writing—original draft preparation, S.M.P.; writing—review and editing, V.D.Z., T.A. and M.B.; visualization, T.A.; supervision, V.D.Z.; project administration, V.D.Z.; funding acquisition, V.D.Z. All authors have read and agreed to the published version of the manuscript.

Funding: This research was funded by the Nova Scotia Department of Agriculture project DEV21-023 "Remediation of Pb polluted soils in Nova Scotia" awarded to V.D. Jeliazkov (Zheljazkov).

Data Availability Statement: Data is contained within the article.

Acknowledgments: We appreciate the help from Paul McNeil and Stephanie Butler from Faculty of Agriculture, Dalhousie University in Truro, Nova Scotia, Canada.

Conflicts of Interest: The authors declare no conflict of interest.

Abbreviations

Pb TC—Pb tissues concentration; Pb TU—Pb tissue uptake; Pb SA—soil available PB; P TC—P tissues concentration; P TU—P tissue uptake; P SA—available soil P.

References

1. Li, R.; Deng, H.; Zhang, X.; Wang, J.J.; Awasthi, M.K.; Wang, Q.; Xiao, R.; Zhou, B.; Du, J.; Zhang, Z. High-efficiency removal of Pb(II) and humate by a CeO_2–MoS_2 hybrid magnetic biochar. *Bioresour. Technol.* **2019**, *273*, 335–340. [CrossRef] [PubMed]
2. Li, R.; Wang, J.J.; Gaston, L.A.; Zhou, B.; Li, M.; Xiao, R.; Wang, Q.; Zhang, Z.; Huang, H.; Liang, W.; et al. An overview of carbothermal synthesis of metal–biochar composites for the removal of oxyanion contaminants from aqueous solution. *Carbon* **2018**, *129*, 674–687. [CrossRef]
3. Kushwaha, A.; Hans, N.; Kumar, S.; Rani, R. A critical review on speciation, mobilization and toxicity of lead in soilmicrobe-plant system and bioremediation strategies. *Ecotoxicol. Environ. Saf.* **2018**, *147*, 1035–1045. [CrossRef] [PubMed]
4. Brhane, G.; Dargo, H. Assessment of some heavy metals contamination in some vegetable and canned foods: A Review. *Int. J. Emerg. Trends Sci. Technol.* **2014**, *1*, 1394–1403.
5. Lanier, C.; Bernard, F.; Dumez, S.; Leclercq, J.; Lemière, S.; Vandenbulcke, F.; Nesslany, F.; Platel, A.; Devred, I.; Cuny, D.; et al. Combined effect of Cd and Pb spiked field soils on bioaccumulation, DNA damage, and peroxidase activities in *Trifolium repens*. *Environ. Sci. Pollut. Res.* **2016**, *23*, 1755–1767. [CrossRef]
6. ATSDR. Agency for Toxic Substances & Disease Registry. 2018. Available online: https://www.atsdr.cdc.gov/spl/previous/07list.html (accessed on 1 February 2023).
7. Nasreddine, L.; Parent-Massin, D. Food contamination by metals and pesticides in the European Union. Should we worry? *Toxicol. Lett.* **2002**, *127*, 29–41. [CrossRef]
8. Gogoasa, I.; Oprea, G.; Harmanescu, M.; Trasca, T.I.; Rivis, A.; Gergen, I. Some metals (Fe, Mn, Zn, Cu, Pb, Cd) contents in vegetables from a nonpolluted plain area of Cenad-Banat (Romania). *J. Fac. Food Eng.* **2010**, *9*, 16–24.
9. Kumar, A.; Prasad, M.N.V. Plant-lead interactions: Transport, toxicity, tolerance, and detoxification mechanisms. *Ecotoxicol. Environ. Saf.* **2018**, *166*, 401–418. [CrossRef]
10. Flora, S.J.S. Lead exposure: Health effects, prevention and treatment. *J. Environ. Biol.* **2002**, *23*, 25–41.
11. Nedelescu, M.; Bălălău, D.; Baconi, D.; Jula, M.; Morar, D.; Gligor, A.; Bălălău, C. Preliminary assessment of heavy metals content of vegetables grown in industrial areas in Romania. *Farmacia* **2015**, *63*, 296–300.
12. Boskabady, M.; Marefati, N.; Farkhondeh, T.; Shakeri, F.; Farshbaf, A.; Boskabady, M.H. The effect of environmental lead exposure on human health and the contribution of inflammatory mechanisms, a review. *Environ. Int.* **2018**, *120*, 404–420. [CrossRef] [PubMed]
13. McBride, M.B. Arsenic and Lead Uptake by Vegetable Crops Grown on Historically Contaminated Orchard Soils. *Appl. Environ. Soil Sci.* **2013**, *2013*, 283472. [CrossRef] [PubMed]
14. McBride, M.B.; Shaylera, H.A.; Russell-Anellia, J.M.; Spliethoffb, H.M.; Marquez-Bravo, L.G. Arsenic and lead uptake by vegetable crops grown on an old orchard site amended with compost. *Water Air Soil Pollut.* **2015**, *226*, 265. [CrossRef] [PubMed]
15. Peijnenburg, W.J.G.M.; Jager, T. Monitoring approaches to assess bioaccessibility and bioavailability of metals: Matrix issues. *Ecotoxicol. Environ. Saf.* **2003**, *56*, 63–77. [CrossRef] [PubMed]
16. Chutia, P.; Kato, S.; Kojima, T.; Satokawa, S. Arsenic adsorption from aqueous solution on synthetic zeolite. *J. Hazard. Mater.* **2009**, *162*, 440–447. [CrossRef]

17. Hussain Lahori, A.; Zhang, Z.; Guo, Z.; Mahar, A.; Li, R.; Kumar Awasthi, M.; Ali Sial, T.; Kumbhar, F.; Wang, P.; Shen, F.; et al. Potential use of lime combined with additives on (im)mobilization and phytoavailability of heavy metals from Pb/Zn smelter contaminated soils. *Ecotoxicol. Environ. Saf.* **2017**, *145*, 313–323. [CrossRef]
18. Codling, E.E. Long-term effects of lime, phosphorus, and iron amendments on water-extractable arsenic, lead, and bioaccessible lead from contaminated orchard soils. *Soil Sci.* **2007**, *172*, 811–819. [CrossRef]
19. Ngole, V.M. Using soil heavy metal enrichment and mobility factors to determine potential uptake by vegetables. *Plant Soil Environ.* **2011**, *57*, 75–80. [CrossRef]
20. Guo, J.; Li, Y.; Hu, C.; Zhou, S.; Xu, H.; Zhang, Q.; Wang, G. Ca-containing amendments to reduce the absorption and translocation of Pb in rice plants. *Sci. Total Environ.* **2018**, *637–638*, 971–979. [CrossRef]
21. Mehmood, S.; Saeed, D.A.; Rizwan, M.; Khan, M.N.; Aziz, O.; Bashir, S.; Ibrahim, M.; Ditta, A.; Akmal, M.; Mumtaz, M.A.; et al. Impact of different amendments on biochemical responses of sesame (*Sesamum indicum* L.) plants grown in lead-cadmium contaminated soil. *Plant Physiol. Biochem.* **2018**, *132*, 345–355. [CrossRef]
22. Chisholm, D. Lead, arsenic, and copper content of crops grown on lead arsenate-treated and untreated soils. *Can. J. Plant Sci.* **1972**, *52*, 583–588. [CrossRef]
23. Peryea, F.J.; Kammereck, R. Phosphate enriched movement of arsenic out of lead arsenate contaminated topsoil and through uncontaminated subsoil. *Water Air Soil Pollut.* **1997**, *93*, 243–254. [CrossRef]
24. MacLean, K.S.; Langille, W.M. Arsenic in orchard and potato soils and plant tissue. *Plant Soil.* **1981**, *61*, 413–418. [CrossRef]
25. Hood, E. The Apple Bites Back: Claiming Old Orchards for Residential Development. *Environ. Health Perspect.* **2006**, *114*, A470–A476. [CrossRef]
26. Intawongse, M.; Dean, J.R. Uptake of heavy metals by vegetable plants grown on contaminated soil and their bioavailability in the human gastrointestinal tract. *Food Addit. Contam.* **2006**, *23*, 36–48. [CrossRef]
27. Lima, F.S.; Nascimento, C.W.A.; Silva, F.B.V.; Carvalho, V.G.B.; Filho, M.R.R. Lead concentration and allocation in vegetable crops grown in a soil contaminated by battery residues. *Hortic. Bras.* **2009**, *27*, 362–365. [CrossRef]
28. Nguyen, H.S.; Huong, B.T.L.; Truong, T.A. Research on Heavy Metal Accumulation of Cu, Pb, Cd and Zn of the carrot. *Int. J. Agric. Technol.* **2015**, *11*, 1119–1127.
29. Miretzky, P.; Fernandez-Cirelli, A. Phosphates for Pb immobilization in soils: A review. *Environ. Chem. Lett.* **2008**, *6*, 121–133. [CrossRef]
30. Brady, N.C.; Weil, R.R. *The Nature and Properties of Soil*, 13th ed.; Upper, S.R., Ed.; Prentice Hall: Hoboken, NJ, USA, 2002.
31. Zheljazkov, V.D.; Astatkie, T. Effect of plant species and benomyl on lead concentration and removal from lead-enriched soil. *HortScience* **2011**, *46*, 1604–1607. [CrossRef]
32. Zheljazkov, V.; Warman, P. Application of high Cu compost to swiss chard and basil. *Sci. Total Environ.* **2003**, *302*, 13–26. [CrossRef]
33. Montgomery, D.C. *Design and Analysis of Experiments*, 10th ed.; Wiley: New York, NY, USA, 2020.
34. Urra, J.; Alkorta, I.; Garbisu, C. Potential Benefits and Risks for Soil Health Derived from the Use of Organic Amendments in Agriculture. *Agronomy* **2019**, *9*, 542. [CrossRef]
35. Burducea, M.; Lobiuc, A.; Asandulesa, M.; Zaltariov, M.-F.; Burducea, I.; Popescu, S.M.; Zheljazkov, V.D. Effects of Sewage Sludge Amendments on the Growth and Physiology of Sweet Basil. *Agronomy* **2019**, *9*, 548. [CrossRef]
36. Burducea, M.; Lobiuc, A.; Dirvariu, L.; Oprea, E.; Olaru, S.M.; Teliban, G.-C.; Stoleru, V.; Poghirc, V.A.; Cara, I.G.; Filip, M.; et al. Assessment of the Fertilization Capacity of the Aquaculture Sediment for Wheat Grass as Sustainable Alternative Use. *Plants* **2022**, *11*, 634. [CrossRef]
37. Zhang, B.; Fuqing, S.; John, Y.; Suqin, Y.; Peng, Z. Effects of inorganic and organic soil amendments on yield and grain cadmium content of wheat and corn. *Environ. Eng. Sci.* **2016**, *33*, 11–16. [CrossRef]
38. Burducea, M.; Zheljazkov, V.D.; Dincheva, I.; Lobiuc, A.; Teliban, G.C.; Stoleru, V.; Zamfirache, M.M. Fertilization modifies the essential oil and physiology of basil varieties. *Ind. Crops Prod.* **2018**, *121*, 282–293. [CrossRef]
39. Zheljazkov, V.D.; Warman, P.R. Application of high Cu compost to dill and peppermint. *J. Agr. Food Chem.* **2004**, *52*, 2615–2622. [CrossRef]
40. Sharma, P.; Dubey, R.S. Lead toxicity in plants. *Braz. J. Plant Phy.* **2005**, *17*, 35–52. [CrossRef]
41. Gupta, D.; Nicoloso, F.; Schetinger, M.; Rossato, L.; Pereira, L.; Castro, G.; Srivastava, S.; Tripathi, R. Antioxidant defense mechanism in hydroponically grown *Zea mays* seedlings under moderate lead stress. *J. Hazard. Mater.* **2009**, *172*, 479–484. [CrossRef]
42. Zhou, J.; Zhang, Z.; Zhang, Y.; Wei, Y.; Jiang, Z. Effects of lead stress on the growth, physiology, and cellular structure of privet seedlings. *PLoS ONE* **2018**, *13*, e0191139. [CrossRef]
43. Hamid, Y.; Tang, L.; Yaseen, M.; Hussain, B.; Zehra, A.; Aziz, M.Z.; He, Z.; Yang, X. Comparative efficacy of organic and inorganic amendments for cadmium and lead immobilization in contaminated soil under rice-wheat cropping system. *Chemosphere* **2019**, *214*, 259–268. [CrossRef]
44. Rate, A.W.; Lee, K.M.; French, P.A. Application of biosolids in mineral sands mine rehabilitation: Use of stockpiled topsoil decreases trace element uptake by plants. *Bioresour. Technol.* **2004**, *91*, 223–231. [CrossRef] [PubMed]
45. Lee, S.Z.; Chang, L.; Yang, H.H.; Chen, C.M.; Liu, M.C. Absorption characteristics of lead onto soils. *J. Hazard. Mater.* **1998**, *63*, 37–49. [CrossRef]

46. Blaylock, M.J.; Salt, D.E.; Dushenkov, S.; Zakarova, O.; Gussman, C.; Kapulnik, Y.; Ensley, B.D.; Raskin, I. Enhanced accumulation of Pb in Indian mustard by soil-applied chelating agents. *Environ. Sci. Technol.* **1997**, *31*, 860–865. [CrossRef]
47. Cao, X.D.; Ammar, L.W.; Bing, M.; Yongliang, L.Y. Immobilization of Zn, Cu, and Pb in contaminated soils using phosphate rock and phosphoric acid. *J. Hazard. Mater.* **2009**, *164*, 555–564. [CrossRef]

horticulturae

MDPI

Article

Satureja montana L. Cultivated under Polypropylene Woven Fabric on Clay-Textured Soil in Dry Farming Conditions

Snežana Mrđan [1], Tatjana Marković [1,*], Tihomir Predić [2], Ana Dragumilo [1], Vladimir Filipović [1], Željana Prijić [1], Milan Lukić [1] and Dragoja Radanović [1]

1 Institute for Medicinal Plants Research "Dr Josif Pančić" Belgrade, Tadeuša Košćuška 1, 11000 Belgrade, Serbia
2 Department of Agroecology, Agricultural Institute of Republic of Srpska, Knjaza Miloša 17, 78000 Banja Luka, Bosnia and Herzegovina
* Correspondence: tmarkovic@mocbilja.rs

Abstract: During a five-year field trial established with *Satureja montana* L. under polypropylene woven fabric (PPWF) on clay-textured soil in dry farming conditions in South Banat, Serbia, the influence of a single basal application of compost and mineral fertilizers at different planting densities (3.6 and 5 plants m^{-2}) on the yield was investigated. Single-dose fertilization positively influenced the yields of *S. montana* in both applied fertilization models. In the third production year, the dry herb yield achieved by applying organic fertilizers in dense cultivation (1016 g m^{-1}) was comparable with that obtained in the mineral plot (961 g m^{-1}). Furthermore, the plants were optimally supplied with N, P, and K macronutrients, with equal amounts removed by yield, in both tested fertilization plots. The use of PPWF proved beneficial to cultivated plants in terms of water-use efficiency and weed suppression. However, severe yield loss was observed in the fourth and fifth production years as a result of the extreme rainfall conditions; the excess moisture retained by PPWF applied to heavy clay soil favored the development of soil-borne pathogens. Other mulch materials should be further investigated for the production of *S. montana* on heavy clay soil.

Keywords: winter savory; cultivation; fertilization; planting density; permeable mulch

Citation: Mrđan, S.; Marković, T.; Predić, T.; Dragumilo, A.; Filipović, V.; Prijić, Ž.; Lukić, M.; Radanović, D. *Satureja montana* L. Cultivated under Polypropylene Woven Fabric on Clay-Textured Soil in Dry Farming Conditions. *Horticulturae* **2023**, *9*, 147. https://doi.org/10.3390/horticulturae9020147

Academic Editor: Bastien Christ

Received: 12 December 2022
Revised: 12 January 2023
Accepted: 16 January 2023
Published: 22 January 2023

1. Introduction

Winter savory (*Satureja montana* L.) is a perennial semi-evergreen subshrub with medicinal, spicy, and honey-bearing properties. It belongs to the *Lamiaceae* family. The plants spontaneously grow in sub-Mediterranean areas in dry, sunny, and rocky habitats [1,2]. The genus *Satureja* has around 30 species, 9 of which have been recorded in the central and western Balkans; among them, *S. montana* is considered the most economically important species [3]. The leaves, flowers, and stems are used to make herbal tea and to cure numerous diseases. It is also regarded as a culinary herb due to the distinctive flavor of its leaves [1,4]. The essential oil (EO) of *Satureja* is rich in phytochemicals such as carvacrol, thymol, β-caryophyllene, γ-terpinene, p-cymene, and linalool, which have strong antioxidant, antimicrobial, antiviral, antispasmodic, and antidiarrheal activities [5–8].

Since the natural resources of this medicinal species are being exhausted due to irrational exploitation and increasing demand for its raw plant material [8], it is crucial to start cultivating *S. montana* over a larger area. The production of seedlings is the first step in the cultivation process. Seedlings require special attention and numerous protective measures in the first two years after planting in the field, particularly against weeds, which inhibit growth at the early stage of crop development by competing with the crop for water, nutrients, light, and space below and above ground [9]. Weed control is a major concern, especially for perennial medicinal plant species. Alternative weed management approaches are preferred to the use of herbicides. Mulching is a physical weed-control method that improves agricultural conditions through soil-moisture conservation, by altering soil temperature, and by positively affecting microorganism activity [10–13]. Although the use of mulch is an investment, it has been demonstrated to be justified and to have long-term effectives, taking into account the reduction in labor force input, which is often becoming increasingly expensive and difficult to obtain, and the elimination of pesticide application [14,15]. Furthermore, mulch use is recognized as an effective crop-production cultivation strategy used in the face of climate-change-related extreme weather conditions (increasing air temperatures, relatively little precipitation, and heavy rainfalls) [16,17]. Polypropylene woven fabric (PPWF) is known for its ability to conserve moisture and to prevent weed growth, and it has an advantage over black plastic mulch in that it is permeable to air and water and does not disintegrate when exposed to sunlight [18–20]. Though the use of mulch has many advantages, some studies suggest limitations involving excess moisture conservation underneath the mulch, resulting in restricted oxygen in the root zone on poorly drained soils and creating an environment conducive to the development of diseases and pests [11,21,22].

Most studies on *S. montana* primarily focused on the secondary metabolites and biochemical activity of its raw material, but only a few of them addressed proper growing conditions [23]. Certain agricultural practices, such as the amount and type of fertilizer used, as well as soil properties, have an impact on both the yield and quality. Mulch application on heavy soils has previously been studied on grain [24–26] and vegetable crops [27–29] but, to the best of our knowledge, never on *S. montana*. In heavy soils, clay minerals absorb water molecules and expand in wet conditions, whereas soil shrinkage occurs in dry conditions, resulting in vertical cracking. Water percolates fast through crack networks after rain or irrigation and is then lost at the root zone in deeper subsoil. Consequently, cracks in the soil profile impair the development and distribution of roots, causing yield loss [30,31]. Ameliorative strategies reported in studies for improving clay soil physico-mechanical properties involve applying river sand and organic matter before plowing [32,33]. In some studies, it has been reported that the application of mulch in heavy soils can alter some soil properties, resulting in improved crop productivity [26,34]. In addition, the clay fraction forms the bulk of the colloidal system of the soil, which is responsible for the adsorption of ions and the storage of plant nutrients in the soil in an exchangeable form available to plants. This makes clay one of the most important factors in soil fertility [35].

With a cultivation trial set up on heavy clay soil under PPWF, which does not allow for annual soil tillage, the goal of this study was to discover the advantages and disadvantages of this mulch applied in the dry farming cultivation of *S. montana*. The first two production years showed satisfactory preliminary data with regard to the plants' development [36]. In the following three production years of *S. montana*, the influence of the crop densities (3.6 and 5 plants m^{-2}) and applied fertilization models (mineral and organic) on the yield was further estimated, focusing on the sufficiency of a single basal fertilization applied prior to PPWF set-up. This study also aimed to resolve whether mineral fertilizers might be replaced by organic fertilizers to produce satisfactory yields.

2. Materials and Methods

2.1. Locality

A field trial was established in Pančevo (44°52′20.0″ N; 20°42′04.7″ E), South Banat, Republic of Serbia. This location has a moderate continental climate, characterized by the local southeasterly prevailing wind "Košava" [37]. The average annual temperature is around 11 °C, with summer temperatures ranging from 21 °C to 23 °C and winter temperatures hovering around −2 °C [38]. The average annual precipitation ranges from 580 to 620 mm, with a rainy season in early summer (May and June) and periods with little precipitation (November and March) [39]. The temperatures and precipitation during the five years of the experiment are presented in Figure 1.

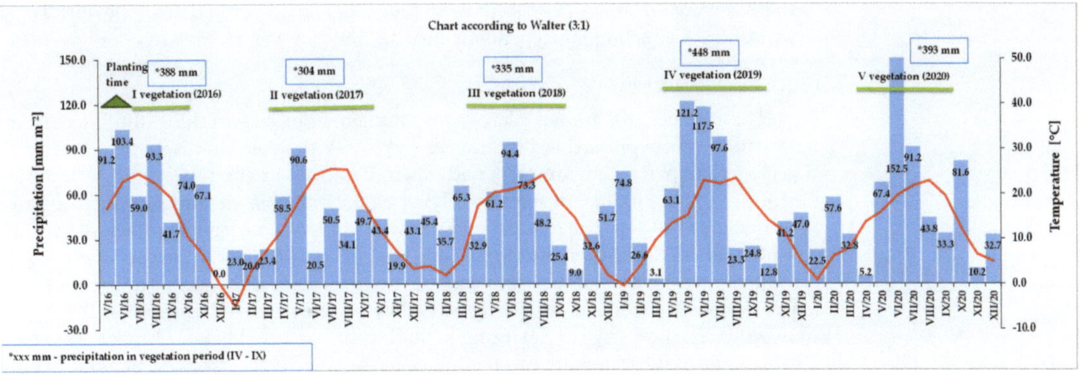

Figure 1. Climatic conditions during five production years of *S. montana* L. in Pančevo, South Banat, Serbia.

According to the calculated five-year average in the vegetation period of the trial duration, from 2016 to 2020, the monthly average precipitations in April, May, June, July, August, and September were 51.5, 86.3, 97.7, 74.3, 48.5, and 35.0 mm, respectively, whereas the monthly average temperatures were 14.2, 17.5, 22.1, 23.2, 23.7, and 18.7 °C, respectively.

The soil type was a variety of Chernozem (leached gleyed), characterized by a less favorable mechanical composition (high clay content in the soil, >42%). Its main properties are given in Table 1.

Table 1. Chemical properties and clay content of the arable soil layer (0–30 cm).

Soil Type	pH		$CaCO_3$ (%)	Humus (%)	Total N (%)	Al Method (mg kg^{-1})		Clay Content (%)
	In H_2O	In KCl				P_2O_5	K_2O	
Chernozem	7.11	5.88	0.42	3.51	0.241	73	445	42.36

2.2. Field Experiment

The field trial was established on 28 May 2016 from experimental fields in Pančevo, South Banat, Serbia. Over a five-year period, the yield of *S. montana* was studied in response to different organic and mineral fertilization models, as well as different crop spacings (70 × 40 and 50 × 40 cm, representing crop densities of 3.6 and 5 plants m^{-2}, respectively).

The fertilizers were applied prior to the soil being tilled; the mineral fertilizer was applied in a single dose of 100 kg ha^{-1} of N, 120 kg ha^{-1} of P_2O_5, and 270 kg ha^{-1} of K_2O, while the organic fertilizer was compost produced in the Production Unit of the Institute for Medicinal Plant Research. The compost had a total content of NPK = 1.4:0.5:1.4% and was applied in a single dose of 20 t ha^{-1}, resulting in calculated nutrient values of 280 kg ha^{-1} of N, 100 kg ha^{-1} of P_2O_5, and 270 kg ha^{-1} of K_2O. Following the fertilizer treatment, 1 mm thick polypropylene woven fabric (PPWF) from GINEGAR Plastic Products Ltd., Migdal HaEmek, Israel, was spread over the subsequently trenched soil and one-year old nursery plants were transplanted.

In the spring of the 2nd year, the mineral plot received additional fertilization with AN (34% N) at a dose of 200 kg ha^{-1}, while the organic plot received additional fertilization with the liquid organic fertilizer BioGrow (NPK 4:6:3) from Biobizz Worldwide SL, Legizamon, Spain, at a dose of 5 L ha^{-1}.

The experimental design was a two-factor block design with four replications. The plots were arranged according to the applied fertilizers, and each fertilization plot had two different crop density patterns. The size of a single plot was 17.1 m^2 (3.8 m × 4.5 m). There were four rows (replications), and each one was composed of nine plants. The independent variables were fertilizers (levels: mineral and organic) and crop density (levels: 3.6 and 5 plants m^{-2}), affecting the dependent variable, the crop yield [36].

2.3. Field Measurements

The number of adopted plants was counted prior to harvest in the 1st year, and adaptation was expressed in percentages (AP%). With regard to disease occurrence, the infected plant parts were sampled, with microscopic observations being made in the 2nd, 3rd, 4th, and 5th production years by relying on the available literature for microorganism identification [40,41]. The disease severity was assessed in each experimental unit, and the numbers of healthy vs. diseased plants were recorded. Prior to harvest in subsequent production years, the number of live plants per treatment was counted and expressed as a percentage of the initial crop density. Each year, plants were harvested at the stage of flowering. The first stage of flowering in the 1st year occurred in late October (28 October 2016), while in the following production years, flowering occurred earlier, with harvests performed between 15 September and 2 October. For each plot, the dry biomass of each plant was measured (yield per single plant) and also calculated and presented as a yield per unit area (yield m^{-2}).

2.4. Analyses of NPK Macronutrients in Aboveground Plant Biomass

After the harvest, the air-dried plant material was milled (M-20, IKA Universal mill, IKA®-Werke GmbH & Co. KG, Staufen, Germany) and stored for chemical analysis. An analysis of the mineral elements was performed after the dry ashing procedure for organic matter destruction. Total nitrogen (N) in the aboveground plant part was determined by alkaline distillation after the standard Kjeldahl digestion procedure. The dry plant material was subjected to wet acid digestion (conc. H_2SO_4 + H_2O_2 + 450 °C) for the P and K analyses. Phosphorus (expressed as P%) was determined using the vanadate-molybdate method (UV-visible spectrophotometer, Helios Alpha, Thermo Spectronic, PROFCONTROL GmbH, Schönwalde-Glien, Germany). Potassium (expressed as K%) was determined by atomic absorption spectrophotometry using the flame technique (AAS Thermo E.C.—SOLAR S4) [42]. The levels of major macronutrients removed from the soil by aboveground biomass (g m^{-1}) were calculated using the data on plant dry biomass yields obtained by the end of the 3rd, 4th, and 5th vegetation seasons and the data on their concentrations in aboveground plant biomass.

2.5. Statistical Analysis

The data obtained in this study were statistically processed by a two-way factorial analysis of variance (a two-way ANOVA) using the software STATISTICA 7.0. The post hoc Fisher's least significant difference (LSD) test ($p < 0.05$) was used to test the differences between treatments.

3. Results

3.1. Crop Density

From planting to harvest in the first production year, 96% and 95% of the total number of plants were observed in the mineral and organic plots, respectively (the initial density, AP%). In the second and third production years, further losses of 8% and 10% occurred in the mineral plot and 5% and 10% occurred in the organic plot, respectively. By the end of the third production year, 86% of the original plants in the mineral and 85% in the organic plot, respectively, were still alive. In the first three years of production, crop density was unaffected by the applied crop densities or fertilizers. However, it rapidly decreased after being severely affected by disease in the fourth and fifth years; the loss was higher in the organic plot, 41% and 57%, respectively, compared to the mineral one, 21% and 34%, respectively, compared to corresponding initial densities, regardless of applied crop densities (Figure 2).

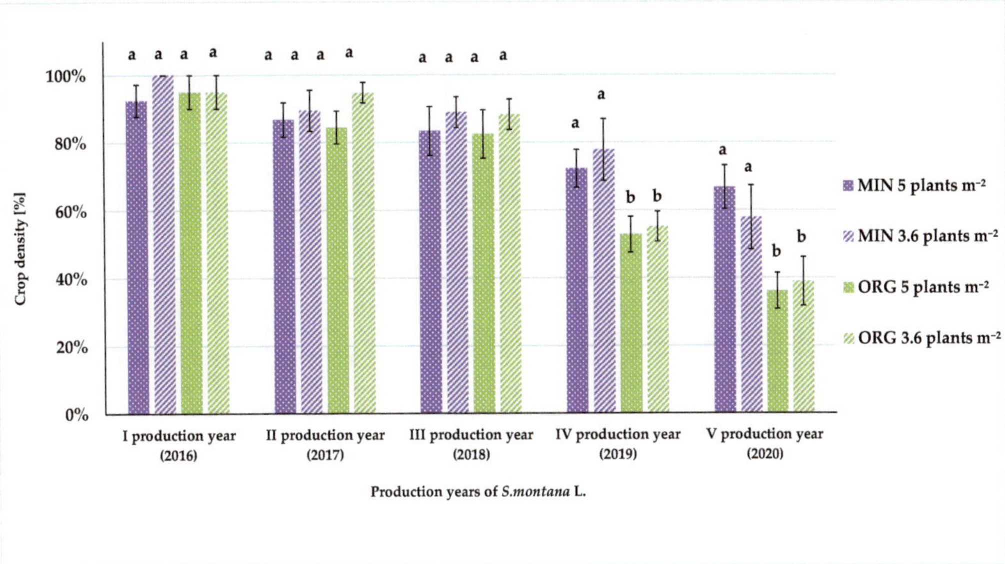

Figure 2. Crop density of *S. montana* prior to harvests in a five-year production period in relation to the initial crop density (%); MIN—mineral plot; ORG—organic plot; lower crop density (3.6 plants m^{-2}); higher crop density (5 plants m^{-2}); means ± standard error followed by the same letter in the same year are not significantly different ($p < 0.05$).

3.2. Yield per Single Plant and Area

In the first production year, neither crop density nor the fertilization model had an effect on yield per single plant (Table 2) or yield per area (Table 3).

Table 2. Average yields of individual *S. montana* plants in relation to crop density and fertilization model over a five-year production period.

Fertilization Model	Crop Density (plants m^{-2})	Yields (g plant^{-1})				
		1st Year	2nd Year	3rd Year	4th Year	5th Year
Mineral	5	72.53 ± 5.19 [a*]	151.06 ± 10.82 [b]	242.03 ± 24.44 [b]	151.88 ± 20.06 [b]	172.9 ± 8.27 [b]
	3.6	81.40 ± 10.56 [a]	186.00 ± 13.27 [ab]	313.77 ± 21.17 [a]	204.43 ± 12.98 [ab]	187.9 ± 28.66 [b]
Organic	5	80.95 ± 7.52 [a]	178.44 ± 16.62 [b]	280.54 ± 19.98 [ab]	230.50 ± 28.55 [a]	178.5 ± 23.17 [b]
	3.6	80.95 ± 12.45 [a]	221.94 ± 13.98 [a]	297.74 ± 19.91 [ab]	251.73 ± 25.09 [a]	274.3 ± 30.97 [a]

* Values are presented as means ± standard error. Values with the same letter in each column showed no statistically significant difference ($p < 0.05$).

Table 3. Average yields of *S. montana* per unit area in relation to crop density and fertilization model in a five-year production period.

Fertilization Model	Crop Density (plants m^{-2})	Yields (g m^{-2})				
		1st Year	2nd Year	3rd Year	4th Year	5th Year
Mineral	5	300.0 ± 49.41 [a*]	755.3 ± 54.11 [ab]	983.9 ± 30.26 [ab]	556.3 ± 50.47 [ab]	570.8 ± 43.46 [a]
	3.6	293.0 ± 37.98 [a]	669.6 ± 47.78 [b]	986.6 ± 34.05 [ab]	557.0 ± 40.15 [ab]	342.7 ± 61.16 [b]
Organic	5	359.4 ± 45.25 [a]	892.3 ± 34.26 [a]	1016.3 ± 70.00 [a]	634.0 ± 79.06 [a]	337.5 ± 82.66 [b]
	3.6	274.5 ± 50.35 [a]	799.0 ± 50.33 [ab]	938.4 ± 44.76 [b]	492.1 ± 65.45 [b]	367.9 ± 72.49 [b]

* Values are presented as means ± standard error. Values with the same letter in each column showed no statistically significant difference ($p < 0.05$).

In the second production year, the effect of crop density was observed, being the highest in the organic plot of lower density (Table 2), and was different from all other treatments. Both organic plots had higher yields compared to the mineral ones (Table 2). Regarding the yields per area, the highest for all treatments was observed in the organic plot with lower crop density, while in the mineral plot, the highest one was observed at higher crop density (Table 3).

In the third production year, alongside the observed lower individual yield in the higher density of the mineral plot, the highest individual plant yield in the lower density of the mineral plot was comparable to the yield of the plants in the organic plot (Table 2), ultimately resulting in a maximum yield in the higher density of the organic plot (Table 3).

Unfortunately, in the fourth and fifth production years, disease caused a rapid decline in crop density (Figure 2). In the fourth year, the plants in the organic plots produced generally larger individual yields than those in mineral plots (Table 2). The average yields per area in the fertilization models differed, although not by much, as plant loss in the fourth year was likewise higher in the organic model (Figure 2) than it was in the conventional model. The yield per area in the organic plot, however, was much higher at a higher density than it was at a lower one (Table 3). Finally, in the fifth year, with crop density averaging 60% in mineral plots and less than 40% in organic plots compared to the initial ones (Figure 2), the highest yield was observed in the higher-density mineral plot (Table 3).

3.3. Content of NPK Nutrients in Yield and Its Removal from the Soil

The average content of N present in the aboveground plant at the time of harvest during the third, fourth, and fifth production years was in the range 1.32–1.78% (Figure 3). The content of N in the yield of *S. montana* decreased from the third to the fifth production years. The implemented planting densities (3.6 and 5 plants m^{-2}) had no effect on the level of N in the plants' biomass. Considering the applied fertilization models, the content of N did not differ in plants produced in organic and mineral plots in the third and fourth years, while in the fifth year, it was higher in plants treated with organic fertilizers than in those treated with mineral fertilizers. Nitrogen removal, by yield, ranged on average from 4.78 to 17.57 g m^{-1}, and it continuously decreased from the third to the fifth years (Figure 4). In general, the removal of N was not affected by the applied fertilizers or planting densities. Even though the N content was higher in plants in the organic plot in the fifth year, the plants in both fertilizer plots removed the same amount of N by the end of the fifth year due to the higher yield in the mineral plot.

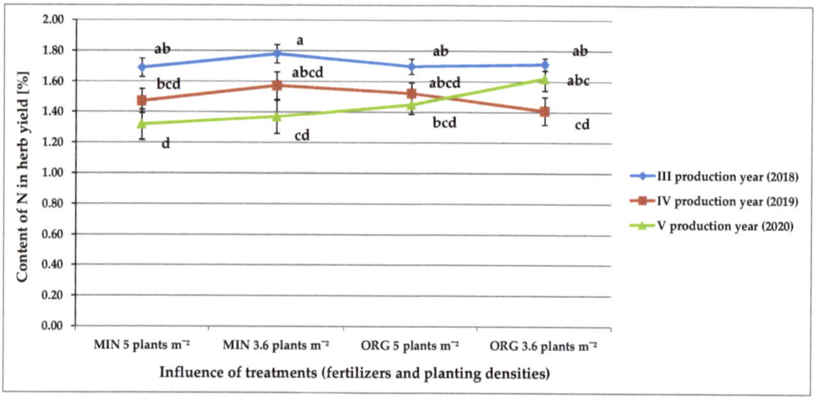

Figure 3. Nitrogen content (N%) in *S. montana* L yield in relation to fertilization and crop density in the third to fifth production years; MIN—mineral plot; ORG—organic plot; lower crop density (3.6 plants m^{-2}), higher crop density (5 plants m^{-2}); means ± standard error followed by the same letter in the same year are not significantly different ($p < 0.05$).

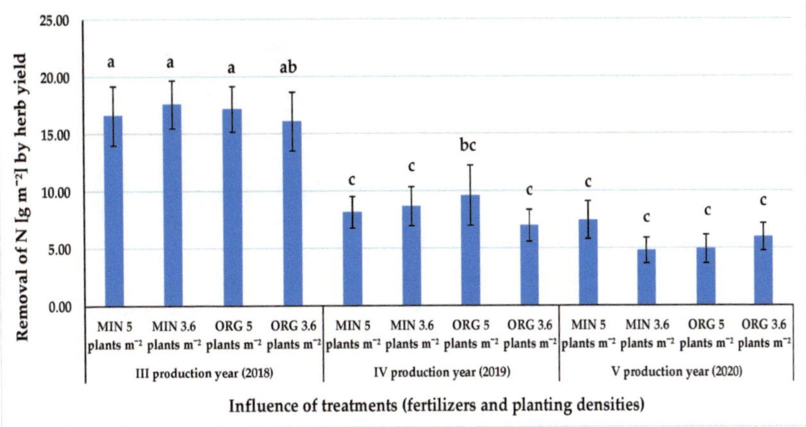

Figure 4. Nitrogen removal (N g m^{-1}), by yield, in relation to crop density and fertilization model in the third to fifth production years; MIN—mineral plot; ORG—organic plot; lower crop density (3.6 plants m^{-2}), higher crop density (5 plants m^{-2}); error bars denote standard error; means followed by the same letter in the same year are not significantly different ($p < 0.05$).

The average content of P in plants right before the harvest, from the third to the fifth production years, ranged from 0.30 to 0.47% and was not affected by fertilizers or planting densities (Figure 5). The average P removal by the plant yield ranged from 1.43 to 4.54 g m^{-1} and was not affected by treatments within the same production year (Figure 6). Though without significance, slightly lower P removal was noticed at MIN × 3.6 plants m^{-2} due to the lowest recorded P levels in plants. Considering K content in plants observed during the third, fourth, and fifth production years, it was not affected by treatments and ranged from 1.47 to 1.52% (Figure 7). As for K removal by plants, it ranged from 5.05 to 15.43 g m^{-1}, with K removal not depending on applied treatments in the third year, whereas in the fourth year, K removal was higher in the mineral plot and ORG × 5 plants m^{-2} with similar removal observed in MIN × 5 plants m^{-2}, in the fifth year. Considering that the content of K was uniform in all three years, the strongest influence on its removal was the decreasing yield from the third to the fifth years (Figure 8).

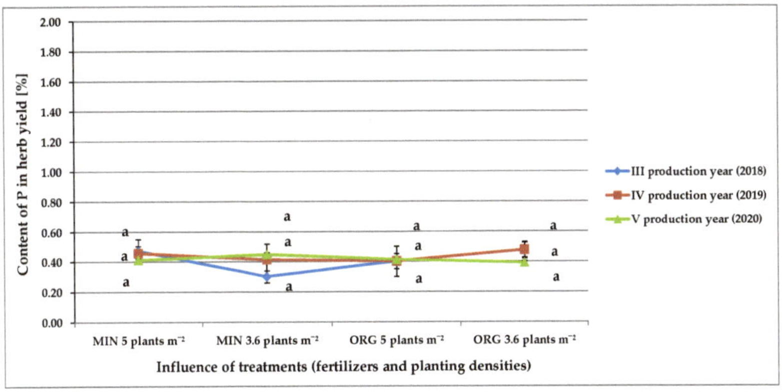

Figure 5. Phosphorus content (P%) in *S. montana* L. yield in relation to fertilization and crop density the third to fifth production years; MIN—mineral plot; ORG—organic plot; lower crop density (3.6 plants m^{-2}), higher crop density (5 plants m^{-2}); means ± standard error followed by the same letter in the same year are not significantly different ($p < 0.05$).

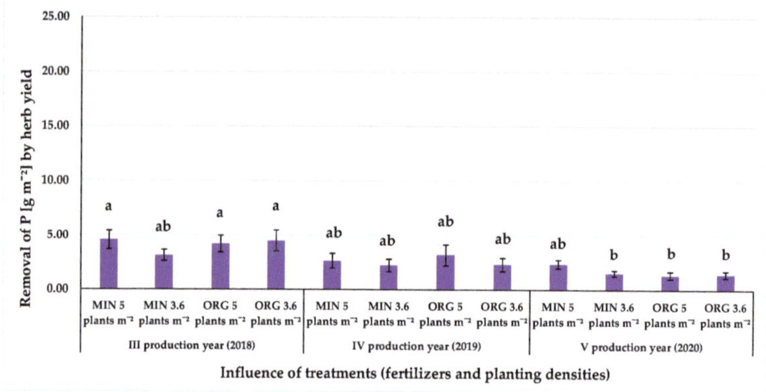

Figure 6. Phosphorus removal (P g m^{-1}), by yield, with regard to crop density and fertilization models in the third to fifth production years; MIN—mineral plot; ORG—organic plot; lower crop density (3.6 plants m^{-2}), higher crop density (5 plants m^{-2}); error bars denote standard error; means followed by the same letter in same year are not significantly different ($p < 0.05$).

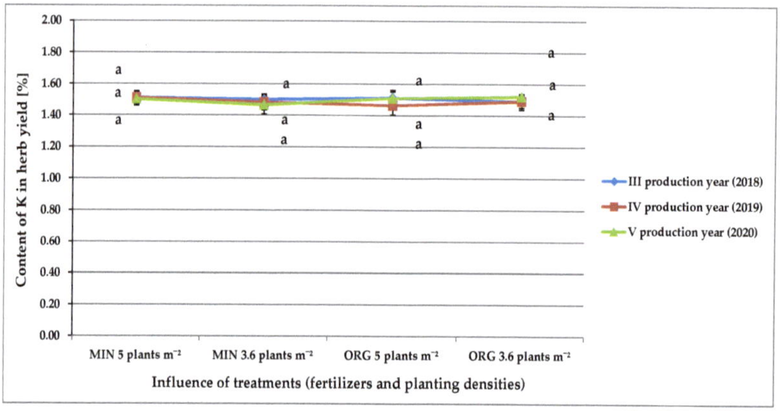

Figure 7. Potassium content (K%) in *S. montana* L. yield in relation to fertilization and crop density in the third to fifth production year; MIN—mineral plot; ORG—organic plot; lower crop density (3.6 plants m^{-2}), higher crop density (5 plants m^{-2}); means ± standard error followed by the same letter in the same year are not significantly different ($p < 0.05$).

Given that *S. montana* was cultivated under a mulch film, when calculating the removal of each macronutrient per unit of area, it was assumed that PPWF covered approximately 70% of the total experimental area (10.000 m^2), with the remaining area (30%) being the intermediate paths.

The average N removal, by yield, in the third, fourth, and fifth years in the mineral plots was 17.08, 8.41, and 6.11 g m^{-1}, respectively, while in organic plots, it was 16.62, 8.28, and 5.43 g m^{-1}, respectively, regardless of crop density. On 7000 m^2, in the third, fifth, and fifth years, it accounted for 120, 59, and 43 kg ha^{-1} in both fertilization plots, respectively.

The average P removal, by yield, in the third, fourth, and fifth years in the mineral plots was 3.83, 2.42, and 1.94 g m^{-1}, respectively, while in the organic plots, it was 4.33, 2.75, and 1.40 g m^{-1}, respectively, regardless of crop density. On 7000 m^2, it accounted for 27, 17, and 14 kg ha^{-1} in the mineral plots and 30, 19, and 10 kg ha^{-1} in the organic plots, respectively.

The average K removal, by yield, in the third, fourth, and fifth years in the mineral plots was 14.86, 8.34, and 6.82 g m^{-1}, respectively, while in the organic plots, it was 14.71, 8.29, and

5.33 g m^{-1}, respectively, regardless of crop density. On 7000 m^2, it accounted for 104, 58, and 48 kg ha^{-1} in the mineral plots and 103, 58, and 37 kg ha^{-1} in the organic plots, respectively.

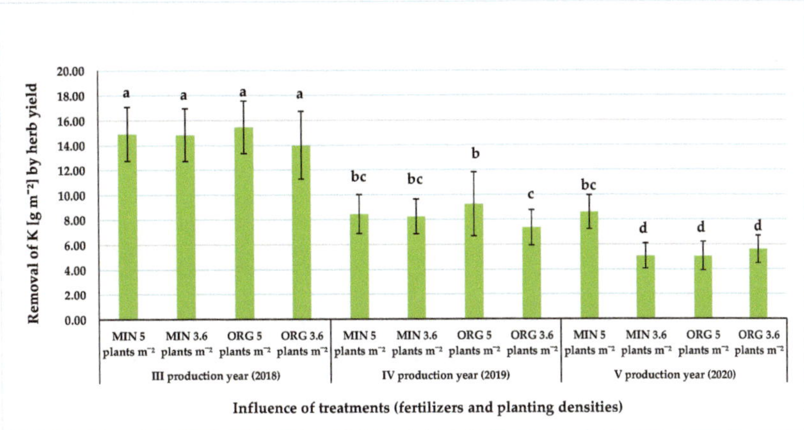

Figure 8. Potassium removal (K g m^{-1}), by yield, with regard to crop density and fertilization models in three experimental years; MIN—mineral plot; ORG—organic plot; lower crop density (3.6 plants m^{-2}), higher crop density (5 plants m^{-2}); error bars denote standard error; means followed by the same letter in same year are not significantly different ($p < 0.05$).

4. Discussion

4.1. Crop Density

In the first two years of production, the development of plants has proven satisfactory, with a denoted average loss of 11% by the end of the second production year in regard to the total number of planted plants, regardless of the applied treatments. Though the harvest in the first year was performed late, followed by a relatively cold winter without snow cover and very low temperatures (Figure 1), plants successfully overwintered and adapted to the temperate continental climate conditions of the South Banat. Higher-crop-density plants began to close inter-row spaces by the end of the second year, whereas lower-crop-density plants occupied only spaces within the rows [36].

The success of mulch in providing favorable growth conditions for plants depends on the appropriate selection of mulch material, soil physico-chemical properties, and the climatic circumstances in which the plant grows. In a previous study conducted with *Arnica montana* [43], black and silver polyethylene plastic (PE) mulch resulted in a higher number of flowers per plant, with a positive influence on rosette diameter and plant height compared with the same crop cultivated without mulching. A higher biomass yield was gained with PE mulch as it elevated the soil temperature and preserved soil moisture, thus favoring initial plant growth. On the other hand, due to high air temperatures and drought, a decrease in root yield has been reported in *Gentiana lutea* L. cultivated under black and silver PE mulch [44].

The positive effect of PPWF in our study was recognized with the securing of unhindered initial growth of plants by suppressing weeds in the first two years of plant production until the full crop closure in the third production year, as the summer period of the second production year was characterized by a lack of precipitation and slightly higher temperatures. Using PPWF in the cultivation of *S. montana* ensured successful weed management and water efficiency by relying on rainfall for water irrigation throughout the entire duration of crop production in dry farming conditions. Furthermore, allowing gas exchange with PPWF avoided the potentially negative effect of heated air accumulation, as seen in *Gentiana lutea* cultivated under PE mulch [44]. In an experiment conducted in the cultivation of *Mentha piperita* L., the use of PPWF proved to be the most effective in weed suppression as compared to many other synthetic mulches [45]. Even though mulch was no longer present from the third to the sixth production years, the weed control achieved by using water–air permeable yet biodegradable film was critical in resulting in high Gentiana root yield in the fifth and sixth production years [44].

Although many studies emphasized the advantages of using mulch in medicinal plant cultivation, a few studies also reported that mulches contribute to the development of soil-borne pathogens. In studies conducted with *Thymus vulgaris* L. [46,47], plastic mulch was not beneficial to the crop as it promoted the development of soil fungal diseases. Disease occurrence on *Nepeta cataria* and *Hypericum perforatum* cultivated under the wool mat was 31% and 46%, respectively, whereas under the oat straw, it was 25% and 11%, respectively [10]. In this study, the number of decayed *S. montana* plants increased as the crop aged. The first signs of plant decline related to the disease (3.2%) were observed by the end of the second year of production [36]. Afterwards, no significant additional loss was observed in the third production year compared to the previous one, as climate conditions favored plant development with satisfactory sums of temperatures and precipitation, resulting in a yield that reached its maximum. Crop density was equally preserved in both fertilization models, with an average of 86%, regardless of applied densities. Nevertheless, in the fourth and fifth production years, severe loss was observed due to disease. It was noted that segments of branches dry up until the whole plant declines. Symptoms were asymmetrically or randomly distributed in individual plants as well as across a field, particularly in the advanced stages of infection. Noticeable symptoms resembled those caused by some soil-borne pathogens, *Verticillium* spp., and/or *Fusarium* spp. [40,41]. Literature data on the diseases and pests of *Satureja* species scarcely exist. However, wilt, either caused by *Verticillium* spp. or *Fusarium* spp., is a disease that may occur during the cultivation of species of the genus *Satureja* [48,49]. For many crops, economic losses due to soil-borne pathogens are estimated to be 50–75% of the attainable yield [50]. To date, no effective fungicides for the permanent suppression of *Verticillium* spp. have been made. The most effective method for controlling diseases caused by soil-borne pathogens has proven to be soil fumigation with chloropicrin and methyl bromide. Nevertheless, due to the potential risks to human health and the environment, many countries have banned the use of these compounds [51]. On the other hand, medicinal plants belong to the group of "minor crops", and a major problem in their cultivation is a lack of registered and available protection preparations in the market. The use of unapproved preparations may result in a law violation as well as a risk to consumer health due to pesticide residues found in herbal material [52]. With all of this in mind, it is easy to understand why the pathogens in our study were so difficult to control once they appeared. Additionally, since the study on *S. montana* was set up under PPWF and full crop density closure occurred in the third production year, soil fumigation was impossible to implement.

Soil-borne pathogens are more dangerous and infectious in conditions with higher soil moisture and lower soil temperature [49]. Similar findings were confirmed while investigating the pathogenicity of *Phytophthora clandestine* and *Pythium irregulare* in *Trifolium subterraneum*, which cause root infection [53,54]. Furthermore, it has been demonstrated that a higher frequency of extreme rain events can saturate the soil, favoring conditions for the development of soil-borne pathogens [55].

In this regard, a significant difference in climatic conditions was observed during the experimental period 2016–2020, which could be explained by the calculated five-year averages of precipitation and temperature shown in Figure 1. In August 2016, an extremely rainy period was recorded, with precipitation being 90% higher than the calculated five-year average, though crop production was not negatively affected. Although a dry summer was observed in 2017, with precipitations being 80% and 30% lower in June and July, respectively, than the calculated five-year average, moisture was conserved under the mulch, which favored crop development. In 2018, except for the low precipitation observed in May, the sums of precipitation and temperatures corresponded to the averages. Nevertheless, an extremely rainy period with lower temperatures was observed in 2019 and 2020 when compared to the calculated five-year averages of precipitation and temperature. Thus, the precipitations in May, June, and July of 2019 were 40%, 20%, and 30% higher, while those in June and July of 2020 were 60% and 20% higher than the calculated five-year average, respectively. Additionally, the temperatures in the indicated periods in 2019 and 2020 were, on average, 1.5 °C and 2.2 °C lower than those of the calculated five-year average, respectively.

There is an assumption that a combination of factors such as PPWF being applied to the soil with high clay content (>42%) and the occurrence of extreme rainfalls contributed to creating favorable conditions (high soil moisture and low soil temperature) for the development of soil-borne pathogens and, thus, the severe plant loss in the fourth and fifth vegetation seasons. Considering the progression of climate change in the future and the appearance of extreme weather conditions, particularly high rainfalls, there is a possibility of reestablishing conditions in the soil that might promote the development of soil-borne pathogens again. In that regard, this should be taken into account when mulching heavy soils.

4.2. Yield

In the first year of production, plants developed uniformly in both fertilized plots, were supplied with enough available nutrients, and were grown in favorable growing conditions, with the applied densities providing them sufficient space for development. PPWF prevented weeds from competing for nutrients and light in the soil and promoted early vegetative growth of the cultivated crops, as observed when investigating the efficiency of two different polypropylene black mulches on the growth of *Rosmarinus officinalis* L., *Lavandula officinalis* L., and *Thymus vulgaris* L. [56].

In the second production year, nutrient availability affected individual plant development in such a way that *S. montana* plants in the lower-density plots were more supplied than those in higher-density plots, as they competed for nutrients in an equal-sized area. It resulted in larger plants for plots with lower densities, occupying an area larger than 0.2 m^2, which was also the amount of space available to the plants in denser formations. In terms of the effect of planting densities, investigations on *S. montana* L., *Satureja sahendica* Bornm, and *Satureja khuzestanica* Jamzad [23,57,58] revealed that plants grew larger in diameter with wider spacing, allowing more branching and lateral growth of the shoots, whereas dense cultivation resulted in higher yields per area. Other studies [59–64] discovered that narrower spacing resulted in higher yields per unit area.

In the second production year, though a 15% higher yield in the organic plot was noted, there was no difference in yield per area for higher densities in either fertilizer plots. Even if the difference was not statistically confirmed, the higher yield in the denser organic plot suggested a trend of increasing yields per area in subsequent years. Similar to our results, in the second production year of *Satureja mutica* cultivated under dry farming conditions [65], treatments involving cattle manure and the highest applied planting density (80,000 plants ha^{-1}) gave the highest dry yield compared to the plot with lower planting densities (40,000 and 26,666 plants ha^{-1}) without any fertilizer. When assuming that mineralization intensifies under mulch due to conserved moisture and optimal temperature for microbial activity, a greater influence of organic fertilization on plant growth over mineral fertilization is observed in the second and third years after fertilizer application [65,66]. Finally, after the plots were closed and full crop density was reached in the third production year, the highest yield (on average, 1016 g m^{-1}) was detected in the denser organic plot compared to the yields in the mineral plots with higher (in average 984 g m^{-1}) and lower (in average 987 g m^{-1}) densities. When calculated on a 7000 m^2 area under the crop (mulch area), it accounted for average yields of 7114 kg ha^{-1} in the higher-density organic plot and 6887 kg ha^{-1} and 6906 kg ha^{-1} for the higher- and lower-density mineral plots, respectively. Thus, it leads to the conclusion that a satisfactory level of nutrition in winter savory for maximum yield could be provided with the application of organic fertilizers. There has been a similar study on fertilizer application reported on another species of the *Satureja* genus. Thus, testing the effects of cow manure (30 t ha^{-1}) and an NPK fertilizer (50:25:25 kg ha^{-1}) on yield in a two-year production of *Satureja macrantha* C.A.Mey [67] showed that plants treated only with cow manure and with a combination NPK and cow manure gave, on average, 22% and 18%, respectively, higher yield compared to the control (2000 kg ha^{-1} yield).

Severe yield loss was observed in the fourth and fifth production years, in which the drawbacks of the applied mulch on clay soil in circumstances of high rainfall and low soil temperature were seen. It was assumed that mulch, during months with a higher precipitation rate, retained excess soil moisture and, with poor drainage in clay soil, contributed to increased soil-borne pathogen development. The potential drawbacks of the applied mulch were described in some studies. Hence, the cultivated *Origanum majorana* L. [68] did not benefit from mulching, since higher yields were gained without mulching (96 kg 100 m^{-2}) than with mulching (80 kg 100 m^{-2}), as plants under the mulch were severely affected by fungal diseases (*Alternaria* sp. and *Botrytis cinerea*). Similarly, it was revealed that mulching was not beneficial for *Thymus vulgaris* L. production [47], because the occurrence of high rainfall and used drip irrigation under PE mulch was attributed to a high incidence of fungal diseases, resulting in the lowest yield (332 kg ha^{-1}) compared to the highest yield (1604 kg ha^{-1}) gained with irrigation and without mulch.

4.3. Content of NPK Nutrients in Herb and Removal, by Yield

The content of the three major plant nutrients, nitrogen (N), phosphorus (P), and potassium (K), in the aboveground plant biomass and their removal from the soil are considered indicators of whether the cultivated plants were properly supplied with the implemented fertilization models in each of the tested planting densities.

Nitrogen (N) is a constituent of a number of organic compounds, such as amino acids, proteins, nucleic acids, and secondary metabolites. Depending on the plant species, organs, and age, the content in plant tissues is in the range of 1–5% (10–50 g kg^{-1}) of total plant dry matter [69]. In our study, observed in dry aerial plant parts, the highest N content, by average, was 1.72%, noted in the third production year, and the average N content decreased by 13% and 16% in the fourth and fifth production years, respectively. The highest average N content of 4% was observed in the first production year of *S. macrantha* L. [70], which was higher than the highest average N content (1.8%) obtained in the treatment MIN $\times$ 3.6 plants m^{-2} in the third year of our trial. During a two-year production period, a similar N content (3.6%) was discovered in the herb of *S. hortensis* L. [[71]. The lower N content present in the plants from our trial when compared to the reported ones could be explained by the difference in the plants' ages, since a higher N content is usually present in younger plants. After the third production year and with the crop aging, the proportion of woody stems in the structure of the aerial part of *S. montana* becomes higher.

The achieved plant yield per area being equal in both the mineral and organic plots in the third production year indicates favorable conditions for the N mineralization of the applied organic fertilizer due to optimal moisture and temperature conditions for microbial activity under mulch. In contrast, in the fourth and fifth years, it was assumed that excess moisture conserved under mulch on heavy clay soil during rainy seasons resulted in conditions conducive to reduced N uptake under anaerobic conditions. In that regard, it was confirmed that clay soils impeded N uptake due to higher saturation levels [72]. Nitrogen-deficient plants are often stunted with narrow leaves. Chlorosis, caused by a nitrogen deficiency, commonly develops in older leaves as N is remobilized to younger leaves. In the field, N-deficient crops appear light green or even yellow [69]. *S. montana* plants in this study were properly supplied with N throughout the study, regardless of treatment, with no signs of N deficiency.

Phosphorus (P) is a structural constituent of nucleic acids, as well as a component of the ATP/ADP system in plant cells. It is also required for carbohydrates to be transported in leaf cells. The most common phosphorus needs for optimal plant growth during the growing season are in the range of 0.3–0.5% (3–5 mg g^{-1}) of total plant dry matter [69]. In our study, the content of P in the aerial part of S. *montana* was in that exact interval

(Figure 4). In the third year, the highest average P content (0.5%) in *S. montana* plants was achieved in treatment MIN x 5 plants m^{-2}. Other fertilization experiments revealed a lower P content in various *Satureja* species. For instance, in the two-year production of *S. mutica* Fisch. & C.A. Mey [73], the highest average P content was 0.24%, while a P content of 0.3% was found in *S. macrantha* C.A.Mey [70], which corresponds to the lowest P content recorded in the treatment MIN × 3.6 plants m^{-2} in the third year. In a two-year production of *S. macrantha* C.A.Mey [67], 0.33% of P was reported in the first year in which the applied mineral fertilizer was reported and 0.38% of P was reported for the combined organic and mineral fertilizer in the second year. When compared to the results in our study, a higher content was determined with the applied mineral fertilizer (0.47% in MIN × 5 plants m^{-2} in the third year), whereas the average P content in our study, with regard to the applied fertilizers, was 0.42%, which coincides with the P content in *S. macrantha* C.A.Mey [67] in the second production year.

Plants appear stunted in P deficiency; there is a reduction in leaf number and expansion; the foliage turns dark green or, in severe cases, reddish-purple; the shoot growth rate is inhibited; and the formation of reproductive organs is inhibited. Flower initiation is delayed; the number of flowers decreases; and seed formation is restricted [69]. Plant species that grow on soils very poor in phosphorus are adapted to lower concentrations of P [74]. In our study, a P content of 0.42%, present in the analyzed plant material taken in the third, fourth, and fifth years, is typical of plant P contents on a dry mass basis, indicating that the plants received adequate P supplementation throughout the trial period and that there were no signs of P deficiency.

Potassium (K) is required by plants in approximately the same amounts as nitrogen (N). The main role of K is osmoregulation, which is important for cell extension and stomata movement. Potassium affects the rate of mass flow-driven solute movement within the plant. It is required for enzyme activation, protein synthesis, photosynthesis, carbohydrate transport, and water regulation in plants. The K requirement for optimal plant growth is 2–5% (20–50 g kg^{-1}) of total plant dry matter [69]. K was present in plants at an average of 1.5% in the analyzed plant material taken in the third, fourth, and fifth years. Higher contents of K were observed in other species of the *Satureja* genus. For example, studies in a two-year cultivation of *S. macrantha* [67,70] reported highest average K contents of 4.3% and 3.5%, respectively, whereas in a two-year production of *S. mutica*, the highest average K content was 3.3% [73].

The uptake of K occurs in the K+ form. As potassium forms no organic compounds within the plant, it remains in the ionic K+ form. In the soil, K is much less mobile than N but more mobile than P [69]. In clay soils, K ions can be fixed in the interlamellar spaces of clay minerals, which reduces their availability to plants [75]. Although the trial was set up on soil with high clay content, according to the soil analysis, the amount of K present in the soil was at optimal levels (Table 1). When K levels are low, plants become stunted and develop poor root systems; K transports from mature leaves to younger leaves. Mature leaves become chlorotic and necrotic, which results in leaf shedding as the deficiency continues. Furthermore, it causes other plant disorders such as decreased turgor; decreased cambium growth rate; and disruption of lignification of vascular bundles, which leads to increased lodging in some plants [69]. Despite this, none of these symptoms were observed on *S. montana* plants throughout the years of cultivation. In addition, although a higher K content was found in other *Satureja* species, in our study, the level of K in the aboveground part at the time of harvest was found to be at a uniform level of approximately 1.5%. That level did not depend on the applied treatments or the climatic differences of the growing seasons (Figure 1). It could be concluded that the plants were sufficiently supplied with K+ during the entire growing period, and it was assumed that the content of 1.5% K+ in the aboveground part is characteristic for *S. montana* in the period of full crop development.

The average N, P, and K removal in the third year coincided with the removal of nutrients by the *Satureja* species dry aerial plant yield reported by Bomme and Nast [76], who also recommended proper dosages of applied fertilizers comparable to those used in this trial, though slightly modified by taking into account soil analysis data. Even though nutrient removal, by yield, decreased dramatically in the fourth and fifth years due to disease-affected yield, it suggested that an appropriate single dose of fertilizer was applied in both the organic and mineral models.

5. Conclusions

According to the obtained results, wider spacing of *S. montana* resulted in higher yields per plant, whereas narrower spacing resulted in higher yields per unit area. In the third production year, the yield achieved with the application of organic fertilizer during dense cultivation was comparable with the yield obtained in the mineral plot, and that year produced the highest yields. According to the results on macronutrient content, the *Satureja* plants were properly supplied with N, P, and K during the entire duration of the trial, regardless of the type of applied fertilizer, and without the manifestation of any deficiencies, whereas an equal level of NPK nutrients removed, by yield, in both tested fertilization plots suggested that a single basal fertilization plus a single additional surface fertilization in spring of the second year were applied in an appropriate amount in both fertilization plots.

Based on these findings, it is possible to conclude that, in situations where surface fertilizer application is no longer feasible after mulching, the use of organic fertilizer at high plant density, in an appropriate single dose, could provide satisfactory nutrition levels for a cultivated crop comparable to mineral-fertilizer nutrition levels. This might be encouraging for further organic agriculture applications in medicinal crop production.

In regard to *S. montana* cultivation on heavy clay soil, PPWF allowed rainwater infiltration and was effective in terms of water-use efficiency and weed suppression. Nevertheless, the occurrence of extreme rainfalls resulted in severe yield loss due to excess moisture preserved under the mulch, creating conditions favorable for the development of soil-borne pathogens. Since extreme rainfalls are anticipated to reappear in the future as a result of climate change, the cultivation of winter savory could be primarily constrained by heavy soil texture. Other mulch materials, such as organic materials for *S. montana* cultivation on heavy clay soil, should be further explored.

Author Contributions: Conceptualization, D.R. and T.M.; methodology, D.R., T.M. and A.D.; software, A.D. and S.M.; validation D.R., T.M. and S.M.; formal analysis, D.R., T.P., A.D., V.F. and S.M.; investigation, D.R., S.M. and T.M.; resources, D.R., T.M., A.D., Ž.P., S.M. and V.F.; data curation, D.R., A.D. and S.M.; writing—original draft preparation, S.M. and D.R.; writing—review and editing, T.M.; visualization, Ž.P. and D.R.; supervision, M.L.; project administration, D.R.; funding acquisition, M.L., D.R. and T.P. All authors have read and agreed to the published version of the manuscript.

Funding: This research was funded by the Ministry of Education, Science and Technological Development of the Republic of Serbia, grant: 451-03-68/2022-14/200003.

Data Availability Statement: Not applicable.

Conflicts of Interest: The authors declare no conflict of interest.

References

1. Rzepa, J.; Sajewicz, M.; Baj, T.; Gorczyca, P.; Włodarek, M.; Głowniak, K.; Waksmundzka-Hajnos, M.; Kowalska, T. A Comparison of Methodical Approaches to Fingerprinting of the Volatile Fraction from Winter Savory (*Satureja montana*). *Chromatogr. Res. Int.* **2012**, *2012*, 596807. [CrossRef]
2. Hudz, N.; Makowicz, E.; Shanaida, M.; Białoń, M.; Jasicka-Misiak, I.; Yezerska, O.; Svydenko, L.; Wieczorek, P.P. Phytochemical Evaluation of Tinctures and Essential Oil Obtained from *Satureja montana* Herb. *Molecules* **2020**, *25*, 4763. [CrossRef] [PubMed]
3. Čopra-Janićijević, A.; Vidic, D.; Maksimović, M. Characterisation of *Satureja montana* L. Essential Oil and Headspace Volatiles. *Nat. Volatiles Essent. Oils* **2020**, *7*, 22–34. [CrossRef]
4. Ćetković, G.; Čanadanović-Brunet, J.; Djilas, S.; Tumbas, V.; Markov, S.; Cvetković, D. Antioxidant Potential, Lipid Peroxidation Inhibition and Antimicrobial Activities of *Satureja montana* L. subsp. kitaibelii Extracts. *Int. J. Mol. Sci.* **2007**, *8*, 1013–1027. [CrossRef]

5. Aćimović, M.; Šovljanski, O.; Pezo, L.; Travičić, V.; Tomić, A.; Zheljazkov, V.D.; Ćetković, G.; Švarc-Gajić, J.; Brezo-Borjan, T.; Sofrenić, I. Variability in Biological Activities of *Satureja montana* subsp. *Montana* and Subsp. *Variegata* Based on Different Extraction Methods. *Antibiotics* **2022**, *11*, 1235. [CrossRef]
6. Đorđević, N.; Karabegović, I.; Cvetković, D.; Šojić, B.; Savić, D.; Danilović, B. Assessment of Chitosan Coating Enriched with Free and Nanoencapsulated *Satureja montana* L. Essential Oil as a Novel Tool for Beef Preservation. *Foods* **2022**, *11*, 2733. [CrossRef]
7. Wesołowska, A.; Grzeszczuk, M.; Jadczak, D. Influence of Harvest Term on the Content of Carvacrol, p-Cymene, Î³-Terpinene and Î²-Caryophyllene in the Essential Oil of *Satureja montana*. *Not. Bot. Horti Agrobot. Cluj-Napoca* **2014**, *42*, 392–397. [CrossRef]
8. Maccelli, A.; Vitanza, L.; Imbriano, A.; Fraschetti, C.; Filippi, A.; Goldoni, P.; Maurizi, L.; Ammendolia, M.G.; Crestoni, M.E.; Fornarini, S. *Satureja montana* L. Essential Oils: Chemical Profiles/Phytochemical Screening, Antimicrobial Activity and O/W NanoEmulsion Formulations. *Pharmaceutics* **2019**, *12*, 7. [CrossRef]
9. Korav, S.; Dhaka, A.K.; Singh, R.; Premaradhya, N.; Reddy, G.C. A Study on Crop Weed Competition in Field Crops. *J. Pharmacogn. Phytochem.* **2018**, *7*, 3235–3240.
10. Duppong, L.M.; Delate, K.; Liebman, M.; Horton, R.; Romero, F.; Kraus, G.; Petrich, J.; Chowdbury, P.K. The Effect of Natural Mulches on Crop Performance, Weed Suppression and Biochemical Constituents of Catnip and St. John's Wort. *Crop Sci.* **2004**, *44*, 861–869. [CrossRef]
11. Patil Shirish, S.; Kelkar Tushar, S.; Bhalerao Satish, A. Mulching: A Soil and Water Conservation Practice. *Res. J. Agric. For. Sci.* **2013**, *1*, 26–29.
12. Matković, A.; Božić, D.; Filipović, V.; Radanović, D.; Vrbničanin, S.; Marković, T. Mulching as a Physical Weed Control Method Applicable in Medicinal Plants Cultivations. *Lek. Sirovine* **2015**, *35*, 37–51. [CrossRef]
13. Pavlović, D.; Vrbničanin, S.; Anđelković, A.; Božić, D.; Rajković, M.; Malidža, G. Non-Chemical Weed Control for Plant Health and Environment: Ecological Integrated Weed Management (EIWM). *Agronomy* **2022**, *12*, 1091. [CrossRef]
14. Li, Q.; Liu, J.; Zhang, L.; Qin, C.; Yu, J.; Surya, N.A. Using Water-Retaining Agent and Mulch to Improve Growth and Yield of Potato under Dry Farming. *Trans. Chin. Soc. Agric. Eng.* **2013**, *29*, 83–90.
15. Yang, F.; Cen, R.; Feng, W.; Liu, J.; Qu, Z.; Miao, Q. Effects of Super-Absorbent Polymer on Soil Remediation and Crop Growth in Arid and Semi-Arid Areas. *Sustainability* **2020**, *12*, 7825. [CrossRef]
16. Lalljee, B. Mulching as a Mitigation Agricultural Technology against Land Degradation in the Wake of Climate Change. *Int. Soil Water Conserv. Res.* **2013**, *1*, 68–74. [CrossRef]
17. Iriany, A.; Chanan, M.; Djoyowasito, G. Organic Mulch Sheet Formulation as an Effort to Help Plants Adapt to Climate Change. *Int. J. Recycl. Org. Waste Agric.* **2018**, *7*, 41–47. [CrossRef]
18. George, S.; Idicula, S.P.; Joseph, K. Polypropylene Woven Fabric: A Good Mulch Material for Young Rubber Plants. *Short Sci. Rep.* **2015**, *43*, 171–177. [CrossRef]
19. Nedunchezhiyan, M.; Sahoo, B.; Pati, K.; Chauhan, V.B.S.; Bansode, V.; Kumar, J.S.; Tripathy, S.; Sahoo, K.; Sunita, K.; Toppo, M. Polypropylene Fabric Ground Cover Effects on Weed Control and Profit in Elephant Foot Yam Cultivation. *Int. J. Curr. Microbiol. App. Sci* **2020**, *9*, 1100–1111. [CrossRef]
20. Prem, M.; Ranjan, P.; Seth, N.; Patle, G.T. Mulching Techniques to Conserve the Soil Water and Advance the Crop Production—A Review. *Curr. World Env.* **2020**, *15*, 10–30. [CrossRef]
21. Kader, M.A.; Senge, M.; Mojid, M.A.; Ito, K. Recent Advances in Mulching Materials and Methods for Modifying Soil Environment. *Soil Tillage Res.* **2017**, *168*, 155–166. [CrossRef]
22. El-Beltagi, H.S.; Basit, A.; Mohamed, H.I.; Ali, I.; Ullah, S.; Kamel, E.A.R.; Shalaby, T.A.; Ramadan, K.M.A.; Alkhateeb, A.A.; Ghazzawy, H.S. Mulching as a Sustainable Water and Soil Saving Practice in Agriculture: A Review. *Agronomy* **2022**, *12*, 1881. [CrossRef]
23. Zawiślak, G.; Nurzyńska-Wierdak, R. Variation in Winter Savory (*Satureja montana* L.) Yield and Essential Oil Production as Affected by Different Plant Density and Number of Harvests. *Acta Sci. Pol. Hortorum Cultus* **2017**, *16*, 159–168. [CrossRef]
24. Unger, P.W.; Jones, O.R. Effect of Soil Water Content and a Growing Season Straw Mulch on Grain Sorghum. *Soil Sci. Soc. Am. J.* **1981**, *45*, 129–134. [CrossRef]
25. Mupangwa, W.; Twomlow, S.; Walker, S.; Hove, L. Effect of Minimum Tillage and Mulching on Maize (*Zea mays* L.) Yield and Water Content of Clayey and Sandy Soils. *Phys. Chem. Earth Parts A/B/C* **2007**, *32*, 1127–1134. [CrossRef]
26. Paul, P.L.C.; Bell, R.W.; Barrett-Lennard, E.G.; Kabir, E. Impact of Rice Straw Mulch on Soil Physical Properties, Sunflower Root Distribution and Yield in a Salt-Affected Clay-Textured Soil. *Agriculture* **2021**, *11*, 264. [CrossRef]
27. Palada, M.C.; Davis, A.M.; Kowalski, J.A.; Crossman, S.M.A. Yield and Irrigation Water Use of Vegetables Grown with Plastic and Straw Mulch in the US Virgin Islands. *Int. Water Irrig.* **2003**, *23*, 1–9.
28. Filipović, V.; Romić, D.; Romić, M.; Borošić, J.; Filipović, L.; Mallmann, F.J.K.; Robinson, D.A. Plastic Mulch and Nitrogen Fertigation in Growing Vegetables Modify Soil Temperature, Water and Nitrate Dynamics: Experimental Results and a Modeling Study. *Agric. Water Manag.* **2016**, *176*, 100–110. [CrossRef]
29. Wang, F.; Liu, Y.; Liang, B.; Liu, J.; Zong, H.; Guo, X.; Wang, X.; Song, N. Variations in Soil Aggregate Distribution and Associated Organic Carbon and Nitrogen Fractions in Long-Term Continuous Vegetable Rotation Soil by Nitrogen Fertilization and Plastic Film Mulching. *Sci. Total Environ.* **2022**, *835*, 155420. [CrossRef]
30. Tuong, T.P.; Cabangon, R.J.; Wopereis, M.C.S. Quantifying Flow Processes during Land Soaking of Cracked Rice Soils. *Soil Sci. Soc. Am. J.* **1996**, *60*, 872–879. [CrossRef]

31. Mokhtari, M.; Dehghani, M. Swell-Shrink Behavior of Expansive Soils, Damage and Control. *Electron. J. Geotech. Eng.* **2012**, *17*, 2673–2682.
32. Boekel, P. The Effect of Organic Matter on the Structure of Clay Soils. *Neth. J. Agric. Sci.* **1963**, *11*, 250–263. [CrossRef]
33. Alekseev, V.V.; Aleksandrov, R.I.; Vasiliev, S.A.; Chuchkalov, S.I. Hydrophysical Aspects of Soil Assessment in Melioration. In *IOP Conference Series: Earth and Environmental Science*; IOP Publishing: Bristol, UK, 2019; Volume 341, p. 12106. [CrossRef]
34. Mulumba, L.N.; Lal, R. Mulching Effects on Selected Soil Physical Properties. *Soil Tillage Res.* **2008**, *98*, 106–111. [CrossRef]
35. Mengel, K.; Kirkby, E.A. The Soil as a Plant Nutrient Medium. In *Principles of Plant Nutrition*; Mengel, K., Kirkby, E.A., Kosegarten, H., Appel, T., Eds.; Springer Science & Business Media: Dordrecht, The Netherlands, 2012; pp. 15–96.
36. Radanović, D.; Matković, A.; Đurović-Pejčev, R.; Marković, T.; Filipović, V.; Mrđan, S.; Vasin, J. Preliminary Results of Winter Savory (*Satureja montana* L.) Cultivated under Permeable Mulch Film in Dry Farming Conditions of South Banat. *Lek. Sirovine* **2018**, *38*, 51–57. [CrossRef]
37. Gavrilov, M.B.; Marković, S.B.; Schaetzl, R.J.; Tošić, I.; Zeeden, C.; Obreht, I.; Sipos, G.; Ruman, A.; Putniković, S.; Emunds, K. Prevailing Surface Winds in Northern Serbia in the Recent and Past Time Periods; Modern-and Past Dust Deposition. *Aeolian Res.* **2018**, *31*, 117–129. [CrossRef]
38. Malinović-Milićević, S.; Mihailović, D.T.; Radovanović, M.M.; Drešković, N. Extreme Precipitation Indices in Vojvodina Region (Serbia). *J. Geogr. Inst. Jovan Cvijić SASA* **2018**, *68*, 1–15. [CrossRef]
39. Tošić, I.; Hrnjak, I.; Gavrilov, M.B.; Unkašević, M.; Marković, S.B.; Lukić, T. Annual and Seasonal Variability of Precipitation in Vojvodina, Serbia. *Theor. Appl. Climatol.* **2014**, *117*, 331–341. [CrossRef]
40. Peaden, R.N. *A Guide for Identification of Verticillium Wilt in Alfalfa*; US Department of Agriculture, Agricultural Research Service: Beltsville, MD, USA, 1984.
41. Nelson, P.E.; Toussoun, T.A.; Marasas, W.F.O. *Fusarium Species: An Illustrated Manual for Identification*; Pennsylvania State University Press: University Park, PA, USA, 1983.
42. Jones, J.B. *Laboratory Guide for Conducting Soil Tests and Plant Analysis*; CRC Press: Boca Raton, FL, USA, 2001.
43. Radanović, D.; Pljevljakušić, D.; Marković, T.; Ristić, M.T. Influence of Fertilization Model and PE Mulch on Yield and Quality of Arnica (*A. montana*) at Dystric Cambisol. *Zemljište I Biljka* **2007**, *56*, 85–95.
44. Radanović, D.; Marković, T.; Vasin, J.; Banjac, D. The Efficiency of Using Different Mulch Films in the Cultivation of Yellow Gentian (*Gentiana lutea* L.) in Serbia. *Ratar. I Povrt./Field Veg. Crops Res.* **2016**, *53*, 30–37. [CrossRef]
45. Matković, A.; Marković, T.; Filipović, V.; Radanović, D.; Vrbničanin, S.; Božić, D. Preliminary Investigation on Efficiency of Muches and Other Mechanical Weeding Methods Applied in *Mentha piperita* L.: Cultivation. *Lek. Sirovine* **2016**, *36*, 61–74. [CrossRef]
46. Collingwood, C.D.; Crossman, S.M.A.; Navarro, A.A. Response of Selected Herbs to Improved Production Practices. In Proceedings of the Caribbean Food Crops Society, 27th Annual Meeting, Roseau, Dominica, 14–20 July 1991.
47. Palada, M.C.; Crossman, S.M.A.; Collingwood, C.D. Irrigation Water Use and Yield of Thyme in the Virgin Islands. In Proceedings of the Caribbean Food Crops Society, 29th Annual Meeting, Martinique, France, 1 July 1993. [CrossRef]
48. Hammer, M.; Junghanns, W. Bohnenkraut (*Satureja hortensis* L.) Und Winter-Bohnenkraut (*S. montana* L.). In *Handbuch des Arznei- und Gewürzpflanzenbaus*; Verein für Arznei- und Gewürzpflanzen Saluplanta e.V: Bernburg, Deitschland, 2012; pp. 265–276.
49. Zimowska, B. Diversity of Fungi Occurring on Savory (*Satureja hortensis* L.). *Herba Pol.* **2010**, *56*, 29–37.
50. Lewis, J.A.; Papavizas, G.C. Biocontrol of Plant Diseases: The Approach for Tomorrow. *Crop Prot.* **1991**, *10*, 95–105. [CrossRef]
51. Falcón-Piñeiro, A.; Remesal, E.; Noguera, M.; Ariza, J.J.; Guillamón, E.; Baños, A.; Navas-Cortes, J.A. Antifungal Activity of Propyl-Propane-Thiosulfinate (PTS) and Propyl-Propane-Thiosulfonate (PTSO) from Allium Cepa against *Verticillium Dahliae*: *In Vitro* and in Planta Assays. *J. Fungi* **2021**, *7*, 736. [CrossRef] [PubMed]
52. Szpyrka, E.; Słowik-Borowiec, M. Consumer Health Risk to Pesticide Residues in *Salvia Officinalis* L. and Its Infusions. *J. Environ. Sci. Health Part B* **2019**, *54*, 14–19. [CrossRef] [PubMed]
53. Wong, D.H.; Sivasithamparam, K.; Barbetti, M.J. Influence of Soil Temperature, Moisture and Other Fungal Root Pathogens on Pathogenicity of Phytophthora Clandestina to Subterranean Clover. *Trans. Br. Mycol. Soc.* **1986**, *86*, 479–482. [CrossRef]
54. You, M.P.; Rensing, K.; Renton, M.; Barbetti, M.J. Modeling Effects of Temperature, Soil, Moisture, Nutrition and Variety as Determinants of Severity of Pythium Damping-off and Root Disease in Subterranean Clover. *Front. Microbiol.* **2017**, *8*, 2223. [CrossRef]
55. Homet, P.; González, M.; Matías, L.; Godoy, O.; Pérez-Ramos, I.M.; García, L.V.; Gómez-Aparicio, L. Exploring Interactive Effects of Climate Change and Exotic Pathogens on Quercus Suber Performance: Damage Caused by Phytophthora Cinnamomi Varies across Contrasting Scenarios of Soil Moisture. *Agric. For. Meteorol.* **2019**, *276*, 107605. [CrossRef]
56. Fontana, E.; Hoeberechts, J.; Nicola, S. Effect of Mulching on Medicinal and Aromatic Plants in Organic Farm Guest Houses. In Proceedings of the I International Symposium on the Labiatae: Advances in Production, Biotechnology and Utilisation, Acta Horticulturae, Sanremo, Italy, 30 November 2006; 723, pp. 405–410. [CrossRef]
57. Abbaszadeh, B.; Sefidkon, F.; Layegh Haghighi, M.; Karegar Hajiabadi, E. The Effect of Planting Time and Planting Density on Yield and Essential Oil of Satureja Sahendica Bornm. *J. Med. Plants By-Prod.* **2014**, *3*, 141–146.
58. Hadian, J.; Hekmati, M.; Ghorbanpour, M. Agromorphological Variations and Essential Oil Production of *Satureja Khuzestanica* Jamzad under Different Planting Densities. *J. Essent. Oil Bear. Plants* **2016**, *19*, 1102–1110. [CrossRef]
59. Shalaby, A.S.; Razin, A.M. Dense Cultivation and Fertilization for Higher Yield of Thyme (*Thymus vulgaris* L.). *J. Agron. Crop Sci.* **1992**, *168*, 243–248. [CrossRef]

60. Singh, M.; Shivaraj, B.; Sridhara, S. Effect of Plant Spacing and Nitrogen Levels on Growth, Herb and Oil Yields of Lemongrass (Cymbopogon Flexuosus (Steud.) Wats. Var. I Cauvery). *J. Agron. Crop Sci.* **1996**, *177*, 101–105. [CrossRef]
61. Singh, M. Effects of Plant Spacing, Fertilizer, Modified Urea Material and Irrigation Regime on Herbage, Oil Yield and Oil Quality of Rosemary in Semi-Arid Tropical Conditions. *J. Hortic. Sci. Biotechnol.* **2004**, *79*, 411–415. [CrossRef]
62. Kumar, V.; Sood, M. Effect of Transplanting Time, Spacing and Fertilizers on Herbage and Oil Yield of *Mentha piperita* L. *Int. J. Farm Sci.* **2011**, *1*, 68–74. [CrossRef]
63. Kumar, R.; Sood, S.; Sharma, S.; Kasana, R.C.; Pathania, V.L.; Singh, B.; Singh, R.D. Effect of Plant Spacing and Organic Mulch on Growth, Yield and Quality of Natural Sweetener Plant Stevia and Soil Fertility in Western Himalayas. *Int. J. Plant Prod.* **2014**, *8*, 311–334. [CrossRef]
64. Saki, A.; Mozafari, H.; Asl, K.K.; Sani, B.; Mirza, M. Plant Yield, Antioxidant Capacity and Essential Oil Quality of Satureja Mutica Supplied with Cattle Manure and Wheat Straw in Different Plant Densities. *Commun. Soil Sci. Plant Anal.* **2019**, *50*, 2683–2693. [CrossRef]
65. Bandopadhyay, S.; Martin-Closas, L.; Pelacho, A.M.; DeBruyn, J.M. Biodegradable Plastic Mulch Films: Impacts on Soil Microbial Communities and Ecosystem Functions. *Front. Microbiol.* **2018**, *9*, 819. [CrossRef]
66. Bahadur, S.; Pradhan, S.; Verma, S.; Maurya, R.; Verma, S.K. Role of Plastic Mulch in Soil Health and Crop Productivity. *Clim. Chang. Its Implic. Crop Prod. Food Secur.* **2018**, 338–344.
67. Baniyaghoub, M.; Mozafari, H.; Karimzadeh, K.; Rajabzadeh, F.; Azimi, R. The Changes in Yield and Chemical Profile of Essential Oil and Leaf Minerals of *Satureja Macrantha* CA Mey. Under Combined Manure and NPK Fertilizer. *J. Med. Plants By-Prod.* **2021**, *10*, 141–148. [CrossRef]
68. Hälvä, S. Studies on Production Techniques of Some Herb Plants: I Effect of Agryl P17 Mulching on Herb Yield and Volatile Oils of Basil (*Ocimum basilicum* L.) and Marjoram (*Origanum majorana* L.). *Agric. Food Sci.* **1987**, *59*, 31–36. [CrossRef]
69. Hawkesford, M.; Horst, W.; Kichey, T.; Lambers, H.; Schjoerring, J.; Skrumsager Møller, I.; White, P. Functions of Macronutrients. In *Marschner's Mineral Nutrition of Higher Plants*; Marschner, P., Ed.; Elsevier: Amsterdam, The Netherlands, 2012; pp. 135–189.
70. Bakhtiari, M.; Mozafari, H.; Asl, K.K.; Sani, B.; Mirza, M. Bio-Organic and Inorganic Fertilizers Modify Leaf Nutrients, Essential Oil Properties, and Antioxidant Capacity in Medic Savory (*Satureja Macrantha* L.). *J. Biol. Res. della Soc. Ital. di Biol. Sper.* **2020**, *93*, 8477. [CrossRef]
71. Skubij, N.; Dzida, K.; Jarosz, Z.; Pitura, K.; Jaroszuk-Sierocińska, M. Nutritional Value of Savory Herb (*Satureja hortensis* L.) and Plant Response to Variable Mineral Nutrition Conditions in Various Phases of Development. *Plants* **2020**, *9*, 706. [CrossRef] [PubMed]
72. Torbert, H.A.; Wood, C.W. Effects of Soil Compaction and Water-filled Pore Space on Soil Microbial Activity and N Losses. *Commun. Soil Sci. Plant Anal.* **1992**, *23*, 1321–1331. [CrossRef]
73. Saki, A.; Mozafari, H.; Karimzadeh Asl, K.; Sani, B.; Mirza, M. Physiochemical Properties and Leaf Nutrients of *Satureja Mutica* Fisch & CA Mey Treated with Cattle Manure at Different Plant Densities under Dryland Farming System. *J. Med. Plants By-Prod.* **2020**, *11*, 51–57. [CrossRef]
74. Lambers, H.; Brundrett, M.C.; Raven, J.A.; Hopper, S.D. Plant Mineral Nutrition in Ancient Landscapes: High Plant Species Diversity on Infertile Soils Is Linked to Functional Diversity for Nutritional Strategies. *Plant Soil* **2011**, *348*, 7–27.
75. Barré, P.; Montagnier, C.; Chenu, C.; Abbadie, L.; Velde, B. Clay Minerals as a Soil Potassium Reservoir: Observation and Quantification through X-Ray Diffraction. *Plant Soil* **2008**, *302*, 213–220. [CrossRef]
76. Bomme, U.; Nast, D. Nährstoffentzug Und Ordnungsgemäße Düngung Im Feldanbau von Heil-Und Gewürzpflanzen. *Z. Arznei-Und Gewürzpflanzen* **1998**, *3*, 82–90.

horticulturae

Article

Essential Oil Content and Composition of the Chamomile Inflorescences (*Matricaria recutita* L.) Belonging to Central Albania

Ivan Salamon [1,*], Alban Ibraliu [2] and Maryna Kryvtsova [3]

1 Department of Ecology, Faculty of Humanities and Natural Science, University of Presov, 01, 17th November St., SK-081 16 Presov, Slovakia
2 Department of Agriculture Sciences, Faculty of Agriculture and Environment, Agricultural University of Tirana, 1029, Koder Kamez, AL-1021 Tirana, Albania
3 Department of Genetics, Plant Physiology and Microbiology, Biological Faculty, Uzhhorod National University, 32, Voloshyn St., UA-88000 Uzhhorod, Ukraine
* Correspondence: ivan.salamon@unipo.sk; Tel.: +421-917984060

Abstract: The chamomile drug (*Chamomillae anthodium*) is widely known and has been used as a medicine for ages. Today, the drug is officially registered in the European Pharmacopoeia. Despite the economic importance of the chamomile *Matricaria recutita* L., little is known about the extent and nature of the essential oil variability and composition of this species in Albania. Therefore, information about the extent of the uses of various gene pools is extremely valuable for the development of future chamomile cultivation and breeding programs. This study aimed to analyze the differences among 29 populations in different sites in central Albania. The quantities of essential oils ranged from a low of 0.10 ± 0.01% in Fier and Tirana to a high of 0.75 ± 0.05% in Linzë. The yield of volatile oils depended on the geography, altitude, and other factors, including stress influences on the site of plant population growth. This fact was confirmed by various subclimatic characteristics obtained from individual localities. Essential oil extracted from chamomile inflorescences was recorded to have between 23 and 43 chemical components. It was found that /-/-α-bisabololoxides B and A were the major constituents in 25 samples, with only 4 having dominant /-/-α-bisabolol. The uniquely determined chemical type of the chamomile wild populations in Albania was chemical type B (/-/-α-bisabololoxide B > /-/-α-bisabololoxide A > /-/-α-bisabolol). Based on the study of chamomile's pharmacodynamic properties, the sesquiterpenes /-/-α-bisabolol and chamazulene are considered to be the most valuable constituents. Consequently, a very intensive improvement breeding program must begin, with emphasis on drug yield, polyploidization, essential oil quantity, and both component contents in the oil.

Keywords: chamomile; chemical type; essential oil; GC-MS; sesquiterpenes; central Albania

Citation: Salamon, I.; Ibraliu, A.; Kryvtsova, M. Essential Oil Content and Composition of the Chamomile Inflorescences (*Matricaria recutita* L.) Belonging to Central Albania. *Horticulturae* **2023**, *9*, 47. https://doi.org/10.3390/horticulturae9010047

Academic Editor: Othmane Merah

Received: 1 December 2022
Revised: 14 December 2022
Accepted: 24 December 2022
Published: 2 January 2023

1. Introduction

Chamomile, *Matricaria recutita* L., of the *Asteraceae* family, is of great importance as a raw material in pharmaceutical, cosmetic, and food industries [1]. The species is native to eastern and southern Europe as well as parts of western Asia and now occurs throughout almost all of Europe as well as Turkey and the Caucasus region (Georgia), in addition to parts of Iran and Afghanistan [2]. The material of commerce is obtained, for the most part, from farms in Egypt, Germany, Argentina, Poland, and, to a lesser extent, Chile [3], the Czech Republic, Slovakia, Spain, and several of the Balkan countries (Bosnia and Herzegovina, Bulgaria, Croatia, and Serbia). It is also wild-collected for commercial trade in Hungary, as well as in Albania, Bulgaria, Croatia, Kosovo, and North Macedonia [4,5].

In Albania, chamomile plants are commonly consumed as a tea [6]. They have a long history as an effective herbal remedy used in folk medicine to treat fevers, common colds, cough stomach ailments, and gastrointestinal disorders, and their constituents show antiseptic, anti-inflammatory, and antispasmodic properties. They are usually drunk as an infusion but can also be added to bathwater to soothe dermatitis. The chamomile drug in pharmaceutic products is commonly used in the form of macerates, extracts, infusions, and essential oils. Similar raw materials are added to a variety of cosmetic preparations such as creams, soaps, skin lotions, bath ingredients, hair shampoos, etc. [7].

The geographical distribution of ubiquitous chamomile populations in the Balkan country involves mainly grassy sites, uncultivated areas, along roads, and in house yards; they are less cultivated as a commercial medicinal plant. They reach 1200 m in altitude, ranging from fields to alpine valleys. They occur in sun-loving (heliophilous) or medium shade-loving (mesoskiophilous) conditions. The species is commonly found in rich base soils, more or less rich in nutrients with a basic or slightly acidic pH, and very dry and cool soils [8].

This complex study presents the yield variation of chamomile essential oils and the sesquiterpene compositions (/-/-α-bisabololoxide B, /-/-α-bisabololoxide A, /-/-α-bisabolol, chamazulene, and en-yn-dicycloethers) of autochthonous populations in central Albania. Despite the economic importance of the chamomile, *Matricaria recutita* L., little is known about the extent and nature of the essential oil variability and composition of this species in Albania. Therefore, information about the extent of uses of various gene pools is extremely valuable for the development of future chamomile cultivation and breeding programs.

2. Material and Methods

2.1. Collection of the Plant Material

Inflorescences were collected from 29 localities in central Albania in period from 2017 to 2019 (Table 1, Figure 1). The typical biotopes and plant formations of this plant species were dry lawns, forest margins, footpath margins, house yards, countryside, and around ruins. It is principally associated with forest margins (*Origanetalia*) and invasive oak forest (*Quercus pubescenti-petraeae*) [9].

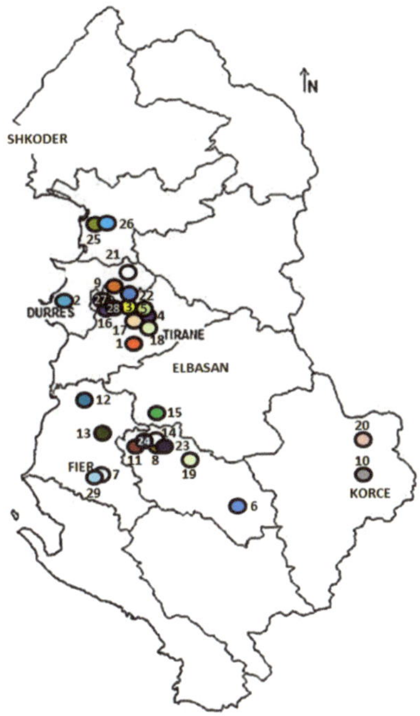

Figure 1. Albania (www.d-maps.com) with its prefectures and the localities, where chamomile flower samples were collected: 1. Baldushk, 2. Rruga Gjiri i Lazit, 3. Zall Her, 4. Dajti, 5. Linzë, 6. Skrapar: Melove, 7. Verbas, 8. Kuçovë, 9. Krujë: Larushk, 10. Korçë, 11. Poshnje: Berat, 12. Karbunarë Lushnjë,

13. Karbunarë e Vogël, 14. Kuçovë: Rruga S. Zuber, 15. Kuçovë: Rruga Kozare, 16. Gjokaj, 17. Tirana: Botanic Garden, 18. Tirana: Grand Park, 19. Skrapar: Tomorri Mount., 20. Sheqeras, 21. Krujë: Droja, 22. Tapizȅ, 23. Kuçovë: Parku i Qytetit, 24. Poshnje, 25. Lezhȅ, 26. Lezhȅ: Kolsh, 27. Fushȅ Prezȅ: Palaq, 28. Fushȅ Prezȅ, 29. Fier.

Table 1. The basic characters of selected sites in Albania with the occurrence of chamomile populations.

	Locality	Geographical Latitude	Terrestrial Longitude	Altitude [m]	Aspect	Slope
01	Baldushk	N 40°38′43′′	E 20°59′38′′	250	south-west	30°
02	Rruga Gjiri i Lazit	N 41°37′17′′	E 19°60′56′′	27	north-west	10°
03	Zall Her	N 41°38′92′′	E 19°82′59′′	115	north-west	0°
04	Dajti	N 41°21′57′′	E 19°55′32′′	450	north-west	15°
05	Linzȅ	N 41°35′34′′	E 19°88′35′′	400	north-west	12°
06	Skrapar: Melove	N 40°58′85′′	E 20°28′75′′	879	South	0°
07	Verbas	N 40°70′16′′	E 19°61′59′′	22	south-west	7°
08	Kuçovë	N 40°80′44′′	E 19°91′34′′	44	south-west	5°
09	Krujë: Larushk	N 41°43′68′′	E 19°73′27′′	33	north-west	17°
10	Korçë	N 40°61′58′′	E 20°77′72′′	870	south-east	27°
11	Poshnje Berat	N 40°78′29′′	E 19°84′43′′	26	south-west	30°
12	Karbunarë Lushnjë	N 40°92′60′′	E 19°76′82′′	99	south-west	0°
13	Karbunarë e Vogël	N 40°91′92′′	E 19°71′89′′	16	south-west	0°
14	Kuçovë: Rruga S. Zuber	N 40°80′13′′	E 19°91′49′′	48	south-west	16°
15	Kuçovë: Rruga Kozare	N 40°86′78′′	E 19°92′10′′	37	south-west	2°
16	Gjokaj	N 41°35′87′′	E 19°67′38′′	88	north-west	1°
17	Tirana: Botanic Garden	N 41°32′68′′	E 19°81′87′′	200	north-west	13°
18	Tirana: Grand Park	N 41°31′14′′	E 19°82′94′′	154	north-west	21°
19	Skrapar: Tomorri Mount.	N 40°66′12′′	E 20°17′01′′	1708	South	30°
20	Sheqeras	N 40°74′53′′	E 20°77′07′′	815	south-east	30°
21	Krujë: Droja	N 41°51′11′′	E 19°79′23′′	543	north-west	30°
22	Tapizȅ	N 41°41′61′′	E 19°75′99′′	55	north-west	31°
23	Kuçovë: Parku i Qytetit	N 40°80′45′′	E 19°91′44′′	46	south-west	10°
24	Poshnje	N 40°78′29′′	E 19°84′43′′	26	south-west	30°
25	Lezhȅ	N 41°77′65′′	E 19°64′74′′	0	north-west	8°
26	Lezhȅ Kolsh	N 41°77′87′′	E 19°68′85′′	336	north-west	30°
27	Fushȅ Prezȅ: Palaq	N 41°40′24′′	E 19°66′92′′	52	north-west	9°
28	Fushȅ Prezȅ	N 41°40′24′′	E 19°66′92′′	52	north-west	4°
29	Fier	N 40°72′51′′	E 19°55′82′′	20	south-east	0°

The flower heads were separated and dried in a sheltered, open air area at a temperature below 32 °C for 14 to 20 days with low humidity of 2 to 5%. The moisture content of the flower tissue was lowered to 12% to prevent mold infection. The dried materials were cleaned, packed, labelled, and stored in a clean and dry place until further use.

2.2. Chamomile Oil Isolation

Each sample of dried flower with a weight of 20 g was ground in a blender. The essential oil from this raw material was prepared using hydro-distillation (2 h) in Clevenger-type apparatus according to the European Pharmacopoeia (Ph. Eur.) 10th Edition [10]. Hexane was used as a collecting solvent. The essential oils were stored under N_2 at +4 °C in a dark space before their GC-FID and GC-MS analyses.

2.3. GC-FID Analysis

The analysis of the chamomile essential oils was carried out using a Vega Series Carloerba gas chromatograph, connected to a Spectrophysics SP 4270 integrator. The following operating conditions were used: column: DB5, 30 m × 0.32 mm inner diameter (i.d.); film thickness: 0.25 mm; carrier gas: nitrogen, adjusted to a flux of 1 mL per min; and injection and FID-detector temperatures: 220 and 250 °C, respectively. Components were identified using their GC retention times, and the resulting values were comparable to those of the literature. Oil component standards for comparison were supplied by Extrasynthese, s.a., Genay, France.

2.4. GC/MS Analysis

GC/MS analysis was carried out on a Varian 450-GC connected with a Varian 220-MS. The separation was achieved using a Factor Four TM: Capillary Column VF 5 ms (30 m × 0.25 mm i.d., 0.25 µm film thickness). Injector type 1177 was heated at temperature 220 °C. Injection mode was split less (1 µL of a 1:1000 n-hexane/diethyl ether solution). Helium was used as a carrier gas at a constant column flow rate of 1.2 mL per min. Column temperature was programmed: initial temperature was 50 °C for 10 min, then up to 100 °C at 3 °C per min, isothermal condition for 5 min and then continued to heating at 150 °C at 10 °C per min. Total time for analysis of one sample was completed in 59.97 min. Identification of components was performed by comparison of their mass spectra with those stored in NIST 02 mass spectra from the literature [11] and a home-made library, as well as on comparison of their retention indices with the standards.

2.5. Statistical Analysis

Statistical analysis was performed using confidence intervals with a significance level $p < 0.05$ using calculation through the mean, standard deviation, and standard error.

The differences between chamomile populations for the mean values of the essential oil constituents were carried out using ANOVA analysis. All statistics data were calculated employing the SAS JMP Statistical Discovery and a dendrogram (ward method) [12], and a relationship diagram (chamomile populations x the essential oil constituents) was carried out.

3. Results

3.1. The Dependence of Essential Chamomile Oil Yield on Their Habitats

The chamomile essential oil hydro-distilled from the fresh dried flower samples from all localities was pastel bluish and later turned greenish yellow. The oil from the Albanian locations of Zall Her (locality 03), Dajti (locality 04), Kuçovë (locality 08), and Linzë (locality 05) had a deeper blue color. The yield ranged from 0.10 ± 0.01% (v/w) in Fier (locality 29) and the Botanical Garden in Tirana (locality 17) to 0.75 ± 0.05% (v/w) on a dry-weight basis in Linzë (locality 05) (Table 2). The quantity of extractible substances using 60% ethanol did not correspond to the amount of essential oil isolated from the flower heads (Table 2).

Table 2. Chamomile inflorescence's yield of essential oil in % (*v*/*w* – expressed as a dry weight).

	Habitat/Locality	Extractible Substances using 60% Ethanol [%]	Essential Oil Yield [%]
01	Baldushk	24.0 ± 0.5	0.40 ± 0.05
02	Rruga Gjiri i Lazit	23.5 ± 0.5	0.40 ± 0.05
03	Zall Her	33.0 ± 0.5	0.45 ± 0.05
04	Dajti	31.5 ± 0.5	0.50 ± 0.06
05	Linzȅ	33.0 ± 0.5	0.75 ± 0.05
06	Skrapar: Melove	29.0 ± 1.0	0.60 ± 0.02
07	Verbas	35.0 ± 1.0	0.28 ± 0.02
08	Kuçovë	23.0 ± 1.0	0.20 ± 0.02
09	Krujë: Larushk	28.0 ± 1.0	0.30 ± 0.02
10	Korçë	34.0 ± 1.5	0.22 ± 0.02
11	Poshnje: Berat	34.0 ± 1.0	0.52 ± 0.03
12	Karbunarë Lushnjë	25.0 ± 0.5	0.35 ± 0.02
13	Karbunarë e Vogël	38.0 ± 1.5	0.37 ± 0.02
14	Kuçovë: Rruga S. Zuber	31.0 ± 1.0	0.53 ± 0.02
15	Kuçovë: Rruga Kozare	34.0 ± 1.0	0.40 ± 0.02
16	Gjokaj	28.0 ± 0.5	0.20 ± 0.02
17	Tirana: Botanic Garden	21.0 ± 0.5	0.10 ± 0.01
18	Tirana: Grand Park	27.5 ± 0.5	0.15 ± 0.01
19	Skrapar: Tomorri Mount.	29.0 ± 1.0	0.40 ± 0.02
20	Sheqeras	29.0 ± 1.0	0.20 ± 0.02
21	Krujë: Droja	35.0 ± 1.0	0.34 ± 0.02
22	Tapizȅ	33.0 ± 1.0	0.40 ± 0.02
23	Kuçovë: Parku i Qytetit	35.0 ± 1.0	0.44 ± 0.03
24	Poshnje	24.0 ± 1.0	0.35 ± 0.02
25	Lezhȅ	36.0 ± 1.5	0.36 ± 0.02
26	Lezhȅ Kolsh	39.0 ± 1.0	0.42 ± 0.03
27	Fushȅ Prezȅ: Palaq	31.0 ± 1.0	0.40 ± 0.02
28	Fushȅ Prezȅ	39.0 ± 1.0	0.26 ± 0.02
29	Fier	27.0 ± 1.0	0.10 ± 0.01

3.2. Qualitative and Quantitative Composition of Essential Oils

The total chemical profile of the tested essential oils of three selected samples from the localities of Poshnje Berat (locality 11), Krujë Larushk (locality 09), and Lezhȅ Kolsh (locality 26) are shown in Figure 2, whereas the qualitative and quantitative compositions of the isolated essential oils are shown in Table 3.

However, the quantitative composition of the individual components in the isolated essential oils differed. Essential oil extracted from the chamomile flower sample from the Poshnje Berat locality was recorded to contain 43 chemical components. It was found that /-/-α-bisabololoxide B was the major constituent (45.47 ± 3.5%), followed by /-/-α-bisabolol (17.01 ± 1.0%), cis-, trans-en-yn-dicycloether (10.98 ± 1.0%), /-/-α-bisabololoxide A (10.33 ± 1.0%), and chamazulene (0.63 ± 0.1%). The essential oil composition from Krujë: Larushk had 23 chemical constituents determined, with dominant /-/-α-bisabololoxide A (34.95 ± 3.0%) and /-/-α-bisabololoxide B (26.62 ± 2.0%), followed by /-/-α-bisabolol

(16.27 ± 1.0%), cis-, trans-en-yn-dicycloether (5.16 ± 0.7%), and chamazulene (3.71 ± 0.2%). The content of the most abundant components (/-/-α-bisabolol and chamazulene) was highest in the oil isolated from the Lezhȅ: Kolsh (26) locality (39.69 ± 2.0% and 6.16 ± 1.0%, respectively; Table 3). In addition, pre-treatment resulted in a higher content of cis-, trans-en-yn-dicycloether (17.24 ± 1.5%), /-/-α-bisabololoxide A (16.78 ± 1.0%), and /-/-α-bisabololoxide B (10.24 ± 1.0%) from 27 identified secondary metabolites.

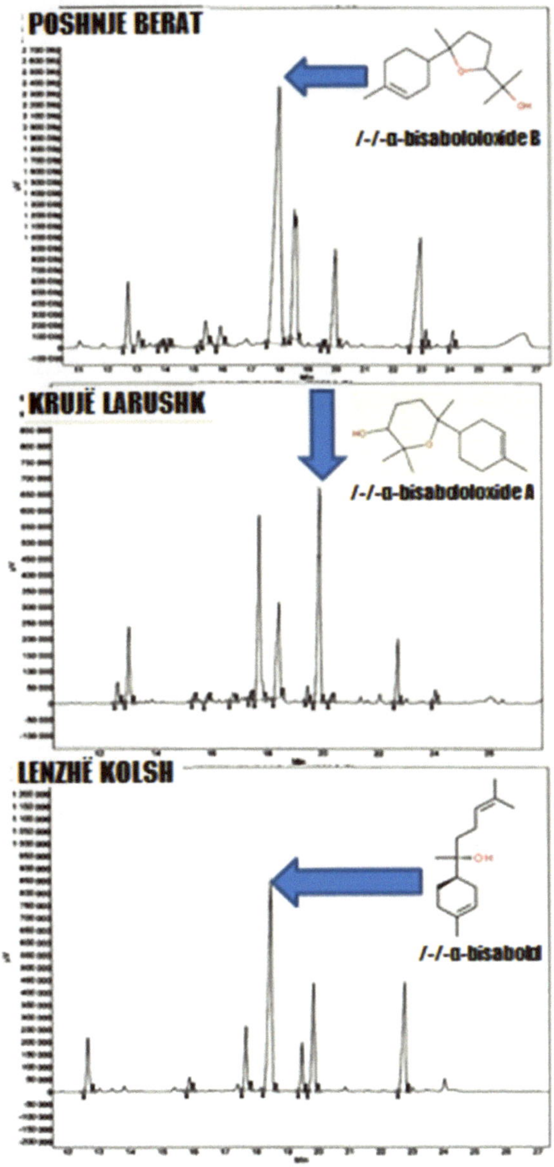

Figure 2. The differences in the chromatographic profiles of chamomile essential oil from three Albanian localities: Poshnje Berat, Krujë Larushk and Lezhȅ Kolsh.

Table 3. Main components of the essential oil isolated from chamomile inflorescences belonging to central Albania.

	Localities	Basic Composition of Essential Oil in %					
		Fa	Bo	Ch	BoA	BoB	*c-,t*-Dc
01	Baldushk	3.5 ± 0.5	11.5 ± 1.0	1.0 ± 0.1	12.7 ± 1.0	29.5 ± 2.0	7.0 ± 1.0
02	Rruga Gjiri i Lazit	4.0 ± 0.5	12.5 ± 1.0	4.0 ± 0.5	20.5 ± 2.5	26.0 ± 2.0	9.0 ± 1.5
03	Zall Her	8.2 ± 1.5	13.5 ± 1.0	11.0 ± 0.5	11.5 ± 1.5	30.5 ± 2.5	8.5 ± 1.0
04	Dajti	8.2 ± 1.5	15.5 ± 2.0	7.5 ± 1.5	12.0 ± 1.5	32.0 ± 3.0	5.5 ± 0.5
05	Linzë	8.2 ± 1.5	16.2 ± 2.0	5.0 ± 0.7	14.0 ± 1.5	31.0 ± 3.0	5.0 ± 0.5
06	Skrapar: Melove	7.0 ± 1.0	16.0 ± 1.0	1.5 ± 0.2	16.0 ± 1.0	28.0 ± 2.0	12.0 ± 2.0
07	Verbas	11.0 ± 1.0	16.0 ± 1.0	1.2 ± 0.2	14.1 ± 1.0	16.0 ± 1.0	16.0 ± 2.5
08	Kuçovë	2.5 ± 0.5	16.0 ± 1.0	10.0 ± 1.0	12.5 ± 0.5	29.0 ± 2.0	8.0 ± 0.5
09	Krujë Larushk	2.0 ± 0.2	16.1 ± 1.0	1.6 ± 0.2	32.0 ± 3.0	24.0 ± 2.0	7.5 ± 1.5
10	Korçë	3.0 ± 0.5	17.0 ± 1.0	0.8 ± 0.1	27.0 ± 2.0	32.0 ± 3.0	16.0 ± 2.0
11	Poshnje Berat	3.0 ± 0.5	16.0 ± 1.0	0.3 ± 0.1	10.0 ± 1.0	46.0 ± 3.5	12.0 ± 1.5
12	Karbunarë Lushnjë	6.0 ± 1.0	16.5 ± 1.0	1.5 ± 0.1	9.5 ± 1.0	26.7 ± 1.5	10.0 ± 1.0
13	Karbunarë e Vogël	5.1 ± 1.0	18.0 ± 1.0	0.7 ± 0.2	24.0 ± 2.0	26.0 ± 1.5	6.0 ± 0.5
14	Kuçovë: Rruga S. Zuber	5.1 ± 1.0	19.0 ± 1.0	2.0 ± 0.2	18.0 ± 1.0	29.0 ± 2.0	10.0 ± 1.0
15	Kuçovë: Rruga Kozare	4.0 ± 0.5	18.0 ± 1.0	1.4 ± 0.2	22.0 ± 2.0	36.0 ± 2.0	12.0 ± 2.0
16	Gjokaj	5.7 ± 1.0	18.5 ± 1.0	0.5 ± 0.1	14.2 ± 2.0	32.7 ± 2.0	9.0 ± 1.5
17	Tirana: Botanic Garden	6.5 ± 1.0	24.0 ± 1.0	1.5 ± 0.4	8.0 ± 1.0	19.7 ± 2.0	4.5 ± 0.5
18	Tirana: Grand Park	4.5 ± 1.0	25.0 ± 3.0	1.5 ± 0.4	6.0 ± 1.0	25.0 ± 3.0	5.0 ± 0.5
19	Skrapar: Tomorri Mount.	4.0 ± 1.0	20.0 ± 3.0	1.0 ± 0.2	13.0 ± 1.0	33.0 ± 3.0	17.0 ± 2.5
20	Sheqeras	7.0 ± 1.0	20.0 ± 3.0	0.5 ± 0.1	20.0 ± 2.0	20.0 ± 2.0	10.0 ± 1.0
21	Krujë: Droja	9.0 ± 1.0	21.0 ± 2.0	3.0 ± 0.1	16.0 ± 1.0	26.0 ± 2.0	17.0 ± 2.0
22	Tapizë	6.0 ± 1.0	22.0 ± 2.0	1.5 ± 0.2	15.0 ± 1.0	32.0 ± 2.0	15.0 ± 2.0
23	Kuçovë: Parku i Qytetit	13.0 ± 1.0	24.0 ± 2.0	1.0 ± 0.2	18.0 ± 1.0	27.0 ± 2.0	15.0 ± 1.5
24	Poshnje	6.0 ± 1.0	30.0 ± 2.0	1.2 ± 0.2	6.0 ± 1.0	37.0 ± 2.0	7.0 ± 1.0
25	Lezhë	4.0 ± 0.5	36.0 ± 2.0	5.0 ± 0.2	22.0 ± 2.0	19.0 ± 1.0	16.0 ± 2.0
26	Lezhë Kolsh	7.0 ± 0.5	42.0 ± 2.0	6.0 ± 1.0	11.0 ± 1.0	18.0 ± 1.0	19.0 ± 2.5
27	Fushë Prezë: Palaq	5.0 ± 1.0	42.0 ± 2.0	0.4 ± 1.0	4.0 ± 1.0	20.0 ± 2.0	14.0 ± 1.0
28	Fushë Prezë	9.0 ± 1.0	40.0 ± 2.0	1.0 ± 1.0	10.0 ± 1.0	23.0 ± 2.0	9.0 ± 1.0
29	Fier	4.0 ± 1.0	42.0 ± 2.0	1.0 ± 1.0	11.0 ± 1.0	20.0 ± 2.0	12.0 ± 1.5

Fa—trans-β-farnesene; Bo—/-/-α-bisabolol; Ch—chamazulene; BoA—/-/-α-bisabololoxide A; BoB—/-/-α-bisabololoxide B; *c-,t*-Dc—cis-, trans-en-yn-dicycloether.

3.3. Chamomile Chemical Types

The qualitative–quantitative essential oil diversity of wild-growing chamomile populations is shown in Table 3. The frequency of the population of /-/-α-bisabololoxide B content is very high (from 16.0 ± 1.0 to 45.47 ± 3.5%) in the regions located at Baldush (locality 01), Rruga Gjirit i Lazit (locality 02), Zall Her (locality 03), Dajti (locality 04), Linzë (locality 05), Skrapar: Melove (locality 06), Verbas (locality 07), Kucovë (locality 08), Krujë: Larushk (locality 09), Korçë (locality 10), Poshnje: Berat (locality 11), Karbunarë Lushnjë (locality 12), Karbunarë e Vogël (locality 13), Kuçovë: Rruga S. Zuber (locality 14), Kuçovë: Rruga Kozare (locality 15), Gjokaj (locality 16), Tirana: Botanic Garden and Grand Park (localities 17, 18), Skrapar: Tomorri Mountains (locality 19), Sheqeras (locality 20), Krujë:

Droja (locality 21), Tapizë (22), Kuçovë: Parku i Qytetit (locality 23), and Poshnje (locality 24). The chamomile plants with a high /-/-α- bisabolol content were very rare. The range of this oxide was from 6.0 ± 1.0 to 32.0± 3.0%. The frequency of the high /-/-α-bisabolol content (from 40.0 ± 2.0 to 42.0 ± 2.0%) was in the localities of Lezhë: Kolsh (locality 26), Fier (locality 29), and Fushë Prezë (localities 27, 28). However, the accumulations of chamazulene content (from 1.0 ± 0.1 to 11.0 ± 0.5%) were affected by the conditions of the year, and regional differences were established.

As a result, four basic types of chamomile A, B, C, and D are recognized, according to the qualitative and quantitative composition of the essential oils (Table 4 and Figure 2), which are characterized as follows: type A (/-/-α-bisabololoxide A is the dominant component), type B (/-/-α-bisabololoxide B is the dominant component), type C (typically with the highest /-/-α-bisabolol content), and type D (/-/-α-bisabolol and /-/-α-bisabolol oxide A and B present in 1:1 ratio approximately).

Table 4. The marked differences in chemical composition of chamomile essential oil from tree localities in central Albania [%].

Component as % of Essential Oil *	GC-MS t_R (min.) **	Kovat's Index	Localities of Chamomile Flower Collection		
			Poshnje Berat	Krujë Larushk	Lezhë Kolsh
trans-2-hexanal	9.26	842	1.15	1.06	——
tricyclene	10.41	879	0.03	0.96	——
α-pinene	14.81	929	0.03	0.19	——
camphene	15.23	942	0.02	0.36	——
sabinene	15.53	973	0.02	——	——
β-pinene	15.85	984	0.03	——	——
β-myrcene	16.15	989	0.09	——	——
p-cymene	16.55	1032	0.04	——	——
(ε,ε)-matricaria esther	19.07	1491	1.27	1.63	0.60
γ-muurolene	19.24	1508	0.07	——	——
β-farnesene	19.55	1513	3.09	1.15	2.63
α-muurolene	19.89	1521	0.25	——	——
α-bulnesene	20.05	1528	0.11	——	0.06
γ-cadinene	20.23	1539	0.09	——	0.22
calamenene	20.65	1543	0.07	——	0.02
α-acoradiene	20.81	1554	0.13	——	——
cadina-1,4-diene	20.99	1558	0.08	——	——
δ-cadinene	23.22	1563	0.02	——	——
α-amorphen	23.49	1564	0.05	——	——
α-calacorene	24.07	1567	0.09	——	——
trans-nerolidol	24.89	1573	0.77	——	——
epiglobulol	25.48	1584	0.52	——	0.15
junenol	26.07	1590	0.09	——	0.06
spatulenol	26.79	1602	0.15	0.72	0.08
β-eudesnol	26.97	1603	0.10	——	0.21
globulol	27.15	1611	0.06	——	0.11
tremetone	27.34	1615	0.33	——	0.34

Table 4. *Cont.*

Component as % of Essential Oil *	GC-MS t_R (min.) **	Kovat's Index	Localities of Chamomile Flower Collection		
			Poshnje Berat	Krujë Larushk	Lezhë Kolsh
caryophyllene oxide	27.9	1617	0.04	0.02	0.06
viridiflorene	28.06	1626	0.48	0.18	0.08
dillapiole	28.52	1641	0.72	0.77	0.19
cubebol	29.4	1652	0.38	0.25	0.10
β-bisabolole	29.85	1660	0.30	0.28	1.01
τ-muurolol	30.61	1663	1.64	0.51	0.99
α-bisabololoxide B	31.28	1679	**45.47**	26.62	10.24
α-bisabolonoxide A	31.69	1684	0.26	0.29	1.08
valerianol	32.05	1689	2.20	2.13	0.05
cadalene	33.05	1702	0.12	0.70	0.06
α-bisabolol	33.59	1711	17.11	16.27	**39.69**
chamazulene	43.4	1773	0.63	3.71	6.16
α-bisabololoxide A	44.98	1781	10.33	**34.95**	16.78
guaiazulene	46.13	1802	0.18	1.09	1.91
cis-en-yn-dicycloether	59.00	1927	10.82	5.02	17.18
trans-en-yn-dicycloether	59.03	1942	0.16	0.14	0.06
Total [%]			100.00	100.00	100.00

Note: *—data are expressed as area in % of the 100.00 % of all identified peaks; **—retention times.

Analyzing the chamomile samples collected from 29 regions in Albania, 2 main groups were separated: populations accumulating /-/-α-bisabololoxide B, and populations characterized by the presence of /-/-α-bisabolol.

3.4. Hierarchical Cluster Analysis

In order to provide additional insights into the chemotypes of chamomile essential oils, we carried out a hierarchical cluster analysis based on the constituents. The dendrogram of the analysis is shown in Figure 3. Based on this analysis, there are two different confirmed chemotypes: chemotype B, dominated by /-/-α-bisabololoxide B and chemotype C, dominated by /-/-α-bisabolol.

The average Euclidean distance among the populations calculated based on the essential oil constituents was 41.6, ranging from 8.9 (between P3 and P4) to 79.1 (between P1 and P5) (Figure 3).

The results obtained from the cluster analysis show the existence of a high variability within the essential oils and differences among the following chamomile populations collected in different areas in Albania (Figure 3). From the 29 populations submitted to multivariate analysis, two well-defined groups of essential oils were differentiated using cluster analysis. Two subclusters can be observed: the first subset contains five populations collected in Fer, Fishe Preze, Palaq, Fushe Preze, Kolsh: Lezhe, and Lezhe, and the second subset includes the samples collected in the other areas.

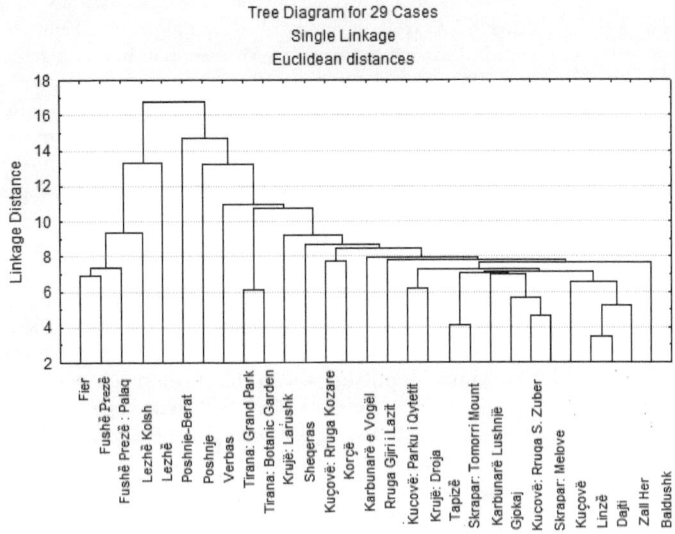

Figure 3. Dendrogram of relationships among Albanian chamomile populations and qualitative–quantitative characteristics of essential oils.

4. Discussion

The chamomile inflorescences of diploid plants usually contain from 0.4 to 1.0% essential oils. While the quantity of essential oil in the flowers from different localities may be genetically coded, the differences in production at the 29 test sites indicated considerable interaction between the plants and the environment [13]. These observations confirm previous studies in Eastern Slovakia [14] and Iran [2]. In both of these later cases, the range in the essential oil content from 0.4 to 0.8% occurred but was less than the accepted 0.9 to 1.0% variation accepted for chamomile flower drugs of variety certifications.

In a natural ecosystem, chamomile species separated on the bases of their secondary products contribute to the development of the modification of the existing ecological balance [15]. However, for a full understanding of the biosynthetic processes of these plant populations, a detailed analysis of the geographical occurrences and their chemotypes are required.

The compounds of essential oils are clearly genetically determined to a higher degree than their contents. The oil content is more strongly influenced by environmental factors and shows considerable variation, even within a relatively small area [14].

In 1972 and 1973, Schilcher [16,17] found that according to the oil composition, chamomile could be differentiated into chemical types based on the contents of /-/-α-bisabololoxide A and B, /-/-α-bisabolol (type A: /-/-α-bisabololoxide B > /-/-α-bisabololoxide B > /-/-α-bisabolol; type B: /-/-α-bisabololoxide A > /-/-α-bisabololoxide B > /-/-α-bisabolol; type C: /-/-α-bisabolol > /-/-α-bisabololoxide B > /-/-α-bisabololoxide A; and type D: /-/-α-bisabololoxide B ≈ /-/-α-bisabololoxide A ≈ /-/-α-bisabolol).

With regard to the sesquiterpene alcohols, the wild plants of this species mostly show bisabololoxides. The plant populations rich in /-/α- bisabolol could be found endemically in Catalonia (Spain) and, to a smaller extent, in other populations. This fact was confirmed by the study of small populations on the islands of Malta, Cyprus, and Crimea [18].

Analyses of the essential oils obtained from Hungarian, Czech, Slovak, German, and Spanish-grown chamomile flowers by Honcariv and Repcak [19] revealed that the chemical composition of chamomile oil varied extremely widely. They fully confirmed the existence of chemotypes in the chamomile species.

The chemical diversity of chamomile populations was justified in the Eastern Slovakian Lowland as well. It was proved by the investigations of Salamon (2004) [20] analyzing populations for the presence of essential oil and its composition during the period of 1995–1998.

In the chamomile samples from Attica in Greece, after essential oil hydro-distillation and GC/MS identification, 21 components were determined in 1993. The major constituents were β-farnesene (21.2 and 5.6%) and /-/-α-bisabololoxide A (16.0 and 9.0%) [21].

In Italy [22], the compounds of the oil from tubular and ligulae florets and from the receptacle were (E)-β-farnesene (14.4–17.1%), spathulenol (4.4–12.6%), α-bisabolone oxide A (9.2–11.2%), chamazulene (8.4–13.7%), α-bisabolol oxide A (4.9–11.6%), and cis-en-yn-bicycloether (2.7–13.4%). The bisabololoxide chemotype was proven.

The chemotype C chamomile populations were located in the northern, southwestern and southern parts of Bulgaria. The ones from the northwestern and southwestern regions had an average of 39% /-/-α-bisabolol and blue oil with 2.7% chamazulene content. Chamomile populations with the chemotype A essential oil were found in the central regions of North Bulgaria. They had an average of 34% /-/-α-bisabololoxide A and blue essential oil with the highest chamazulene content of 7.7% for all chemotypes. Chemotype D chamomile, with almost equal amounts of (-)-α-bisabolol and (-)-bisabolol oxide A and B, was discovered on one single site near to the city of Kjustendil [23].

The essential oil composition of the Hungarian populations can be divided into two groups: the α-bisabolol chemotype (where the ratio of this component is between 45 and 58%) and the bisabololoxide A chemotype (where the ratio of bisabololoxide A varied between 34 and 43%) [24].

In Poland [25], the presence of 22 compounds was identified in the essential oil obtained from the inflorescences of wild-growing chamomile, among which sesquiterpene compounds had the highest proportions: the dominant compounds were /-/-α-bisabololoxide A (31.70%), /-/-α-bisabololoxide B (17.09%), and /-/-α-bisabolnoxide A (15.73%); and chamazulene was also found in a large amount (15.58%). Among the other compounds, the presence of β-farnesene was found in small quantities (4.89%).

In total, 13 Brazilian chamomile populations from different geographic origins, agricultural practices, and harvest times, grown in the Santa Catarina and Paraná states in Brazil were analyzed [26]. Eleven samples were classified as chemotype B (rich in /-/-α-bisabololoxide B), and the remaining two were classified as chemotype A (rich in /-/-α-bisabololoxide A). The predominance of chemotype B was observed concerning the agricultural practices (organic or conventional) or geographic origin (Santa Catarina or Paraná).

Finally, there is the latest study of chamomile chemical profiles from plants collected from semi-wild populations in Latvia [27]. All samples were characterized with a high quality, reaching the thresholds of European Pharmacopoeia for essential oil content with high levels of /-/-α-bisabololoxide A.

This scientific study of the chamomile populations was complemented by a research on the variability in Albania, focusing on morphological traits [27,28]. To characterize and identify the different chemotypes of the Albanian populations and their essential oil composition using cluster analysis, it is possible to compare with different results and the dendrogram of the *Matricaria recutita* populations in Iran [29].

Thus, chamomile with a particular chemical composition is used as a drug as it shows specific pharmacological activity. As efficient methods for determining the drug's constituents and effectiveness have been developed, the content of /-/-α-bisabolol and its oxides in the flowers has become an important indicator of the raw material's quality and value.

5. Conclusions

The cultivation of chamomile is practiced all over the world; however, the collection of the wild population has not lost its importance, especially in European countries in the east and south. In Albania, the blossom of wild chamomile starts in the second half of May and continues until the beginning of July. Much of the Albanian production of chamomile is harvested from wild or unselected populations and is hand-harvested by low-cost labor. The essential oil contents from the localities of Zall Her, Dajti, Kuçovë, and Linzë had a deeper blue color. The yield ranged from 0.10 ± 0.01% to 0.75 ± 0.05% on a dry-weight basis, which is an insufficient amount. Analyses of the essential oil obtained from Albanian wild-grown chamomile reveal that the chemical composition of the oil varied extremely widely with the dominant chemotype B. As a result, four basic chemical types (A, B, C, and D) were recognized, according to the quality of the product. The country, as the major supplier of this raw material for the world market, needs to initial intensive improvement programs to produce chamomile with high levels of oils with a defined chemical composition placing significance on /-/-α-bisabolol and chamazulene.

Author Contributions: I.S. processed the experimental data, performed the biochemical–biological analysis, and drafted the manuscript. A.I. carried out almost all of the technical details and performed numerical calculations for the suggested experiment with plant population's collection in Albania. M.K. devised the project of Thymus biodiversity, the main conceptual ideas, and proof outline. All authors discussed the results and commented on the manuscript. All authors have read and agreed to the published version of the manuscript.

Funding: This research received no external funding.

Institutional Review Board Statement: Not applicable for studies not involving humans or animals.

Informed Consent Statement: Not applicable for studies not involving humans.

Data Availability Statement: The data are openly available in a public repository that issues datasets with DOIs.

Conflicts of Interest: The authors declare that the research was conducted in the absence of any commercial or financial relationships that could be construed as a potential conflict of interest.

References

1. Salamon, I. Chamomile Production. *Focus Herbs–A Mag. Dissem. Herb. Knowl.* **1992**, *2*, 6–11.
2. Salamon, I.; Ghanavati, M.; Abrahimpour, F. Potential of Medicinal Plant Production in Iran and Variability of Chamomile (*Matricaria recutita* L.) Essential Oil Quality. *J. Essent. Oil Bear. Plants* **2010**, *13*, 638–643. [CrossRef]
3. Weldt, E. Chamomile in Chile Cultivation and Industrialization. In *1st International Symposium on Chamomile Research, Development and Production: Program and Abstract Book, Presov, Slovak Republic, 7–10 June 2006*; International Society for Horticultural Science: Leuven, Belgium, 2007; p. 27.
4. Salamon, I. Ecobiology of the Chamomile [*Matricaria recutita* (L.) Raushert]. *Hortic. Bras.* **1994**, *12*, 226–229.
5. Chauhan, R.; Singh, S.; Kumar, V.; Kumar, A.; Kumari, A.; Rathore, S.; Kumar, R.; Singh, S. A Comprehensive Review on Biology, Genetic Improvement, Agro and Process Technology of German Chamomile (*Matricaria chamomilla* L.). *Plants* **2021**, *11*, 29. [CrossRef] [PubMed]
6. Asllani, U. *Essential Oils of Medicinal and Aromatic Plants in Albanian Area (in Albanian)*; Toena Publishing House: Tiranë, Albania, 2004; 147p.
7. Papadhopuli, G. *Medicinal and Aromatic Plants of Albania (in Albanian)*; OMBRA GVG: Durres, Albania, 1987; 231p.
8. Pieroni, A. Traditional uses of wild food plants, medicinal plants, and domestic remedies in Albanian, Aromanian and Macedonian villages in South-Eastern Albania. *J. Herb. Med.* **2017**, *9*, 81–90. [CrossRef]
9. Dring, J.; Hoda, P.; Mersinllar, M.; Mullaj, A.; Pignatti, S.; Rodwell, J. Plant Communities of Albania—A Preliminary Overview. *Ann. Bot.* **2002**, *2*, 7–30. [CrossRef]
10. *European Pharmacopoeia (Ph. Eur.)*, 10th ed.; EDQM Council of Europe: Strasbourg, France, 2022; 3856p.
11. Adams, R.P. *Identification of Essential Oil Components by Gas Chromatography/Mass Spectrometry*; Allured Publishing Corporation: Carol Steam, IL, USA, 1995; ISBN 978-1-932633-21-4.
12. Kezik, J.; Hill, M. *Data Management Essentials Using SAS and JMP*; Cambridge University Press: Cambridge, MA, USA, 2016; pp. 122–132. [CrossRef]
13. Salamon, I. Effect of the internal and external factors on yield and qualitative-quantitative characteristics of chamomile essential oil. *Acta Hortic.* **2007**, 45–65. [CrossRef]
14. Salamon, I. The Slovak gene pool of German chamomile (*Matricaria recutita* L.) and comparison in its parameters. *Hortic. Sci.* **2011**, *31*, 70–75. [CrossRef]

15. Das, M. *Chamomile–Medicinal, Biochemical, and Agricultural Aspects*; CRC Press-Taylo&Francis Group: New York, NY, USA, 2014; 268p; ISBN 9781466577602.
16. Schilcher, H. Neuere Erkenntnisse bei der Qualitätsbeurteilung von Kamillenblüten bzw. Kamillenöl. Teil 2: Qualitative Beurteilung des ätherischen Öles in Flores Chamomillae. Aufteilung der Handelskamille in vier bzw. Fünf Chemische Typen. *Planta Medica* **1973**, *23*, 132–144. [CrossRef] [PubMed]
17. Schilcher, H. Neuere Erkenntnisse bei der Qualitätsbeurteilung von Kamillenblüten bzw. Kamillenöl. *Dtsch. Apoth. Ztg.* **1972**, *112*, 1474–1478.
18. Salamon, I. Myth of the Spain (/-/-α-Bisabolol) Chemotype. In *1st International Symposium on Chamomile Research, Development and Production: Program and Abstract Book, Presov, Slovak Republic, 7–10 June 2006*; International Society for Horticultural Science: Leuven, Belgium, 2007; p. 43.
19. Honcariv, R.; Repcak, M. Chemotypes of *Matricaria chamomilla* L. *Herba Pol.* **1979**, *25*, 261–267.
20. Papazoglou, T.; Anastassaki, T.; Demetzos, C.; Loukis, A. Composition of the Essential Oil of Wild *Chamomilla recutita* (L.)Rausch. Grown in Greece. *J. Essent. Oil Res.* **2011**, *10*, 635. [CrossRef]
21. Trilini, B.; Pagiotti, R.; Meghini, L.; Pintore, G. Essential Oil Composition of Ligulate and Tubular Flow-ers and Receptacle from Wild *Chamomilla recutita* (L.)Rausch. Grown in Italy. *J. Essent. Oil Res.* **2006**, *18*, 42–45. [CrossRef]
22. Nikolova, A.M.; Jankulov, J.K.; Kozhuharova, K.S. Chemical Composition of Chamomile Populations in Bulgaria. *Int. J. Appl. Eng. Res.* **2011**, *6*, 1961–1967. [CrossRef]
23. Gosztola, B.; Németh, E.; Kozak, A.; Sárosi, S.Z.; Szabó, K. Comparative Evaluation of Hungarian Chamomile (*Matricaria recutita* L.) Populations. *Acta Hortic.* **2007**, *749*, 157–162. [CrossRef]
24. Nurzynska-Wierdak, R. The Essential oil *Chamomilla recutita* (L.)Rausch. Cultivated and Wild Growing in Poland. Annales Universitatis Mariae Curie-Sklodowska, Sectio. *Pharmacia* **2011**, *24*, 197–206. [CrossRef]
25. Petronilho, S.; Maraschin, M.; Delgadillo, I.; Coimbra, M.A.; Rocha, S.M. Sesquiterpenic composition of the inflorescences of Brazilian chamomile (*Matricaria recutita* L.): Impact of the agricultural practices. *Ind. Crop. Prod.* **2011**, *34*, 1482–1490. [CrossRef]
26. Mežaka, I.; Kronberga, A.; Nakurte, I.; Taškova, I.; Jakovels, D.; Primavera, A. Genetic, chemical and morphological variability of chamomile (*Chamomilla recutita* L.) populations of Latvia. *Ind. Crop. Prod.* **2020**, *154*, 112614. [CrossRef]
27. Zhuri, N.; Gixhari, B.; Ibraliu, A. Variability of Chamomile (*Matricaria chamomilla* L.) Populations as a Valuable Medicinal Plant in Albania Evaluated by Morphological Traits. *Albanian J. Agric. Sci.* **2019**, *4*, 83–89.
28. Zhuri, N. Diversity of Cultivated Chamomile (*Matricaria Chamomilla* L.) Populations in Different Areas in Albania. *Acta Sci. Agric.* **2022**, *140*, 1–14. [CrossRef]
29. Piri, E.; Sourestani, M.M.; Khaleghi, E.; Mottaghipisheh, J.; Zomborszki, Z.P.; Hohmann, J.; Csupor, D. Chemo-Diversity and Antiradical Potential of Twelve *Matricaria chamomilla* L. Populations from Iran: Proof of Ecological Effects. *Molecules* **2019**, *24*, 1315. [CrossRef] [PubMed]

horticulturae

MDPI

Article

Chemotypes of Species of the Genus *Thymus* L. in Carpathians Region of Ukraine—Their Essential Oil Qualitative and Quantitative Characteristics and Antimicrobial Activity

Maryna Kryvtsova [1,*], Myroslava Hrytsyna [2], Ivan Salamon [3], Maria Skybitska [4] and Olha Novykevuch [2]

1 Department of Genetic, Plant Physiology and Microbiology, Faculty of Biology, Voloshina, 32, 88000 Uzhghorod, Ukraine
2 Faculty of Social Development and Health, Stepan Gzhytskyj National University of Veterinary Medicine and Biotechnologies of Lviv, Pekarska Str., 50, 79010 Lviv, Ukraine
3 Department of Ecology, Faculty of Humanities and Natural Sciences, University of Presov, 01, 17th November Str., SK-081 16 Presov, Slovakia
4 Botanical Garden of Ivan Franko National University of Lviv, Cheremshyny Str., 44, 79000 Lviv, Ukraine
* Correspondence: maryna.krivcova@uzhnu.edu.ua; Tel.: +380-502785497

Abstract: The study of the R&D in this paper is to determine the range of essential oils (EOs) in the raw materials of species of the genus Thymus of the natural flora in the Carpathian region and their antimicrobial activity. It was found that the component range of EO in species of the genus Thymus depends on the microclimatic conditions of the population. The range of essential oils in the raw material of *Th. Serpyllum* and *Th. Pulegoides* is 7–9 mL and *Th. Marschallianus* is 3.5 mL. The research found that the plants *Th. Serpyllum* and *Th. Pulegoides* that grow in sunny habitats have an aromatic mono- and bicyclic monoterpenoid chemotype (K/α-T-neol/ G/p-C/B), with total dominance of carvacrol and p-cymene. The populations of *Th. Serpyllum*, which grow on the edges of sparse pine forests, and populations of *Th. Pullegioide*, with denser plant cover and which grow in meadows, have an acyclic and bicyclic monoterpene chemotype (G/α-T-neol/B/K). Plants that grow in the communities of meadow-steppe vegetation have the following chemotypes: *Th. Serpyllum*—L/K/G with 63% of linalool, *Th. Pullegioides*—G/α-T-neol/L/B, and *Th. Marschallianus*—α-T-neol/K/L/α-T-nen/G/B. Of these, the dominant chemotypes are α-terpineol and carvacrol (28:6.5%). A wide spectrum of antimicrobial activity was registered in samples of *Th. Pullegoides* and *Th. Serpyllum* that have an aromatic-monoterpenoid chemotype. Essential oils of *Th. Pullegoides* were dominated by carvacrol, and p-cymene had the highest fungicidal action (41.00 ± 1.0%). Plant populations of *Th. Pullegoides* and *Th. Serpyllum* with the aromatic-monoterpenoid chemotype are suggested by R&D to be of use in the pharmaceutical industry. They have high contents of natural components, which are effective in a wide spectrum of antimicrobial activity. The EO of *Th. Marschallianus* had the lowest influence on the inhibition of bacterial and fungal reproduction.

Keywords: chemotype; essential oils (EOs); biochemical content; antimicrobial activity; biodiversity

Citation: Kryvtsova, M.; Hrytsyna, M.; Salamon, I.; Skybitska, M.; Novykevuch, O. Chemotypes of Species of the Genus *Thymus* L. in Carpathians Region of Ukraine—Their Essential Oil Qualitative and Quantitative Characteristics and Antimicrobial Activity. *Horticulturae* **2022**, *8*, 1218. https://doi.org/10.3390/horticulturae8121218

Academic Editor: Riccardo Motti

Received: 27 October 2022
Accepted: 15 December 2022
Published: 19 December 2022

1. Introduction

Species of the genus *Thyme (Thymus* L.) are rich in essential oils, exhibit antimicrobial and antioxidant properties, and are widely used in ethnic and traditional medicine, aromatherapy, cosmetology, and cooking. Thyme has been used as a spice since ancient times in many countries of Southern Europe (Mediterranean) and Asia.

The investigation of the quantitative content and component composition of essential oils, depending on the environmental conditions of growth, and their antimicrobial and antifungal activities, depending on the chemotype of the population, are urgent issues today.

The component composition of essential oils depends, first of all, on geographical distribution and climatic conditions [1]. There is regularity in nature: species that grow in xerothermic conditions, in addition to a specific morphological and anatomical structure, have the ability to accumulate substances of secondary synthesis in greater quantities. It was found that in regions with a warm climate, species of the genus *Thymus* are dominated by a chemotype with a predominant content of aromatic monoterpenes (p-cymene, thymol, and carvacrol) [2–5], while in countries with a cool climate, the chemotype is monoterpenes and their oxygen derivatives and sesquiterpenes.

Thus, six chemotypes of *Th. Pulegioides* have been described in the Mediterranean, thymol, carvacrol, geraniol, linalool, fenchone, and alpha-terpenyl acetate, of which phenolic chemotypes predominate [2]. In the aerial organs of *Th. Serpyllum*, collected in the Kumaon and Uttarakhand regions of the Himalayas (India), 29 essential oil components were identified, of which approximately two-thirds of the composition was thymol, r-cymene, thymol methyl ester, and carvacrol methyl ester [3,4]. Thymol (32.2% and 27.4%) was the main compound in oils *Th. Vulgaris* and *Th. Pulegioides* in different thyme species grown in the Botanical Garden in Romania, with carvacrol (25.8%) in the oil of *Th. Serpyllum* and terpinyl acetate (47.6%) in *Thymus Glabrescens* oil [5].

In a temperate climate, the chemotype of monoterpene hydrocarbons is formed. In Lithuania, the essential oils of *Th. Serpyllum* ssp. *serpyllum* were combined into five monoterpene chemotypes, namely 1,8-cineole, germacrene B, (E)-β-ocimene, α-cadinol, and cis-p-ment-2-en-1-ol [6]. Herba *Th. Serpyllum* grown in Estonia had an essential oil yield of 0.6–4.4 and 1.9–8.2 mL/kg^{-1}. The following chemotypes of acyclic (myrcene) and bicyclic (borneol) monoterpenes and acyclic (E)-nerolidol) and bicyclic sesquiterpenes (caryophyllene oxide) are distinguished in it. Thymol and carvacrol, which are mentioned in the *European Pharmacopoeia* as the main components, are not the main components of the essential oil grown in Estonia [7]. The composition of the essential oil of *Th. camphoratus* grown in the Lviv Botanical Garden was dominated by bicyclic monoterpenoids—borneol and fenhol [8].

In the composition of essential oils of the raw material of *Th. Pulegioides* collected in Croatia were acyclic monoterpenoids, geraniol and linalool, followed by aromatic thymol, γ-terpinene, thymol methyl ester, and borneol, the amount of which changed during development. Due to the high content of geraniol in the essential oils of *Th. Pulegioides,* it may be interesting as an aromatic material for the perfume industry or for food flavoring [9].

The content of essential oils depends on the altitude above sea level [10]. In the composition of oil from the population of *Th. Serpyllum* in the Altai Mountains, located at an altitude of 150 m above sea level, monoterpenes predominated; from the population located at 500–700 m above sea level, half of the essential oil (EO) composition was aromatic oils—carvacrol, *p*-cymol, and other monoterpenes—γ-terpinene, and 1,8-cineole. Similar results were obtained from populations of *Th. Pulegioides* in Monte Pisani (Santagallo, Tuscany, 650 m.a.s.l.). Two populations had a thymol chemotype, and the third, with a typical lemon smell, contained more geraniol, with an essential oil level of 1.05 g [11].

The accumulation and qualitative composition of essential oils is affected by the method of their distillation [12]. Volatile substances of *Th. Serpyllum* were obtained by hydrodistillation (HD), simultaneous hydrodistillation and extraction (SDE), and static gas chromatographic mass spectrometry (HS) analysis. Monoterpenes were the most dominant in all three techniques (84.8–94.2%). The essential oil profiles obtained by HD and SDE were similar, with oxygenated monoterpenes being the most common (up to 75.4%). HS was dominated by monoterpene hydrocarbons (94.2%). The main compounds were α-terpinyl acetate (HD and SDE) and myrcene (HS).

In addition to essential oils of different component compositions, the synthesis of which depends on temperature conditions, because p-cymene and γ-terpinene are precursors of thymol, *Th. Serpyllum* contains various phenolic compounds. Fourteen phenolic compounds were detected in it, five of which were phenolic acids and nine of which were flavan-3-ols [13]. Tested samples of *Th. Serpyllum* contained rosmarinic acid (4–23 mg/g) and luteolin-7-*O*-β-D-glucuronide, which were the main components, while protocatechinic, 3-*O*-caffeylquinic, and caffeic acids (0.19–0.62 mg/g) were secondary components [14].

In the raw material of *Th. Marschallianus*, flavonoids were found: luteolin, quercetin, apigenin, and their derivatives, as well as rosmarinic acid and methyl rosmarinate. Significantly higher amounts of flavonoids were detected in the sample collected in the spontaneous flora, while for rosmarinic acid, a higher amount was detected in the cultivated sample. Both samples showed promising antibacterial activity, especially against Gram-positive organisms [15]. The most characteristic components of *Th. Marschallianus* were rosmarinic acid, protocatechuic acid, luteolin 7-*O*-glucoside, and apigenin 7-*O*-glucuronide, and GC-MS analysis of the volatile fraction revealed the presence of thymol and carvacrol. In addition, sesquiterpenoids, fatty acids, and their ethers were detected [16].

When analyzing random amplified polymorphic DNA (RAPD) of thymol and sesquiterpene chemotype plants, *Th. Pulegioides* revealed a partial correlation between molecular and chemical evaluations [17].

The *European Pharmacopoeia 8.0*, 2018 [18], includes raw material of *Thymi herba* species *Thymus Vulgaris* or *Th. Zygis*. In the *State Pharmacopoeia of Ukraine*, 2014 [19], in addition to these species, which are cultivated in the south of Ukraine, is the herb *Serpylli herba* of the natural flora species *Th. Serpyllum*. However, this name is often used in addition to *Th. Serpyllum* and other types of natural flora—*Th. Pulegioides* and *Th. Marshallianus*. For this reason, confusion often arises when preparing medicinal plant raw materials.

The aim of the R&D in this paper is to determine the range of essential oils and their qualitative and quantitative characteristics in the raw materials of genus Thymus and its species, which are growing in the natural flora of the Carpathian region in Ukraine. The study of their antimicrobial and antifungal activities was an important part of the experimental works.

2. Materials and Methods

2.1. Collection of the Plant Material

The object of the research was the medicinal plant material (MPM) of species of the genus Thymus (*Thymus* L.): pine thyme (*Th. Serpyllum* L.), which is included in the *State Pharmacopoeia of Ukraine*, and the non-official species flea thyme (*Th. Pulegioides* L.) and Marshall's thyme (*Th. Marschallianus Willd.*), the raw materials of which are used under the name *Herba serpylli*. *Th. Pulegioides* and *Th. Marschallianus* are promising essential oil medicinal plants, but they are often confused or considered to be admixtures of the morphologically close species *Thymus Serpyllum*, due to their morphological similarity.

The material for research (MPM) was selected in the Lviv region. Raw material of *Th. Serpyllum* was collected in three places of growth in Roztochchja, in the Yavoriv National Park (see the Herbarium at the website: https://yavorivskyi-park.in.ua/en/ (accessed on 16 December 2022)): on the crystalline outcrops of the Bila skelja near the village of Lelehivka (population No 1.1), in communities of psamophilous vegetation on the outskirts of the village of Strach (population No 1.2), and in a pine forest meadow near the village of Trostyanets, Mykolaiv district (population No 1.3). Raw material of *Th. Pulegioides* was collected in communities of grass-herbaceous vegetation in three places of growth: in the eastern outskirts of the city of Pustomyty (population No 2.1), and two in the Yavoriv National Park near the village of Yameljnya (population No 2.2) and the village of Vereshchytsja in the Havrylove tract in the communities of meadow-steppe vegetation (population No 2.3). Individuals of *Th. Marschallianus* were collected on Pishchana Gora (Lion Mountain) in the "Znesinnja" RLP, Lviv (population No 3.1) (Figure 1) (Atlas of the Lviv region: https://geoknigi.com/load.php?id=3 (accessed on 16 December 2022)) [20].

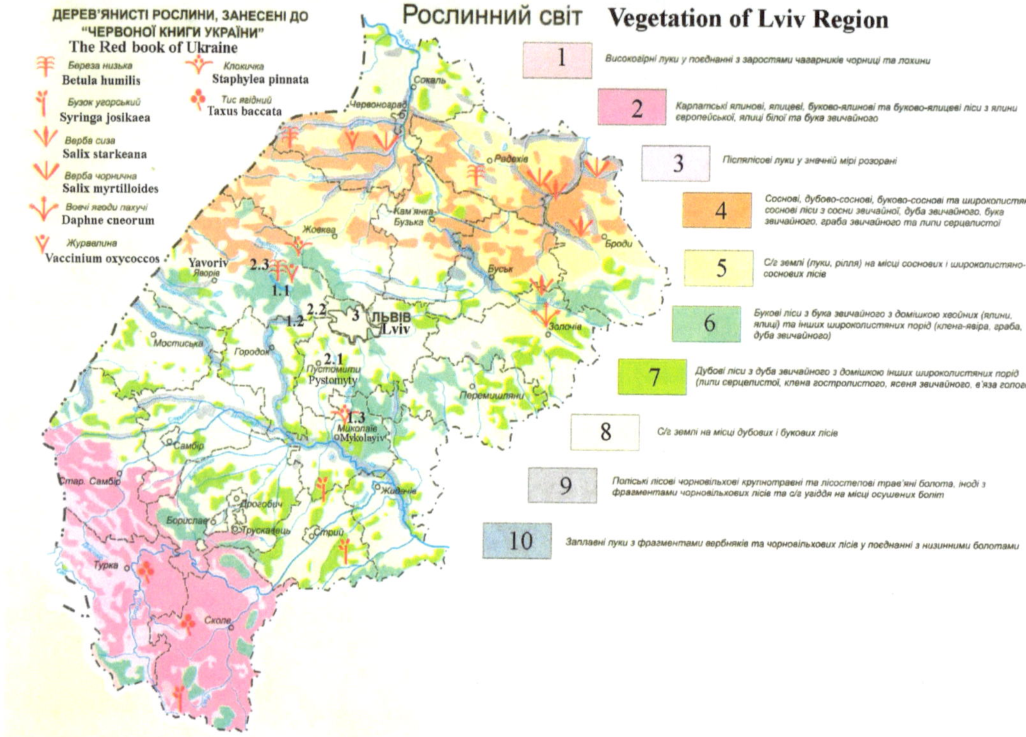

Figure 1. The distribution of species of the genus Thymus in vegetation of Lviv Region. *Th. Serpyllum*: 1.1—population No 1.1; 1.2—population No 1.2; 1.3—population No 1.3. *Th. Pulegioides*: 2.1—population No 2.1; 2.2—population No 2.2; 2.3—population No 2.3. *Th. Marschallianus*: 3—population No 3. Legend: 1 High-altitude meadows combined with thickets of huckleberries and blueberry bushes. 2 Carpathian spruce, fir, beech–spruce, and beech–fir forests from European spruce, white fir, and ordinary beech. 3 Post-forest meadows are largely plowed. 4 Pine, oak–pine, beech–pine, and broadleaf-pine forests of ordinary pine, ordinary oak, ordinary beech, ordinary hornbeam, and heart-leaved linden. 5 Agricultural land (meadow, arable land) in the place of pine and broadleaf-pine forests. 6 Beech forests from ordinary beech with an admixture of conifers (spruce and fir) and *other broad-leaved species (sycamore maple, hornbeam, and ordinary oak)*. 7 Oak forests from ordinary oak with admixture of other broad-leaved species (heart-*leaved linden, sharp-leaf maple, ordinary ash, and bare elm)*. 8 Agricultural land in the place of oak and beech forests. 9 Polissja forest of black alder coarse-grass and forest-steppe grass swamps, sometimes with fragments of black alder forests and agricultural land in place of drained swamps. 10 Floodplain meadows with fragments of willow and black alder forests in combination with lowland swamps.

2.2. Thymus Oil Distillation

Each sample of dry *Thymus* herb with weight of 20 g was grounded in a blender. The EO from this raw material was prepared by hydrodistillation (2 h) in a Clevenger-type apparatus according to the *European Pharmacopoeia 8.0* [18]. Hexane (1 mL) was used as a collecting solvent. The essential oils were stored under N_2 at + 4 °C in a dark space before their GC-FID/MS analysis.

2.3. Analyses of the Essential Oils

The determination of the qualitative composition of essential oils was carried out by GC/FID gas chromatography using a GC-SCION 456. The following operating conditions were used: column SOLGEL-WAX, 60 m × 0.25 mm i.d., thickness of the film 0.25 µm, carrier gas nitrogen regulated to the current of 1 mL/min, the temperature of injection to the detector FID 220 °C and 250 °C, respectively.

The components were identified according to their retention time (RT) (min), and the obtained values were compared with literature data. Standards for comparison of essential oils were provided by Extrasynthesis Ltd. (Lyon, France) and Sigma-Aldrich (St. Louis, USA). Research results are presented in percentages. The percentage of single chromatographic peaks was measured as the ratio of the single peak area to the total peak area. Detailed descriptions of component determination of all *Thymus* species including with RT and chromatographic profiles were given by Hrytsyna et al., 2020 [8].

Analysis of GC/MS was conducted using gas chromatograph Varian 450-GC connected with Varian 220-MS. The division was made using FactorFourTM: capillary column DB-WAX, 30 m × 0.25 mm i.d., thickness of film 0.25 µm. The injector of 1177 type was heated at a temperature of 220 °C. The injection equaled 1 µL of the solution of 1:1000 n-hexane/diethyl ether. Helium was used as carrier gas during constant velocity of the current in the column of 1.2 mL/min. Temperature of the column was set as follows: the initial temperature for 10 min was 50 °C, then to 100 °C at 3 °C/min; isothermic temperature was held for 5 min, and then continued to 150 °C at 10 °C/min. Total time for the analysis of one sample was 54.97 min. The constituents were identified by comparing their mass spectrum with the spectra kept in NIST 02 (library of software) or mass spectra from the literature (Adams, 2007) [21], and also the comparison of their indices obtained with the standards was performed. Statistical analysis was conducted using reliable intervals at the level of $p < 0.05$ with our calculation through mean statistical deviation and standard error.

2.4. Method for the Testing of the Antimicrobial Activity

The antimicrobial activity of EO was studied at the Microbiological Laboratory of the Department of Genetics, Plant Physiology and Microbiology, Uzhhorod National University. The sensitivity of microorganisms against the EOs was determined by agar diffusion test using typical *Staphylococcus aureus* (ATCC25923), *Escherichia coli* (ATCC25922), *Streptococcus pyogenes* (ATCC19615), and *Candida albicans* (ATCC885-653) [22]. As positive controls were used gentamicin (10 mg/disk) for Gram-negative bacteria, ampicillin (10 mg/disk) for Gram-positive bacteria, and nystatin (100 UI) for *Candida*. As negative control DMSO was used. The bacterium inocula 100 µL in the physiological solution were adjusted to the equivalent of 0.5 McFarland (5×10^8 KFU/mL) standard, and evenly spread on the surface of Muller–Hinton agar (incubated at 37 ± 2 °C for 24 h); yeasts were spread on SDA agar (incubated at 35 ± 2 °C for 48 h). The 20 µL extracts were introduced into wells 6 mm in diameter. The diameters of the inhibition zones were measured in millimeters, including the diameter of the well. The antimicrobial effect was assessed by the presence of the growth inhibition zone. Each antimicrobial assay was performed at least three times.

2.5. Statistical Analyses of the Data

For the results of experiment, we used statistical software ANOVA with the calculation of averages, error, and standard deviation. Differences were considered significant at $p < 0.05$.

3. Results

The *European Pharmacopoeia 8.0* (2018) [18] includes raw material of *Thymi herba*. These are whole leaves and flowers of *Th. Vulgaris* L. or *Th. Zygis* L. with an essential oil content of at least 12 mL/kg, in which the sum of thymol and carvacrol is at least 40 percent of the essential oil. In the *State Pharmacopoeia of Ukraine*, 2014 [19], species that are cultivated in the south of Ukraine and the herb *Serpylli herba* of the natural flora species *Th. Serpyllum* are also included. In the herb *Serpylli herba*, the content of essential oil is not less than 3.0 mL/kg, and in the national article is not less than 1.5 mL/kg.

As a result of the research, it was found that individuals from all three research populations of *Th. Serpyllum* contain fairly large contents of essential oil (Table 1). The largest amount of essential oil (0.9 ± 0.05%, or 9 mL/kg) was from plants from the communities of thermophilic vegetation on the outcrops of crystalline rocks of the Bila skelja (population 1.1). In second place in terms of essential oil content (0.8 ± 0.05%, 8 mL/kg) was the population on the edge of a sparse pine forest on the Torton sandstones (population 1.3). Plants growing in the plant communities of psammophilous vegetation had the lowest content of essential oil—0.7 ± 0.05% (7 mL/kg of dry raw material) (population 1.2) (Table 1). In all research populations of *Th. Pulegioídes*, raw material contained 0.8 ± 0.05% (8 g/kg, calculated on anhydrous raw material), 0.85 ± 0.05%, and 0.9 ± 0.05% essential oil. The content of essential oil in the raw material of Marshall's thyme (*Th. Marschallianus*) was 0.35 ± 0.05% (Table 2). The total number of identified main components of EO ranged from 67.2 to 90.6% (Tables 1 and 2).

Table 1. The component composition of essential oils of *Th. Serpyllum* from different places of plant population growth (x ± SD).

Population		**1.1**		**1.2**		**1.3**	
EO Content (%, *v/w*, Expressed in Dry Weight)		**0.90 ± 0.05**		**0.70 ± 0.05**		**0.80 ± 0.05**	
GC Analysis (%)	**RT (min)**	**1***	**2***	**1***	**2***	**1***	**2***
β-myrcene	12.84	0.3 ± 0.1	0.3 ± 0.1	0.5 ± 0.1	0.5 ± 0.1	0.3 ± 0.1	0.3 ± 0.1
1,8-cineole	14.18	trace **	trace	0.1 ± 0.1	0.1 ± 0.1	trace	trace
limonene	14.74	trace	trace	trace	trace	trace	trace
α-terpinene	16.0	3.0 ± 0.5	2.5 ± 0.5	4.0 ± 0.5	3.5 ± 0.5	4.0 ± 0.5	3.2 ± 0.5
p-cymene	16.77	2.0 ± 0.5	2.0 ± 0.5	10.0 ± 1.0	9.1 ± 1.0	1.6 ± 0.2	1.6 ± 0.2
terpinolene	17.91	trace	trace	trace	trace	trace	trace
α-thujone	22.5	0.1 ± 0.1	0.1 ± 0.1	1.6 ± 0.2	0.6 ± 0.1	0.5 ± 0.1	0.5 ± 0.1
linalool	24.54	63.0 ± 2.0	63.0 ± 2.0	0.4 ± 0.1	0.4 ± 0.1	0.7 ± 0.1	0.7 ± 0.1
bornyl acetate	25.26	0.5 ± 0.1	0.3 ± 0.1	4.5 ± 0.5	3.5 ± 0.5	1.4 ± 0.2	1.0 ± 0.2
β-caryophyllene	26.51	2.5 ± 0.5	2.0 ± 0.5	3.5 ± 0.5	3.0 ± 0.5	4.0 ± 0.5	5.5 ± 0.5
borneol	29.25	1.3 ± 0.2	1.7 ± 0.2	7.5 ± 0.5	7.5 ± 0.5	8.0 ± 0.5	9.5 ± 1.0
α- terpineol	29.92	2.5 ± 0.5	2.2 ± 0.5	20.2 ± 1.0	20.0 ± 1.0	17.0 ± 1.0	18.3 ± 1.0
fenchol	32.09	0.2 ± 0.1	0.2 ± 0.1	2.0 ± 0.5	2.0 ± 0.5	4.0 ± 0.5	3.0 ± 0.5
geraniol	33.19	5.0 ± 0.5	4.5 ± 0.5	15.0 ± 1.0	15.1 ± 1.0	35.0 ± 2.0	35.0 ± 2.0
thymol	42.65	0.6 ± 0.1	0.8 ± 0.1	0.2 ± 0.1	0.2 ± 0.1	0.2 ± 0.1	trace
carvacrol	45.53	6.0 ± 0.5	6.0 ± 0.5	20.0 ± 1.0	18.0 ± 1.0	6.0 ± 0.5	6.5 ± 0.5
Content of the main EOs (%)		87.0	85.6	89.5	83.5	82.7	85.1

Note: 1*—GC-SCION 456 SOLGEL-WAX gas chromatograph with a column length of 60 m; 2*—GC-Carlo Erba VEGA DB-WA gas chromatograph with column length 30 m; ** <0.05%.

Table 2. The component composition of *Thymus* essential oils from different places of growth plant species (x ± SD).

Type		*Th. Pulegoides*						*Th. Marschtallianus*	
Population		2.1		2.2		2.3			
EO Content (%, v/w, Expressed in Dry Weight)		0.90 ± 0.05		0.85 ± 0.05		0.80 ± 0.05		0.35 ± 0.05	
GC Analysis (%)	**RT (min)**	1*	2*	1*	2*	1*	2*	1*	2*
β-myrcene	12.84	0.7 ± 0.1	0.7 ± 0.1	0.2 ± 0.1	0.2 ± 0.1	0.5 ± 0.1	0.5 ± 0.1	3.5 ± 0.5	3.5 ± 0.5
1,8-cineole	14.18	0.1 ± 0.1	0.1 ± 0.1	trace	trace	trace	trace	0.4 ± 0.1	0.3 ± 0.1
limonene	14.74	trace **	trace	trace	trace	trace	trace	0.9 ± 0.1	0.8 ± 0.1
α-terpinene	16.0	8.0 ± 0.5	7.5 ± 0.5	1.6 ± 0.2	1.5 ± 0.2	2.5 ± 0.5	2.0 ± 0.5	4.0 ± 0.5	4.0 ± 0.5
p-cymene	16.77	6.0 ± 0.5	6.0 ± 0.5	3.0 ± 0.5	3.2 ± 0.5	1.6 ± 0.2	1.6 ± 0.2	1.0 ± 0.2	1.0 ± 0.2
terpinolene	17.91	trace	trace	trace	trace	1.8 ± 0.2	2.0 ± 0.5	2.2 ± 0.2	2.4 ± 0.2
α-thujone	22.5	1.2 ± 0.2	1.2 ± 0.2	1.2 ± 0.2	1.0 ± 0.2	1.0 ± 0.2	0.8 ± 0.1	0.8 ± 0.1	0.7 ± 0.1
linalool	24.54	0.6 ± 0.1	0.6 ± 0.1	trace	trace	13.0 ± 1.0	12.0 ± 1.0	4.5 ± 0.5	3.5 ± 0.5
bornyl acetate	25.26	3.0 ± 0.5	3.0 ± 0.5	1.0 ± 0.2	1.0 ± 0.2	0,3 ± 0.1	0.4 ± 0.1	4.0 ± 0.5	3.0 ± 0.5
β-caryophyllene	26.51	2.5 ± 0.5	3.5 ± 0.5	4.0 ± 0.5	5.0 ± 0.5	3.0 ± 0.5	4.0 ± 0.5	2.0 ± 0.2	1.0 ± 0.2
borneol	29.25	7.0 ± 0.5	8.0 ± 0.5	10.0 ± 1.0	11.0 ± 1.0	6.0 ± 0.5	7.5 ± 0.5	2.5 ± 0.5	2.5 ± 0.5
α-terpineol	29.92	13.0 ± 1.0	11.0 ± 1.0	20.0 ± 2.0	21.0 ± 2.0	15.0 ± 1.0	15.0 ± 1.0	28.0 ± 1.0	30.1 ± 1.0
fenchol	32.09	3.5 ± 0.5	3.5 ± 0.5	4.0 ± 0.5	4.0 ± 0.5	2.5 ± 0.5	2.5 ± 0.5	1.2 ± 0.2	1.2 ± 0.2
geraniol	33.19	22.0 ± 2.0	22.0 ± 2.0	33.0 ± 2.0	33.0 ± 2.0	15.0 ± 1.0	15.0 ± 1.0	3.5 ± 0.5	2.5 ± 0.5
thymol	42.65	0.2 ± 0.1	0.2 ± 0.1	0.2 ± 0.1	0.2 ± 0.1	2.0 ± 0.5	2.0 ± 0.5	2.2 ± 0.2	2.2 ± 0.2
carvacrol	45.53	17.0 ± 1.0	18.0 ± 1.0	9.5 ± 0.5	9.5 ± 0.5	3.5 ± 0.5	4.0 ± 0.5	6.5 ± 0.5	9.0 ± 0.5
Content of the main EOs (%)		84.8	85.3	87.7	90.6	67.7	69.3	67.2	67.7

Note: 1*—GC-SCION 456 SOLGEL-WAX gas chromatograph with a column length of 60 m; 2*—GC-Carlo Erba VEGA DB-WA gas chromatograph with column length 30 m, ** <0.05 %.

When analyzing the component composition of essential oils, it turned out that the raw materials of *Th. Serpyllum* collected on the lawn of a pine forest on the Torton sandstones and in communities of psamophilous vegetation on sandy soils in Roztochchja (populations 1.2 and 1.3) are similar in their component composition of EO: geraniol, α-terpineol, borneol, and carvacrol, p-cymene (G/α-T-neol/B/K) in the ratio 35:17:7.5:6% (population 1.3) and K/α-T-neol/G/p-C/B in the ratio 20:20:15:10:9.5 (population 1.2) (Table 1; Figure 2). Other essential oils were less than 1–4%: β-myrcene, α-terpinene, α-thujone, bornyl acetate, β-caryophyllene, fenchol, and tenths of thymol. For 1,8-cineole, limonene, terpinolene, and linalool (population 1.2) there were only traces (<0.05%). Plants *Th. Serpyllum* from Lelehivka, which grew on the outcrops of crystalline rocks of the Bila skelja in the community of meadow-steppe vegetation, differed from other populations. Their raw material contained 63% linalool (an acyclic monoterpene) and only 6% of carvacrol, 5% of geraniol, and 2.5% of α-terpineol (L/K/G) in the composition of essential oils, while there were traces of linalool in raw materials from other populations.

Therefore, all populations are dominated by monoterpene volatile compounds; in particular, for population 1.3, located near a pine forest, the mono- and bicyclic terpenoids-carvacrol (G/α-T-neol/K/B/p-C) chemotype is characteristic. Linalool (L/G/K) chemotypes were found in the plants of population 1.1 on the crystalline outcrops of the Bila skelja overgrown with pine trees. However, in population 1.2, which is located in a sunny meadow in the communities of psammophilous plants, it is aromatic, mono- and bicyclic terpenoids (K/α-T-neol/ G/p-C/B) chemotype, with a significant content of aromatic monoterpenes—carvacrol and p-cymene (30%) (Figure 2). Since in the populations 1.2. and 1.3 about 20% is α-terpineol, which is the genetic precursor of carvacrol and p-cymene (aromatic compounds), then we can say that high summer temperatures are not enough for their complete transformation [23].

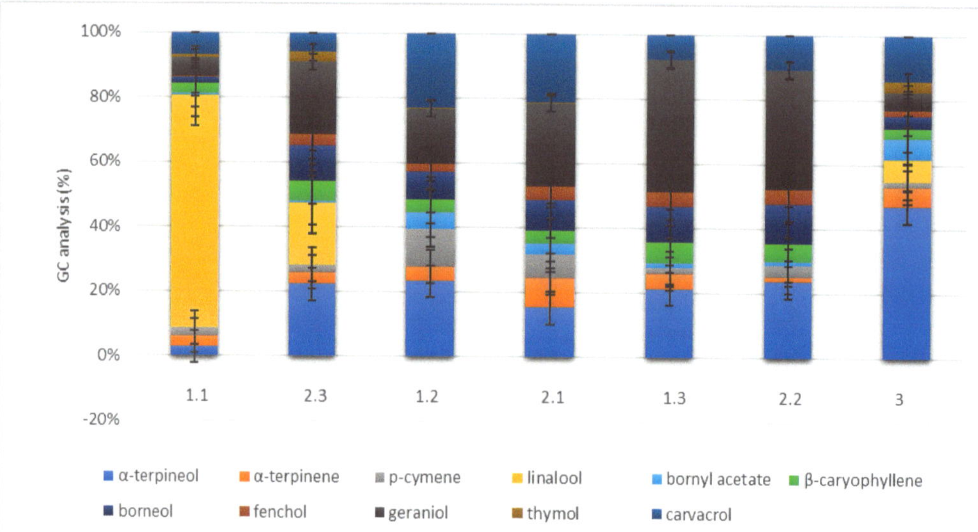

Figure 2. The component composition of essential oils of species of the genus *Thymus* under different growth conditions. Legend: 1.1—population of the *Th. Serpyllum* in the community of meadow-steppe vegetation on the crystalline outcrops of the Bila skelja near the village of Lelehivka in Roztochchja (Yavoriv National Park) (chemotype L/K/G); 1.2—population of the *Th. Serpyllum* in a community of psamophilous vegetation on the outskirts of the village of Strach (chemotype K/α-T-neol/G /p-C/B); 1.3—population of the *Th. Serpyllum* in a community of psamophilous vegetation in a pine forest meadow near the village of Trostyanets, Mykolaiv district (chemotype G/α-T-neol /B/K); 2.1—population of the *Th. Pulegioides* of grass-herbaceous vegetation on the eastern outskirts of the city of Pustomyty (chemotype K/G/α-T-neol/α-T-nen/B/p-C); 2.2—population of the *Th. Pulegioides* of grass-herbaceous vegetation near the village of Yameljnya (Yavoriv National Park) (G/α-T-neol /B/K chemotype); 2.3—population of the *Th. Pulegioides* in the community of meadow-steppe vegetation in the Havrylove tract near the village of Vereshchytsja (Yavoriv National Park) (G/α-T-neol/L/B); 3—population of the *Th. Marschallianus* collected on Pishchana Gora (Lion Mountain) in the "Znesinnja" RLP, Lviv (α-T-neol/K/L/α-T-nen/G/B-at chemotype).

In another species, *Th. Pullegoides*, raw materials were selected from plants growing in populations of the meadow-grass vegetation (Table 2). Monoterpenes and their oxygen derivatives dominate the EO composition of raw materials from all populations. Thus, the raw material collected in population 2.1 from the sunny eastern outskirts of the city of Pustomyty had the highest content of aromatic essential oils (carvacrol and p-cymene, 17 + 6%; geraniol content, 22%; and α-terpineol, 13%). The plants had high contents of α-terpinene (8%) and borneol (7%) (chemotype K/G/α-T-neol/α-T-nen/B/p-C). The EO contents of bornyl acetate, β-caryophyllene, α-thujone, β-myrcene, borneol, and fenchol ranged from 4 to less than 1%, while thymol was 0.2% and only in the population from Vereshchytsja, 2%. Traces of such essential oils as 1,8-cineole, limonene, terpinolene, and linalool were detected. Raw material of *Th. Serpillum* had a similar chemotype with a large percentage of aromatic compounds from the solar psammophilous population 1.2 (Figure 2).

In plants from population 2.2. from the outskirts of the village of Yameljnya, there were more monocyclic alcohols; in particular, there was twice as much geraniol (33%) and α-terpineol (20%) compared to carvacrol (9.5%) and bicyclic monoterpene borneol (10%), and much less α-terpinene and p-cymene (G/α-T-neol/B/K chemotype) (Table 2). All other oils were approximately similar in quantitative content. Population 2.2, where individuals of *Th. pullegoides* compete with forbs, was determined by the geranial-α-terpineol-borneol-carvacrol chemotype, similar to the population of *Th. Serpillum* from the edge of the pine forest of Trostyanets village (population 1.3) (Figure 2).

Plant raw material of *Th. Pulegoides*, collected on the edge of the forest in a community of meadow-steppe vegetation in the Havrylove tract near the village of Vereshchytsja, had 15% of geraniol and α-terpineol and 13% of linalool, traces of which were present in other populations, and 6% of borneol. Aromatic monoterpenes were only 5.5% of the composition (carvacrol and thymol). The monoterpenoid (geranial-α-terpineol-linalool-borneol) (G/α-T-neol/L/B) chemotype is characteristic of the plant material from this population. In the population of *Th. serpillum* from the Bila skelja, the greater half of the EO composition was linalool (Figure 2).

Therefore, individuals of *Th. Pulegoides* from populations 2.1. and 2.2 are also characterized by the geranial-α-terpineol-borneol-carvacrol chemotype. The most aromatic essential oils of p-cymene and carvacrol are in population 2.1—aromatic-geraniol-α-terpineol-α-terpinene-borneol (G/K/α-T-neol/α-T-n/B/p-C) chemotype in the ratio of 22:17:13:8:7:6%. Individuals of population 2.2 have a geranial-α-terpineol-borneol-carvacrol (G/α-T-neol/B/K) chemotype in the ratio of 33:20:9.5:10%, whereas population 1.3 from Vereshchytsja has geraniol-α-terpineol-linalool-borneol (G/α-T-neol/ L/B) in the ratio of 15:15:13:6% (Figure 2).

The content of essential oils in raw material of *Th. Marschallianus* growing on Pishchana Gora ("Znesinnja") was the lowest (0.35 ± 0.05%), about 2–2.5 times compared to other species. The qualitative composition was dominated by α-terpineol (28%) and carvacrol (6.5%), while linalool, α-terpinene, geraniol, bornyl acetate, which were abundant in other species, accounted for only 3–4.5% each (α-T-neol/K/L/α-T-nen/G/B-at chemotype) (Table 2; Figure 2).

A comprehensive comparison of the qualitative and quantitative compositions of the essential oil of individual species of the genus *Thymus* in the Carpathians region of Ukraine is illustrated in Figure 2. These are the first results from Ukraine that determine *Thymus* chemotypes that can be used for tests of therapeutic use.

The research shows that the highest levels of antimicrobial activity were shown by the samples of *Th. pullegoides* (population 2.1) and *Th. Serpyllum* (population 1.2), which had an aromatic-monoterpenoid chemotype, and population 2.2 with a mono-bicyclic-terpenoid-carvacrol chemotype (Tables 3 and 4). At the same time, a wide range of activity of these samples should be noted for both Gram-positive and Gram-negative bacteria, as well as microscopic fungi of the genus *Candida*. It was found that essential oils had the most pronounced antimicrobial effect on microscopic fungi of the genus *Candida*, as the zones of growth retardation in the analysis of all samples ranged from 11.66 ± 0.58 to 41.00 ± 1.00. The most pronounced antimycotic effect was characteristic for raw materials obtained from population 2.1., *Th. pullegoides*, with an aromatic-monoterpenoid chemotype. High activity against *S. pyogenes* was characteristic of the sample of *Th. Marschtallianus* and for the sample from population 1.3, *Th. Serpyllum*, with a monoterpenoid chemotype. Regarding *E. coli*, a sample from population 1.2, *Th. Serpyllum*, had an aromatic-monoterpenoid chemotype.

Table 3. The influence of EO from raw materials of *Th. Serpyllum* from different habitats on typical strains of microorganisms: inhibition zones, mm, M ± m.

	Population		
	1.1	**1.2**	**1.3**
Test microorganisms	Inhibition zone, mm, M ± m		
Staphylococcus aureus ATCC 25923	11.17 ± 0.29	13.17 ± 0.29	11.50 ± 0.50
Escherichia coli ATCC 25922	15.66 ± 0.58	19.83 ± 0.76	9.66 ± 0.58
Streptococcus pyogenes ATCC 19615	11.67 ± 0.58	14.00 ± 0.25	18.50 ± 0.87
Candida albicans ATCC 885-653	11.17 ± 0.29	29.66 ± 1.52	22.66 ± 0.58

Table 4. Antimicrobial and fungicidal action of raw materials of *species of the genus Thymus* from different habitats: inhibition zones, mm, M ± m.

	Th. Pulegoides (Population)			***Th. Marschtallianus***
	2.1	**2.2.**	**2.3**	**3**
Test microorganisms	Inhibition zone, mm, M ± m			
Staphylococcus aureus ATCC 25923	14.50 ± 0.50	13.50 ± 0.50	13.50 ± 0.50	8.67 ± 0.58
Escherichia coli ATCC 25922	14.50 ± 0.50	15.33 ± 0.58	10.83 ± 0.76	0
Streptococcus pyogenes ATCC 19615	15.83 ± 0.29	13.33 ± 0.58	15.50 ± 0.50	18.83 ± 0.76
Candida albicans ATCC 885-653	41.00 ± 1.00	29.00 ± 1.00	11.66 ± 0.58	0

Therefore, a wide spectrum of antimicrobial activity was recorded in samples of *Th. pullegoides* and *Th. Serpyllum* with an aromatic-monoterpenoid chemotype.

Extracts made from plants of population 2.2, *Th. Pulegoides*, which was dominated by monoterpenoid essential oils, which are the precursors of aromatic ones, had slightly lower antimicrobial activity and almost half as much fungicidal activity (29.00 ± 1.00%). Raw materials obtained from population 1.2., *Th. Serpyllum*, with a carvacrol-geraniol chemotype, had a similar fungicidal action (29.66 ± 1.52%) (Tables 3 and 4; Figure 3). All other populations with the monoterpenoid-bicyclic chemotype had moderate antimicrobial activity.

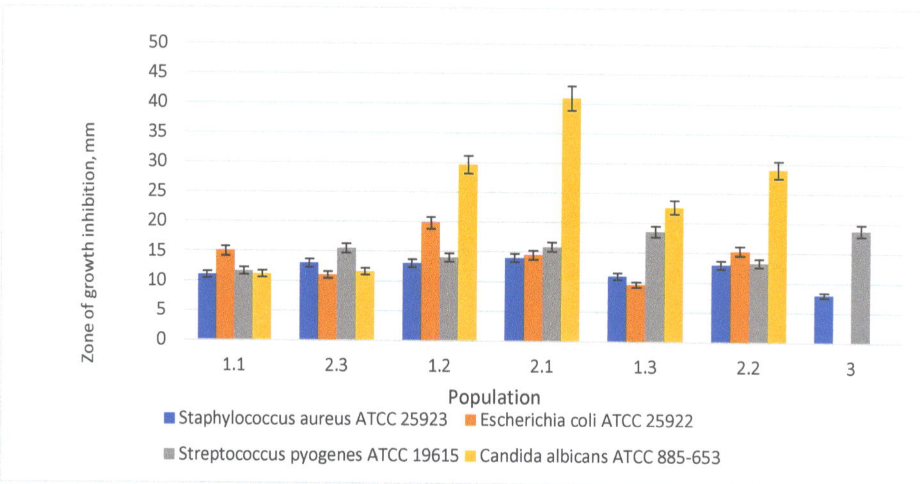

Figure 3. Antimicrobial and fungicidal action of raw materials of species of the genus *Thymus* from different habitats, M ± m. Legend: notations are similar as in Figure 2.

Plants with acyclic monoterpene chemotypes, linalool *Th. Pulegoides* and linalool *Th. Serpyllum*, had low antimicrobial and especially fungicidal activity.

The antimicrobial action on *S. aureus* ATCC 25923 and *S. pyogenes* ATCC 19615 was similar to that of other populations, while on *Escherichia coli* ATCC 25922 it was somewhat lower, and there was a completely low fungicidal effect on *Candida albicans* ATCC 885-653.

The essential oils of *Th. Marschallianus* had the least antimicrobial action, while they had no influence on *C. albicans* ATCC 885-653 and *E.coli* ATCC 25922. Perhaps this can be explained by the low content of geraniol (3.5%), which is abundant in other species and aromatic monoterpenes.

4. Discussion

In the literature, considerable attention is paid not only to the quantitative content, but also to the component composition of essential oils and its dependence on environmental conditions, as well as to the antimicrobial and fungicidal properties of MPM with different component compositions of essential oils.

Therefore, the component composition of essential oils in species of the genus *Thymus* depends, first of all, on the microclimatic conditions of the population. Therefore, in populations of *Th. Serpyllum* growing shaded by a pine forest, or *Th. Pulegoides* in the formed cereal communities and shaded by cereal grasses, characteristic mono- and bicyclic aromatic chemotypes G/α-T-neol/K/B/p-C (*Th. Serpyllum*) and G/α-Teol/B/K (*Th. Pulegoides*) are present, with a slight content of carvacrol (Figure 2: 1.3; 2.2).

Populations in *Th. Serpyllum* and *Th. Pulegoides*, which grow on sunny lawns, have an aromatic mono- and bicyclic monoterpenoids chemotype: K/α-T-neol/G/p-C/B in the first species and G/K/α-T-neol/α-T-nen/B/p-C in *Th. Pulegoides*, with the total dominance of aromatic monoterpenes (carvacrol and p-cymene) (Figure 2: 1.2; 2.1).

In *Th. Serpyllum* plants growing on the crystalline outcrops of the Bila skelja there are linalool (L/G/K) chemotypes. A high content of linalool is found in the population of *Th. Pulegoides* from Vereshchytsja (Havrylove tract) (G/α-T-neol/L/B) (Figure 2: 1.1; 2.3). Both species grow in a group of unique meadow-steppe vegetation. In populations of *Th. Pulegioides* on limestone meadows in Germany, 60% of the research individuals also had the linalyl acetate chemotype, and 20% had the β-caryophyllene-germacrene D-β-bisabolene chemotype [24].

The content of essential oil in the raw material of Marshall's thyme (*Th. Marschallianus*) collected in the communities of strongly changed meadow-steppe vegetation on Pishchana Gora was 0.35 ± 0.05%, and α-terpineol dominated the component composition (Figure 2). *Th. pulegioides*, which grows in meadow coenoses with the dominance of meadow cereals and a high Easter load, has a stable content of essential oils.

Raw material obtained from the population of *Th. Serpyllum* and *Th. Pulegioides* growing in sunny open places (population 1.2 and 2.1) has an aromatic-monoterpenoid chemotype and strong antimicrobial and fungicidal action. This can be explained by the fact that the precursor of carvacrol is α-terpineol, and their transformation depends on the amount of photosynthetically active solar radiation. In addition, the amount of EO in *Th. Pulegioides* decreased in cold rainy weather and increased with the growth of photosynthetically active solar radiation from 0.72% to 0.98% [23,25]. The percentage of carvacrol varied between 16.88 and 29.29%, p-cymene from 5.54–11.33%, and γ-terpinene from 20.60–24.43%. Moreover, the amount of carvacrol grew with the increase in photosynthetically active solar radiation during flowering in July, and the highest content was at the stage of fruiting. The percentage of carvacrol precursors—p-cymene and γ-terpinene—decreased. The p-cymene content showed a significant positive relationship with temperature and precipitation during the April–July period and July temperature. The best time to harvest this type of thyme, both in terms of essential oil yield and phenol content, is during or immediately after full bloom. Research conducted in Kazakhstan showed that the highest content of essential oils—0.4–1.4%—was in the aerial part of *Th. Serpyllum* at the stage of full flowering, and the plants had a phenolic chemotype [10]. In the Left Bank Forest-Steppe, the content of essential oil in the raw material of *Th. Pallasianus* changes during the vegetation period of plants from traces to 0.09%, in *Th. Marcshallianus* from 0.35 to 1.3%, in *Th. Serpyllum* from 0.11 to 0.97%, in *Th. Dimorphus* from 0.24 to 0.37%, and in *Th. Pulegioides* from 0.25 to 0.86% [26].

In addition, *Th. Serpyllum* grows in favorable, thermophilic conditions, which contributes to the synthesis of essential oils in the plant organs. Tymchenko et al., 2007 [27], when studying the resource potential of *Th. Serpyllum* in the Volynj region, came to the conclusion that the reason for the decrease in the resources of species of the genus *Thymus* is the Easter load. The species is important for raw materials mainly in areas with poorly developed grass cover, which are not used for pastures, or in areas with disturbed vegetation cover. As the level of turfing of coenoses increases, the development of populations of *Th. Serpyllum* is suppressed. Formation of populations of *Th. Serpyllum* and *Th. polessicus* in the conditions of the Sumy region is a complicated growth under the tent of a pine forest, and in the local growth there are only generative plants [28]. Only the population formed under the conditions of the *Elytrigioso (repenae)—thymosum* (*serpyllae*) communities is complete in terms of its ontogenetic structure. Morphometric indicators also depend on growth conditions [29]. In our research, the raw material of both species, collected on the edge of a pine forest (population 1.3; 2.2), had monoterpenoids (geranial-α-terpineol-borneol-carvacrol) with a lower antimicrobial and fungicidal action (Figure 3).

The component composition of EOs, as research has shown, depends on geographical distribution and climatic conditions. The content of essential oil from raw materials collected on the territory of the Left Bank Forest-Steppe of Ukraine differed from our results [26]. In the phase of mass flowering for *Th. marcshallianus* it is 0.9%, compared to 0.35% in the Lviv district; for *Th. Serpyllum*, on the contrary, it is 0.26% compared to 0.9 in the Lviv district. For *Th. Pulegioides* the contents are the same: 0.8% in both regions. Such a difference in the content of essential oils is obviously due to the fact that *Th. Serpyllum* in the Lviv district grows in its characteristic populations on pine sands at the edges of pine forests, while the steppe species *Th. marcshallianus* has a three times lower content of essential oil than on the Left Bank, which is due both to its growth on sandy soils in a place with a strong anthropogenic load and to the fact that the investigated region is the extreme limit of distribution of this species.

Chemotypes of monoterpene hydrocarbons and their oxygen derivatives and aromatic-monoterpenoid chemotypes that were isolated for MPM of the species investigated by us of the *Thymus genus* (*Th. Serpyllum, Th. Pulegioides L.* and *Th. Marschallianus* Willd.) are also characteristic of the temperate climate of northern European countries [6,7]. Thus, in *Th. Serpyllum L. ssp. serpyllum var. serpyllum* investigated in the Vilnius region (Eastern Lithuania) [30], the following chemotypes were distinguished: 1,8-cineol/caryophyllene oxide and chemotype with a boron carbon skeleton (camphor + borneol + bornyl acetate). The composition of *Th. Pulegioides* EO was dominated by citral-geraniol with a lemon smell and carvacrol chemotypes [31]. The essential oils of plants with an "ether" odor containing 50–70% α-terpenyl acetate belong to the alpha-terpenyl acetate chemotype, which has not previously been observed in the species *Th. Pulegioides* [32]. In further investigations, it was found that only 35% of the population of *Th. Pulegioides* in the Vilnius region (Lithuania) had the α-terpinyl acetate chemotype [33].

In the south of Ukraine, in the Kherson region, two forms of *Th. Serpyllum* had an essential oil content of 1.1% and 0.67%, and the dominant components were thymol (40.70% and 40.29%) and γ-terpinene (12.88 and 23.31%) [34].

When grown in culture, *Th. Pulegioides* collected from various natural environments in Lithuania had the following chemotypes: geraniol/geranium/neral (G/G/N), thymol (T), linalool (L), carvacrol/γ-terpinene/p-cymene (C/γT/pC), and thymol/carvacrol/γ-terpinene/p-cymene (T/C/γT/pC). It was found that with a sudden change in environmental conditions, two types of *Th. Pulegioides* can be distinguished: plants that retain their chemical composition of essential oils and plants that significantly change their chemical composition of essential oils [35].

The composition of the essential oil is stable and determined genetically [36]. When studying in Lithuania the content of essential oils in *Thymus* ×*Oblongifolius Opiz*, which is a hybrid of *Th. Serpyllum* L. and *Th. Pulegioides* L., eight chemotypes were found: 1-octen-3-ol, (Z)-β-ocimene, geraniol, myrcene, 1,8-cineole, α-terpinyl acetate, germacrene B, and geraniol/germacrene B. It is known that hybrids, in addition to intermediate morphological features, have a diverse chemical composition. When grown in culture for four years, *Th.* × *oblongifolius* had the same oil composition.

EOs are widely used not only as part of herbal preparations and in cosmetology, but also as aromatic ingredients in a variety of food products, beverages, and confectionery [37].

The antibacterial activity of essential oils makes thyme a potential natural preservative in the food industry [38].

The antimicrobial action of EOs is determined by the chemotype of the population. Our research also confirmed this hypothesis. Thus, raw materials collected from populations 1.2 and 2.1 that had an aromatic-monoterpene chemotype had a stronger antimicrobial action, and medicinal plant raw materials *Th. Pulegoides*, whose EO were dominated by carvacrol and p-cymene, had the highest fungicidal activity (41.00 ± 1.00%) against *Candida albicans* ATCC 885-653 in all samples (Figure 3). However, populations 1.3 and 2.2 with mono- and acyclic monoterpenoid essential oils had a slightly lower antimicrobial action and almost half the fungicidal action (29.00 ± 1.00%). Plants with the linalool chemotype of both species (populations 1.1 and 2.3) had low antimicrobial and especially fungicidal activity. EOs of *Thymus Marschallianus* had the least antimicrobial activity, while they had no influence on *C. albicans ATCC 885-653* and *E. coli ATCC 25922*.

EOs of samples *Th. Pulegioides* collected in Campania and Sicily, in which 36–39% have phenols, in particular thymol, showed antibacterial activity against eight selected Gram+ and Gram− microorganisms: *S. aureus* (ATTC 25923), *S. faecalis* (ATTC 29212), *B. subtilis* (ATCC 6633), *B. cereus* (PCI 213), *P. mirabilis* (ATCC 12453), *E. coli* (ATCC 25922), *S. typhi Ty2* (ATCC 19430), and *P. aeruginosa* (ATCC 27853) [2].

Th. Pulegioides EO of the thymol chemotype (T) had the strongest anti-Pseudomonad activity against seven phytopathogenic *Pseudomonas* species isolated from vegetables: it inhibited the growth of all tested bacteria even up to 84.4% compared to the control. Linalool chemotype (L) essential oils inhibited bacterial growth less than chemotype T, and the geraniol/geranium/neral G/G/N chemotype was the least effective [39].

It was found that the phytotoxic action of *Th. Pulegioides* EO with the α-terpinyl acetate chemotype has a stronger influence on monocotyledonous than dicotyledonous plants. It also showed high antimicrobial activity against fungi and dermatophytes, but less activity against bacteria and *Candida* yeast. Therefore, *T. pulegioides* with the α-terpinyl acetate chemotype may be a potential compound for the development of preventive measures and/or medicines against mycosis [33].

All species of the genus *Thymus* investigated in Romania—*Th. Vulgaris* and *Th. Pulegioides* (thymol chemotype), *Th. Serpyllum* (carvacrol), and *Th. Glabrescens* (terpinyl acetate chemotype)—showed complete inhibition of the growth of *P. aeruginosa*, *Listeria inacua*, and *S. pyogenes* [5].

EOs with high contents of carvacrol and thymol showed antifungal activity against *Candida* (seven clinical isolates and four ATCC type strains), *Aspergillus* (five clinical isolates and two Colección Española de Cultivos Tipo (CECT) and two ATCC type strains) and five clinical dermatophyte strains. The results showed that the EO of *Th. Pulegioides* showed significant activity against fungi, due to the formation of lesions in the cytoplasmic membrane and a significant decrease in ergosterol content [40].

During this experimental work, it was found that the EO of lemon thyme (*Th.* × *Citriodorus* (Pers.) Schreb. var. "Silver Queen") has a pronounced antibacterial action against *S. aureus* ATCC 25923 (diameters of growth retardation were 14.60 ± 1.52 mm) and fungicidal activity of *Candida albicans* ATCC 885–6530 (29.30 ± 2.82 mm). The antibacterial activity of thyme EO was found against *E. coli* (19.60 ± 1.85 mm). It was experimentally proven that it had no bactericidal influence on the test strain of *P. aeruginosa* [41].

Productivity (fresh and dry leaves and branches) of *Th. Serpyllum* in the highlands of Baluchistan was 113 kg/ha^{-1}. The EO content was 0.58% and 0.87% in fresh and dry plant samples. Three different concentrations of leaf extracts (0.5 g, 1 g, and 1.5 g) were used for three different time periods (24 h, 48 h, and 72 h). The action of 0.5 g of the *Th. Serpyllum* extract for 24 h against the causative agent of *E. coli* was the most effective [42].

Th. Serpyllum EO had an antimicrobial action on anaerobic bacteria, lactic acid bacteria, and enterobacteria causing spoilage of chicken thigh meat for 2 weeks. The EO slowed down the growth of the colonies of these bacteria. *Pseudomonas* spp. were detected only in the control group stored in air for 12 and 16 days. The results of these investigations indicate the possibility of using EOs of *Th. Serpyllum* as natural food preservatives and potential sources of antimicrobial ingredients for the food industry [43].

EO of *Th. Serpyllum* L. and its components—thymol and total phenol extracted from plants—had an influence on the growth and production of mycotoxins of *Aspergillus ochraceus*, *A. carbonarius*, and *A. niger.* The minimum inhibitory concentration (MIC) determined for the essential oil and thymol and the selected concentration of phenol inhibited fungal growth and ochratoxin A biosynthesis by more than 60%, depending on the conditions and duration of incubation with fungi. The EO showed the strongest inhibitory effect due to the synergistic or cumulative influence of its components [44].

Th. Marschallianus and its components showed antioxidant activity and effect against *Helicobacter pylori* ATCC 43504 with an MIC (minimum inhibitory concentration) = 0.625 mg/mL [16].

5. Conclusions

It has been proven that the quantitative and qualitative composition of EOs depends on the amount of solar radiation and microclimatic and orographic conditions of the population. Different chemical composition of EOs from plant raw materials of three Thymus species from the Carpathians region of Ukraine were recorded in this research study. There are important differences in all monitored characteristics. Plant raw materials from different habitats of the searched species have a high range of EO contents. The plant populations of *Th. serpyllum* and *Th. Pulegioides*, located in sunny meadow conditions, have an aromatic mono- and bicyclic monoterpenoid chemotype, with significant contents of the aromatic monoterpenes carvacrol and p-cymene. Populations of *Th. Serpyllum*, that is shaded by a pine forest and *Th. Pulegoides*, that is shaded by cereal grasses, has mono- and bicyclic monoterpenoid chemotypes. Plant raw materials *Th. Serpyllum* and *Th. Pullegioides* selected from meadow-steppe vegetation on sandy soils has monoterpenoid chemotype with hight contents linalool. *Th. Marschallianus* were reported to have the lowest EO and α-terpineol dominated the component composition. Antimicrobial activity and a high content of natural components of *Th. pullegoides* and Th. serpyllum with an aromatic-monoterpenoid chemotype determines the promise of their use as raw materials in the pharmaceutical industry.

Author Contributions: M.K. and M.H. processed the experimental data, performed the biochemical and biological analyses, and drafted the manuscript. I.S. devised the project of *Thymus* biodiversity, the main conceptual ideas, and the proof outline. M.S. and O.N. worked out almost all of the technical details and performed the numerical calculations for the suggested experiment with the plant population collection in the Ukraine Carpathians. All authors discussed the results and commented on the manuscript. All authors have read and agreed to the published version of the manuscript.

Funding: This research received no external funding.

Institutional Review Board Statement: Not applicable for studies not involving humans or animals.

Informed Consent Statement: Not applicable for studies not involving humans.

Data Availability Statement: Data openly available in a public repository that issues datasets with DOIs.

Conflicts of Interest: The authors declare that the research was conducted in the absence of any commercial or financial relationships that could be construed as a potential conflict of interest.

References

1. Najar, B.; Pistelli, L.; Ferri, B.; Angelini, L.G.; Tavarini, S. Crop Yield and Essential Oil Composition of Two *Thymus vulgaris* Chemotypes along Three Years of Organic Cultivation in a Hilly Area of Central Italy. *Molecules* **2021**, *26*, 5109. [CrossRef] [PubMed]
2. De Martino, L.; Bruno, M.; Formisano, C.; De Feo, V.; Napolitano, F.; Rosselli, S.; Senatore, F. Chemical Composition and Antimicrobial Activity of the Essential Oils from Two Species of Thymus Growing Wild in Southern Italy. *Molecules* **2009**, *14*, 4614–4624. [CrossRef] [PubMed]

3. Verma, R.S.; Lu, R.; Chanotiya, C.S.; Verma, R.K.; Singh, A.; Yadav, A.; Chauhan, A.; Yadav, A.K.; Singh, A.K. Essential oil composition of *Thymus serpyllum* cultivated in the Kumaon region of western Himalaya, India. *Nat. Prod. Commun.* **2009**, *4*, 1934578X0900400723. [CrossRef]
4. Mohan, M.; Seth, R.; Singh, P.; Lohani, H.; Gupta, S. Composition of the Volatiles of *Hyssopus officinalis* (L.) and *Thymus serpyllum* (L.) from Uttarakhand Himalaya. *Natl. Acad. Sci. Lett.* **2012**, *35*, 445–448. [CrossRef]
5. Varga, E.; Bardocz, A.; Belak, A.; Maraz, A.; Boros, B.; Felinger, A.; Böszörményi, A.; Horváth, G. Antimicrobial activity and chemical composition of thyme essential oils and the polyphenolic content of different *Thymus* extracts. *Farmacia* **2015**, *6*, 357–361.
6. Ložionė, K.; Venskutonis, R.P. Chemical Composition of the Essential Oil of *Thymus serpyllum* L. ssp. *serpyllum* Growing Wild in Lithuania. *J. Essent. Oil Res.* **2006**, *18*, 206–211. [CrossRef]
7. Paaver, U.; Orav, A.; Arak, E.; Mäeorg, U.; Raal, A. Phytochemical analysis of the essential oil of *Thymus serpyllum* L. growing wild in Estonia. *Nat. Prod. Res.* **2008**, *22*, 108–115. [CrossRef]
8. Hrytsyna, M.R.; Kryvtsova, M.V.; Salamon, I.; Skybitska, M.I. Promising ex situ essential oil from *Thymus camphoratus* (*Lamiaceae*). *Regul. Mech. Biosyst.* **2020**, *11*, 310–314. [CrossRef]
9. Ani, R.; Josip, M. Essential Oil and Glycosidically Bound Volatiles of *Thymus pulegioides* L. growing Wild in Croatia. *Croat. Chem. Acta* **2008**, *81*, 599–606.
10. Kirillov, V.; Stikhareva, T.; Mukanov, B.; Chebotko, N.; Ryazantsev, O.; Atazhanova, G.; Adekenov, S. Composition of the Essential Oil of *Thymus serpyllum* L. from Northern Kazakhstan. *J. Essent. Oil Bear. Plants* **2016**, *19*, 212–222. [CrossRef]
11. Sárosi, S.; Bernáth, J.; Bertoli, A.; Pistelli, L.; Benvenuti, S. Essential oil polymorphism of *Thymus pulegioides* collected in Monti Pisani, Italy. *Acta Hortic.* **2012**, *955*, 59–64. [CrossRef]
12. Nikolić, B.; Matović, M.; Mladenović, K.; Todosijević, M.; Stanković, J.; Đorđević, I.; Marin, D.P.; Tešević, V. Volatiles of *Thymus serpyllum* Obtained by Three Different Methods. *Nat. Prod. Commun.* **2019**, *14*. [CrossRef]
13. Mariconda, A.; Vassallo, A.; Bonomo, M.G.; Calabrone, L.; Salzano, G.; Claps, M.; Sinicropi, M.S.; Capasso, A.; Saturnino, C. Herbal formulations of *Thymus Serpyllum* L. and *Hypericum Perforatum* L. from Southern Italy: Preparationand Chemical Characterization. *PharmacologyOnLine* **2020**, *1*, 1–10. Available online: https://pharmacologyonline.silae.it/front/archives_2020_1 (accessed on 16 December 2022).
14. Milevskaya, V.; Temerdashev, Z.A.; Butyl'Skaya, T.S.; Kiseleva, N.V. Determination of phenolic compounds in medicinal plants from the *Lamiaceae* family. *J. Anal. Chem.* **2017**, *72*, 342–348. [CrossRef]
15. Niculae, M.; Hanganu, D.; Oniga, I.; Benedec, D.; Ielciu, I.; Giupana, R.; Sandru, C.D.; Ciocârlan, N.; Spinu, M. Phytochemical Profile and Antimicrobial Potential of Extracts Obtained from *Thymus marschallianus* Willd. *Molecules* **2019**, *24*, 3101. [CrossRef] [PubMed]
16. Zhumakanova, B.; Korona-Głowniak, I.; Skalicka-Woźniak, K.; Ludwiczuk, A.; Baj, T.; Wojtanowski, K.; Józefczyk, A.; Zhaparkulova, K.; Sakipova, Z.; Malm, A. Phytochemical Fingerprinting and In Vitro Antimicrobial and Antioxidant Activity of the Aerial Parts of *Thymus marschallianus* Willd. and *Thymus seravschanicus* Klokov Growing Widely in Southern Kazakhstan. *Molecules* **2021**, *26*, 3193. [CrossRef]
17. Pluhár, Z.; Kocsis, M.; Kuczmog, A.; Csete, S.; Simkó, H.; Sárosi, S.; Molnár, P.; Horváth, G. Essential oil composition and preliminary molecular study of four Hungarian *Thymus* species. *Acta Biol. Hung.* **2012**, *63*, 81–96. [CrossRef]
18. Council of Europe European Directorate for the Quality of Medicines & Healthcare. *European Pharmacopoeia 8.0: Published in Accordance with the Convention on the Elaboration of a European Pharmacopoeia*, 8th ed.; European Directorate for the Quality of Medicines & Healthcare Council of Europe: Strasbourg, France, 14 February 2013.
19. State Pharmacopoeia of Ukraine. In *State enterprise "Ukrainian Scientific Pharmacopoeial Center for Quality of Medicines"*, 2nd ed.; Suppl. 3, State Enterprise "Ukrainian Scientific Pharmacopoeial Center for the Quality of Medicines": Kharkiv, Ukraine, 2014; Volume 3, p. 732.
20. Atlas of the Lviv Region. Available online: https://geoknigi.com/view_map.php?id=28 (accessed on 16 December 2022).
21. Adams, R.P. *Identification of Essential Oil Components by Gas Chromatography/Mass Spectrometry*, 4th ed.; Allured Publishing Corporation: Carol Stream, IL, USA, 2007; p. 804.
22. Balouiri, M.; Sadiki, M.; Ibnsouda, S.K. Methods for in vitro evaluating antimicrobial activity: A review. *J. Pharm. Anal.* **2016**, *6*, 71–79. [CrossRef]
23. Vaičiulytė, V.; Butkienė, R.; Ložienė, K. Effects of meteorological conditions and plant growth stage on the accumulation of carvacrol and its precursors in *Thymus pulegioides*. *Phytochemistry* **2016**, *128*, 20–26. [CrossRef]
24. Wester, P.; Möseler, B.M.; Knöss, W. Intra-population terpene polymorphism of *Thymus pulegioides* L.: Evidence for seven chemotypes in a German limestone grassland. *Biochem. Syst. Ecol.* **2020**, *93*, 104173. [CrossRef]
25. Senatore, F. Influence of Harvesting Time on Yield and Composition of the Essential Oil of a Thyme (*Thymus pulegioides* L.) Growing Wild in Campania (Southern Italy). *J. Agric. Food Chem.* **1996**, *44*, 1327–1332. [CrossRef]
26. Gluschenko, L.A. Ecological, cenotic and resource characteristic of the species of genus *Thymus* L. at the territory of Left-bank Forest-Steppe—Manuscript. Thesis on a degree of the candidate of the biological sciences by specialty 03.00.05–botany. M.M.. Ph.D. Thesis, Gryshko National Botanical Garden of the National Ukrainian Academy of Sciences, Kyiv, Ukraine, 2005.
27. Tymchenko, I.À.; Minarchenko, V.M.; Glushchenko, L.A.; Ànishchenko, Ì.M.; Gurinovich, N.V. Resources monitoring of *Thymus* L. in Ukraine. *Ukr. Botan. J.* **2007**, *64*, 78–87.

28. Penkovska, L. Analysis of Ontogenic Structure of *Thymus Serpyllum* L. Emend. Mill. and *Thymus x Polessicus* Klokov (Lamiaceae) Coenopopulations Under the Conditions of the Ympil District, Sumy Region (Ukraine). *Lesya Ukr. East. Eur. Natl. Univ. Sci. Bulletin. Series Biol. Sci.* **2019**, *3*, 38–44. [CrossRef]
29. Penkovska, L. Dimensional features of *Thymus serpyllum* L. Emend. Mill. and *Thymus x polessicus* Klokov (*Lamiaceae*) in the conditions of different phytocenoses in the Shostka geobotanical district, Sumy region. *Bull. Cherkasy University. Series Biol. Sci.* **2020**, *1*, 53–61. [CrossRef]
30. Mockute, D.; Bernotiene, G. 1,8-Cineole—Caryophyllene Oxide Chemotype of Essential Oil of *Thymus serpyllum* L. Growing Wild in Vilnius (Lithuania). *J. Essent. Oil Res.* **2004**, *16*, 236–238. [CrossRef]
31. Mockute, D.; Bernotiene, G. The Main Citral–Geraniol and Carvacrol Chemotypes of the Essential Oil of *Thymus pulegioides* L. Growing Wild in Vilnius District (Lithuania). *J. Agric. Food Chem.* **1999**, *47*, 3787–3790. [CrossRef] [PubMed]
32. Mockute, D.; Bernotiene, G. The α-terpenyl acetate chemotype of essential oil of *Thymus pulegioides* L. *Biochem. Syst. Ecol.* **2001**, *29*, 69–76. [CrossRef]
33. Vaičiulytė, V.; Ložienė, K.; Švedienė, J.; Raudonienė, V.; Paškevičius, A. α-Terpinyl Acetate: Occurrence in Essential Oils Bearing *Thymus Pulegioides*, Phytotoxicity, and Antimicrobial Effects. *Molecules* **2021**, *26*, 1065. [CrossRef]
34. Svydenko, L.V.; Hlushchenko, L.A. Component composition of essential oil in the forms of species creeping thyme (*Thymus serpylum* L.) and broad-leaved thyme (*Thymus pulgioides* L.) in Kherson region. *Agroecol. J.* **2016**, *2*, 129–134.
35. Ložienė, K.; Venskutonis, P. Influence of environmental and genetic factors on the stability of essential oil composition of *Thymus pulegioides*. *Biochem. Syst. Ecol.* **2005**, *33*, 517–525. [CrossRef]
36. StakelienĖ, V.; Ložienė, K. Gynodioecy in *Thymus pulegioides* L., *T. serpyllum* L., and their hybrid T. × *oblongifolius* Opiz (*Lamiaceae*): Flower size dimorphism, female frequency, and effect of environmental factors. *Plant Biosyst. Int. J. Deal. Asp. Plant Biol.* **2013**, *148*, 49–57. [CrossRef]
37. Al-Fekaiki, D.F.; Niamah, A.K.; Al-Sahlany, S.T.G. Extraction and identification of essential oil from *Cinnamomum zeylanicum* barks and study the antibacterial activity. *J. Microbiol. Biotechnol. Food Sci.* **2017**, *7*, 312–316. [CrossRef]
38. Hao, R.; Roy, K.; Pan, J.; Shah, B.R.; Mraz, J. Critical review on the use of essential oils against spoilage in chilled stored fish: A quantitative meta-analyses. *Trends Food Sci. Technol.* **2021**, *111*, 175–190. [CrossRef]
39. Tekorienė, R.; Ložienė, K. Disinfecting capacity of essential oil of *Thymus pulegioides* L. (lamiaceae) chemotypes against phytopathogenic Pseudomonas species. *Acta Aliment.* **2012**, *41*, 257–264. [CrossRef]
40. Pinto, E.; Pina-Vaz, C.; Salgueiro, L.; Gonçalves, M.J.; Costa-De-Oliveira, S.; Cavaleiro, C.; Palmeira, A.; Rodrigues, A.G.; de Oliveira, J.M. Antifungal activity of the essential oil of *Thymus pulegioides* on Candida, Aspergillus and dermatophyte species. *J. Med. Microbiol.* **2006**, *55*, 1367–1373. [CrossRef] [PubMed]
41. Steshenko, Y.M.; Mazulin, O.V.; Polishchuk, N.M. Study of the antimicrobial and fungicidal activity of the essential oil *Thymus* × *citriodorus* (Pers.) Schreb. var. "Silver Queen". *Curr. Issues Pharm. Med. Sci. Pr.* **2021**, *14*, 211–214. [CrossRef]
42. Momin, S.; Irfan, S.; Ahmad, S.; Manzoor, M.; Irfan, H. Comparison of antimicrobial activities, herb and essential oil production of *Thymus vulgaris* L. and *T. serpyllum* L. in Balochistan, Pakistan. *Pak. J. Weed Sci. Res.* **2017**, *23*, 119–125.
43. Kačániová, M.; Terentjeva, M.; Arpášová, H.; Hleba, L.; Ivanišová, E.; Petrová, J.; Kántor, A.; Čuboň, J.; Haščík, P. Effect of *Thymus Serpyllum* and *Ocimum Basilicum* Essential Oils on the Shelf-Life of Chicken's Meat during Refrigerated Storage. *Sci. Pap. Anim. Sci. Biotechnol.* **2016**, *49*, 103–108.
44. Sokolić-Mihalak, D.; Frece, J.; Slavica, A.; Delaš, F.; Pavlović, H.; Markov, K. The effects of wild thyme (*Thymus serpyllum* L.) essential oil components against ochratoxin-producing *Aspergilli*. *Arh. Hig. Rada. Toksikol.* **2012**, *63*, 457–462. [CrossRef]

horticulturae

Article

Preliminary Study on the Impact of Non-Thermal Plasma Activated Water on the Quality of *Triticum aestivum* L. cv. Glosa Sprouts

Alexandru Mandici [1], Daniel Eusebiu Cretu [2], Radu Burlica [2,*], Dragos Astanei [2], Oana Beniuga [2], Craita Rosu [3], Denis Constantin Topa [4], Tudor George Aostacioaei [4], Ana Clara Aprotosoaie [1] and Anca Miron [1,*]

1 Department of Pharmacognosy-Phytotherapy, Grigore T. Popa University of Medicine and Pharmacy, 16 Universitatii Str., 700115 Iasi, Romania
2 Department of Electrical Engineering, Gheorghe Asachi Technical University, 71 D. Mangeron Blvd., 700050 Iasi, Romania
3 Department of Experimental and Applied Biology, Institute of Biological Research Iasi, 47 Lascar Catargi Str., 700107 Iasi, Romania
4 Department of Pedotechnics, University of Life Sciences, 3 Mihail Sadoveanu Al., 700490 Iasi, Romania
* Correspondence: radu.burlica@tuiasi.ro (R.B.); anca.miron@umfiasi.ro (A.M.)

Abstract: The present study aimed to investigate the effects of non-thermal plasma-activated water (PAW) on *Triticum aestivum* L. cv. Glosa sprouts. Two types of PAW were generated by exposing distilled water to a high-voltage electric discharge. Wheat caryopses were treated either with PAW1 (25 mg/L NO_3^-, 4 mg/L NO_2^-, and 6 mg/L H_2O_2) or PAW2 (35 mg/L NO_3^-, 5 mg/L NO_2^-, and 7.5 mg/L H_2O_2) for 8 days, with samples being collected at days 1, 2, 3, and 8. The germination rate, growth parameters, protein, photosynthetic pigments, total phenolic contents, antioxidant activity of free and bound phenolic fractions, and activity of antioxidant enzymes were evaluated. Both PAW1 and PAW2 had positive effects on *Triticum aestivum* L. cv. Glosa sprouts. PAW2 had a better impact on sprouts with respect to growth parameters, free phenolic content, and antioxidant activity of the free phenolic fraction (day 3), the accumulation of proteins, and photosynthetic pigments (day 8). In conclusion, exposure to PAW increases the quality of *Triticum aestivum* L. cv. Glosa sprouts. Further optimization of PAW treatment is required for the most favorable impact on the accumulation of biomass and health-promoting compounds in sprouts.

Keywords: wheat sprouts; germination; seedling growth; proteins; photosynthetic pigments; phenolic compounds; antioxidant activity; antioxidant enzymes

Citation: Mandici, A.; Cretu, D.E.; Burlica, R.; Astanei, D.; Beniuga, O.; Rosu, C.; Topa, D.C.; Aostacioaei, T.G.; Aprotosoaie, A.C.; Miron, A. Preliminary Study on the Impact of Non-Thermal Plasma Activated Water on the Quality of *Triticum aestivum* L. cv. Glosa Sprouts. *Horticulturae* **2022**, *8*, 1158. https://doi.org/10.3390/horticulturae8121158

Academic Editors: Ivan Salamon and Charalampos Proestos

Received: 22 October 2022
Accepted: 29 November 2022
Published: 7 December 2022

1. Introduction

A healthy diet improves quality of life and prevents chronic diseases [1]. In this respect, the consumption of whole grains has overwhelming health benefits [2]. Among cereals, whole grain wheat is an excellent source of nutritional and bioactive compounds. Dietary fiber is an important component of the whole grain wheat (11.6–17.0%). Almost half of all dietary fiber is represented by non-starch polysaccharides, with arabinoxylan (70% of wheat bran) as a major component [3]. Whole grain wheat is also an important dietary source of starch (60–70%), composed of rapidly digestible starch (23.9–34.4%), slowly digestible starch (3.4–23.4%), and resistant starch (1.8–16.9%) [4]. The latter acts as a functional fiber undergoing colonic bacterial fermentation with the production of short-chain fatty acids (acetate, propionate, butyrate) which are essential for colon health [5]. Whole grain wheat is also a source of proteins (7–22%) [6] and lipids (2–2.5%, free fatty acids and acylglycerols being predominant) [7]. Phytosterols, sterol glycosides and esters, alkylresorcinols, vitamin E, and carotenoids are important lipid soluble constituents in wheat grains [8,9]. The average total content of tocopherols and tocotrienols in wheat grains was found to vary from 27.6 to 79.7 µg/g, alpha-tocopherol and beta-tocotrienol being the most abundant ones [10]. Vitamin E is a potent antioxidant playing a major role in the protection against atherosclerosis, arthritis, cataracts, and cancer [11], with the cardioprotective potential being mostly attributed to tocotrienols [12]. Whole grain wheat also contains carotenoids, with large variations in contents (0.94–13.6 mg/kg) depending on the cultivar [13]. As antioxidants and vitamin A precursors, carotenoids have numerous health benefits in the treatment of degenerative eye disorders, cardiovascular diseases, Alzheimer's disease, colorectal cancer, type 2 diabetes, and obesity [14]. Whole grain wheat is also a source of gamma aminobutyric acid (GABA) (3.4 mg/100 g dry weight) [15], betaine (291.2 mg/100 g dry weight), and choline (14.4 mg/100 g dry weight) [16]. GABA, a non-protein amino acid, is an inhibitory neurotransmitter with many other physiological effects in humans such as regulation of blood pressure and heart rate and stimulation of insulin secretion [15]. Betaine and choline reduce hyperhomocysteinaemia, a major risk factor for cardiovascular and neurological diseases [17]. Whole grain wheat also contains phenolic acids with strong antioxidant properties, localized mainly in bran [18]. Ferulic acid is the predominant phenolic acid (269.2–744.7 µg/g), followed by *p*-coumaric (10.4–92.4 µg/g), syringic (7.9–29.5 µg/g), and vanillic (5.6–19.5 µg/g) acids [18–22].

Overall, whole grain wheat contains valuable nutritional and bioactive compounds which provide significant health benefits in the treatment of cardiovascular disease, type 2 diabetes, and cancer [9]. The germination process enhances the nutritional and medicinal value of grains. Contents in soluble dietary fiber, proteins, free lipids, vitamins, and polyphenols increase during germination. Moreover, germination activates different enzymes such as alpha-amylase and proteases which degrade starch and proteins, respectively, resulting in an improved digestibility of grains and an enhanced availability of sugars and free amino acids. Germination was also found to increase the antioxidant capacity of sprouts [23–26]. Therefore, sprouted whole grains are important functional foods, being consumed as part of a diet for the prevention of chronic diseases [15].

There are various methods to control germination and enhance the production and accumulation of nutritional and bioactive compounds in sprouts. The most common ones are the exposure to stress (cold/heat, hypoxia) and use of different forms of physical energy such as pressure, light, ultrasounds, magnetic field, or non-thermal plasma [24]. Non-thermal (cold) plasma, usually generated by electrical discharge in a high-velocity gas flow, is a complex mixture of ions, atoms, molecules, free radicals, and electrons. Non-thermal plasma treatment was reported to enhance the seed germination and accumulation of bioactive compounds in sprouts, reduce the germination time required to accumulate the highest content of phytochemicals, promote plant growth, and prevent microbial contamination of plant material [24,27,28]. Non-thermal plasma can be applied not only directly, but also as non-thermal plasma-activated water (PAW), usually produced by spraying water into the plasma zone. The process generates reactive oxygen and nitrogen species, with hydrogen peroxide and nitrates being considered the major species affecting the seed germination, seedling growth, biosynthetic pathways, and stress response in plants. Nitrates are a source of nitrogen which is essential for the synthesis of proteins; the latter are important not only for growth, but also as components of enzymes with different metabolic functions. Reactive oxygen and nitrogen species trigger a defensive response in the plant tissue consisting in an enhanced production and accumulation of phytochemicals with antioxidant properties such as phenolics, vitamin E, and carotenoids [29]. Treatment with PAW is a promising strategy to improve the nutritional and medicinal potential of sprouts and, at the same time, a promising alternative to chemical fertilizers. Two major advantages of using non-thermal PAW, such as low costs of production and environmental safety, need to be emphasized.

In this respect, we have initiated research studies aiming to boost the nutritional and bioactive potential of sprouts by treatment with non-thermal PAW. The present study reports on the effects of non-thermal PAW on *Triticum aestivum* L. cv. Glosa caryopses regarding germination, growth, accumulation of various metabolites, antioxidant capacity, and antioxidant enzyme activity in sprouts (collected at days 1–3) and shoots (collected at day 8).

2. Materials and Methods

2.1. Chemicals

Folin–Ciocalteu's phenol reagent, gallic acid, sodium carbonate, 2,2-diphenyl-1-picrylhydrazyl (DPPH) radical, Bradford reagent, protein standard (2 mg/mL), *o*-dianisidine, and dimethyl sulfoxide (DMSO) were obtained from Sigma-Aldrich (Steinheim, Germany). Nitro blue tetrazolium (NBT) chloride and riboflavin were from Merck KGaA (Darmstadt, Germany). 1,4-Dithiothreitol and ethylenediaminetetraacetic acid (EDTA) were from Scharlau Chemie (Sentmenat, Spain). Acetone, ethyl acetate, and methanol were from Riedel-de Haën (Seelze, Germany). Titanium sulfate solution was purchased from Fisher Chemical (Fisher Scientific, Loughborough, UK). Sodium hypochlorite and sodium hydroxide were provided by FEELXKIM SRL (Tomesti, Romania) and Chimreactiv SRL (Bucharest, Romania), respectively. Ultrapure water was obtained using the SG Water Ultra Clear TWF water purification system (Barsbüttel, Germany).

2.2. Vegetal Material

Wheat caryopses (*Triticum aestivum* L. cv. Glosa) were supplied by the Territorial Inspectorate for Seed and Propagating Material Quality (Iasi, Romania). Prior to the experiments, the caryopses were immersed in 1.25% sodium hypochlorite solution for 5 min and further rinsed with sterile demineralized water until no chlorine odor was detected [30].

2.3. Generation of Non-Thermal Plasma-Activated Water

Table 1 and PAW2, were obtained in the Department of Electrical Engineering, *Gheorghe Asachi* Technical University (Iasi, Romania). The non-thermal plasma reactor for PAW production, the factors affecting the chemical composition of PAW, and the whole process of PAW production have been previously described [31,32]. Briefly, in our study, a high voltage pulsed power supply generated a low power electrical discharge between two cylindrical point-to-point stainless steel electrodes (1.6 mm diameter). Thus, plasma was produced between the high voltage electrode and ground electrode in a 3.1 mm inner diameter glass cylinder, at a gap of 3 mm. Distilled water was obtained using a Biobase WD-A5 water distiller (Shandong, China) and pumped at a flow of 15 mL/min. Air (carrier gas) was generated by a commercial compressor (2.5 atm). A mass flow regulator MASS-VIEW MV-304 (Bronkhorst High-Tech B.V., The Netherlands) kept the air flow constant at 1.5 L/min. PAW was generated by a pulse electrical discharge between the high voltage electrode of the reactor and ground electrode, in a gas flow (Figure 1). Distilled water was mixed with air in a nozzle with two inlets arranged perpendicularly. The mixture was further sprayed into the plasma area in the reaction chamber, resulting in the formation of various reactive species such as hydrogen peroxide (H_2O_2), nitrates (NO_3^-), and nitrites (NO_2^-) [31]. For the experiments described in the present study, PAW was generated at 60 and 100 Hz for 2 ms pulse width, the distilled water and air flows being constant (15 mL/min and 1.5 L/min, respectively). The electrical parameters of the non-thermal plasma reactor for the frequencies used in our study are shown in Table 1. The current and voltage measurements were carried out using a shunt resistance of 100 Ω and a TT-HVP-2739 high voltage probe (divider voltage ratio 1000:1) (Testec Elektronik GmbH, Dreieich, Germany), respectively [31,33].

Table 1. Electrical parameters of the non-thermal plasma reactor.

Frequency (Hz)	I_{max} (mA)	I_{avg} (mA)	U_{max} (V)	U_{avg} (V)	P_{avg} (W)
60	58	2.5	6049	153	3
100	59	2.8	7200	250	4

Imax—maximum discharge current, Iavg—average current, Umax—maximum discharge voltage, Uavg—average discharge voltage, Pavg—discharge average power.

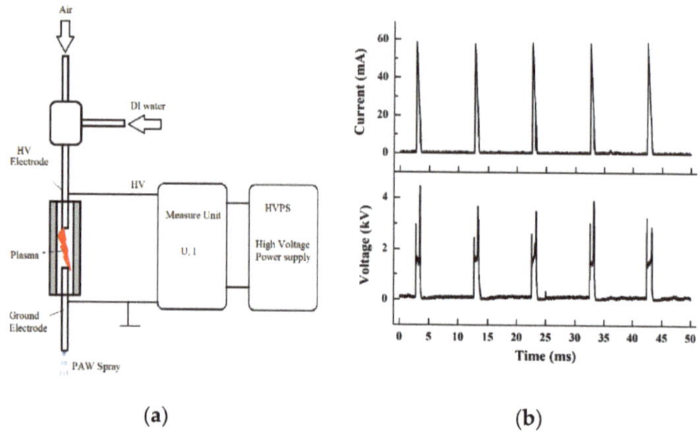

Figure 1. The experimental set-up (**a**) and the current and voltage waveforms (**b**) of the discharge (frequency: 100 Hz, pulse width: 2 ms, distilled water flow: 15 mL/min, air flow: 1.5 L/min).

The current and voltage waveforms were recorded using a WaveSurfer 454 digital oscilloscope (LeCroy, New York, NY, USA) [31,34]; the waveforms corresponding to the frequency of 100 Hz, 2 ms pulse width, distilled water flow of 15 mL/min, and air flow of 1.5 L/min are illustrated in Figure 1. The energy efficiency (EEf), representing the amount of a reactive species (g) produced in PAW per kWh, was calculated as the amount produced (n) divided by the energy consumption (E) for the process, which is n/E, where the value of E is equal to the product of the average power (P_{avg}) and treatment time (t) [31].

2.4. Non-Thermal Plasma-Activated Water: Composition and Analysis

Some of the major chemical reactions occurring in the non-thermal plasma generation area are listed below. As air was used as a working gas to produce plasma, nitrogen oxides were also generated [31,33].

$$H_2O \rightarrow H\cdot + \cdot OH \quad (1)$$

$$H_2O \rightarrow H^+ + e^-_{aq} + \cdot OH \quad (2)$$

$$\cdot OH + \cdot OH \rightarrow H_2O_2 \quad (3)$$

$$N_2 + e^- \rightarrow 2N + e^- \quad (4)$$

$$O_2 + e^- \rightarrow 2O + e^- \quad (5)$$

$$N + O \rightarrow NO \quad (6)$$

$$NO + O \rightarrow NO_2 \quad (7)$$

$$NO_2 + OH \rightarrow HNO_3 \quad (8)$$

$$O_2 + NO_2 + H_2O \rightarrow NO_2^- + NO_3^- + 2H^+ \quad (9)$$

H_2O_2 concentration of PAW was determined spectrophotometrically using titanium sulfate. 1 mL of titanium sulfate was mixed with 2 mL of PAW. A yellow color indicated the presence of H_2O_2 in PAW. The absorbance of the reaction mixture was determined at 410 nm (Shimadzu UVmini-1240 spectrophotometer) [33]. A calibration curve for H_2O_2 was also plotted. The levels of NO_3^- and NO_2^- in PAW were measured using the Visocolor Eco colorimetric test kits (Macherey-Nagel, Germany) and a compact photometer PF-3 (Macherey-Nagel, Germany). The quantification of NO_3^- (Visocolor Eco nitrate colorimetric test kit) involved its reduction to NO_2^- in an acidic medium, followed by the reaction of NO_2^- with an aromatic amine to generate an orange-yellow azo dye. The quantification of NO_2^- (Visocolor Eco nitrite colorimetric test kit) was based on the diazotization of sulfanilamide in an acid solution followed by coupling with naphthylamine to produce a reddish-purple azo dye [33,35,36].

2.5. Treatment and Germination of Wheat Caryopses

Approximately 0.9 kg of disinfected wheat caryopses was divided in three groups which were soaked for 4 h in PAW1, PAW2, and distilled water (control), followed by germination in a Weiss Gallenkamp climatic chamber (Weiss Technik UK Ltd., Loughborough, UK) under controlled conditions (dark, 20 °C, and 65% relative humidity for 3 days). After 3 days, the sprouted caryopses underwent a light/dark cycle (12/12 h) until day 8. PAW1, PAW2, and distilled water were changed at days 2, 4, and 6 with fresh prepared solutions, the wheat caryopses in each group being soaked 2/3 in liquid. Germination rate (%) was determined for the wheat caryopses collected at day 1. Fifty randomly selected wheat caryopses were collected from each group and the germination percentage was calculated as the ratio between the number of germinated caryopses and total number of collected caryopses. Caryopses were considered germinated when the radicle and coleoptile had a minimum length of 1 mm [29]. Growth parameters, superoxide dismutase, catalase, and peroxidase activities were determined in fresh samples immediately after collection. For other assays, samples were stored at −80 °C until analysis.

2.6. Biometric Measurements

Fifty wheat caryopses were randomly collected at days 2, 3, and 8. The lengths of their roots and shoots were measured with a mm scale. The median lengths of roots and shoots were calculated.

2.7. Total Protein Content

Total protein content was determined in the sprouts collected at days 1, 2, and 3, and shoots collected at day 8. In brief, each sample (100 mg) was grinded in 50 mM potassium phosphate buffer (pH = 7) and centrifuged at 10,000 rpm and 4 °C for 10 min (Ortoalresa Digicen 20-R centrifuge, Alvarez Redondo, Madrid, Spain). An aliquot was taken out from supernatant and mixed with Bradford reagent (total volume 1 mL). The absorbance was measured at 595 nm (Specord 210 Plus spectrophotometer, Analytik Jena, Jena, Germany). A standard curve was created by mixing different volumes of protein standard solution with Bradford reagent [37]. The results were expressed as mg protein/g of sample (fresh weight).

2.8. Photosynthetic Pigments Content

Chlorophylls and carotenoids were quantified in the shoots collected at day 8. 0.05 g of shoots were homogenised with 3 mL of acetone and further centrifuged at 10,000 rpm and 4 °C for 15 min (Ortoalresa Digicen 20-R centrifuge, Alvarez Redondo, Madrid, Spain). The final volume was adjusted to 10 mL with acetone. The absorbances were recorded at 662, 645, and 470 nm (Specord 210 Plus spectrophotometer, Analytik Jena, Jena, Germany). The contents, expressed in mg/g of sample (fresh weight), were calculated using the following formulas: chlorophyll a = $(11.24 \times A662) - (2.04 \times A645)$, chlorophyll b = $(20.13 \times A645) - (4.19 \times A662)$, total chlorophylls (chlorophyll a + b) = $(7.045 \times A662) + (18.09 \times A645)$, carotenoids = $[(1000 \times A470) - (1.9 \times \text{chlorophyll a}) - (63.14 \times \text{chlorophyll b})]/214$ [38,39].

2.9. Extraction of Free and Bound Phenolic Fractions

Free and bound phenolic fractions were extracted from the sprouts collected at days 1, 2, and 3, and shoots collected at day 8, according to previously described protocols with minor changes. In brief, free phenolic fraction was obtained as follows: the sample (1 g) was milled and suspended in 5 mL of methanol/ultrapure water/acetic acid mixture (70:29.5:0.5, *v*/*v*/*v*), vortexed for 1 min on ice (Velp Scientifica vortex mixer, Usmate, Italy), ultrasonicated for 40 min at 25 °C (Bandelin Electronic RK31 ultrasonic bath, Bandelin Electronic, Berlin, Germany), and centrifuged at 10,000 rpm for 10 min (Ortoalresa Digicen 20-R centrifuge, Alvarez Redondo, Madrid, Spain). The supernatant was collected, and the extraction was repeated twice. The supernatants were combined and vacuum dried (Büchi R-300 rotary evaporator, Büchi Laboratortechnik AG, Flawil, Switzerland). The residue was stored at −20 °C until further analysis. The pellets of the free phenolics extraction were used to isolate the bound phenolic fraction. Pellets were initially subjected to an alkaline hydrolysis with 4 M sodium hydroxide, followed by ultrasonication for 40 min at 25 °C (Bandelin Electronic RK31 ultrasonic bath, Bandelin Electronic, Berlin, Germany) and overnight rest. After adjusting the pH to a value of 2, each sample was extracted with 3 × 30 mL of ethyl acetate. The combined supernatants were vacuum dried (Büchi R-300 rotary evaporator, Büchi Laboratortechnik AG, Flawil, Switzerland) and stored at −20 °C until further use [26,40].

2.10. Total Phenolic Content

Total phenolic contents in both free and bound phenolic fractions were assessed using the Folin–Ciocalteu assay. The reaction mixture consisted of sample, Folin–Ciocalteu's phenol reagent, 20% sodium carbonate, and ultrapure water in a total volume of 4 mL. After 2 h incubation (dark, room temperature), the absorbance was determined at 765 nm (Specord 210 Plus spectrophotometer, Analytik Jena, Jena, Germany). A calibration curve was plotted using gallic acid [41–43]. The results were expressed as mg of gallic acid equivalents (GAE)/100 g of sample (fresh weight).

2.11. Antioxidant Activity

Antioxidant activity of the free and bound phenolic fractions was assessed by DPPH radical scavenging assay as previously described with minor modifications. Briefly, the sample (0.05 mL) was mixed with DPPH methanolic solution (0.075 mM, 2.95 mL), followed by vigorous shaking. After 30 min, the absorbance of the reaction mixture was determined at 515 nm. Additionally, the absorbance of DPPH solution (2.95 mL) and methanol (0.05 mL) was measured at 515 nm at 0 min (Specord 210 Plus spectrophotometer (Analytik Jena, Jena, Germany). DPPH radical scavenging activity was calculated using the formula: % DPPH scavenging activity = $[(A_{0\,min} - A_{30\,min})/A_{0\,min}] \times 100$ [23,42,44].

2.12. Superoxide Dismutase Activity

Superoxide dismutase (SOD, ***EC*** 1.15.1.1) was quantified in the sprouts collected at day 3 and shoots collected at day 8 according to previously described procedures. In brief, the enzyme was extracted by grinding the vegetal material in 5 mL of TRIS-0.1 M HCl buffer containing 1 mM dithiothreitol and 1 mM EDTA (pH = 7.8) at 4 °C, followed by centrifugation at 10,000 rpm for 15 min (Hettich Mikro 22R centrifuge, Tuttlingen, Germany). The supernatant was used for SOD quantification. The reaction mixture, consisting of 2.55 mL of potassium phosphate buffer (0.067 M, pH = 7.8), 0.2 mL of EDTA (0.1 M, pH = 7.8), 0.1 mL of supernatant, 0.1 mL of NBT (1.5 mM), and 0.05 mL of riboflavin (0.12 mM), was exposed to neon lamp illumination (24 °C, 5 min). The absorbance of the reaction mixture was further determined at 560 nm (Beckman Coulter DU 730 UV/Vis spectrophotometer, Beckman Coulter Inc., San Diego, CA, USA). SOD activity was expressed as U/mg protein/min; one unit (U) represents the amount of SOD causing 50% inhibition of light-induced NBT reduction [45,46].

2.13. Catalase Activity

Similarly, catalase (CAT, ***EC*** 1.11.1.6) activity was determined in the samples collected at days 3 and 8. The vegetal material was grinded in 5 mL of potassium phosphate buffer (0.05 M, pH = 7) and further centrifuged at 4500 rpm for 10 min (Hettich Mikro 22R centrifuge, Tuttlingen, Germany). In total, 1 mL of H_2O_2 (0.059 M) was incubated with 1 mL of phosphate buffer for 4–5 min at 25 °C, followed by the addition of 0.1 mL of supernatant (diluted to cause a decrease in absorbance of 0.03–0.07/min). The decrease in absorbance at 240 nm was measured for 2 min (Beckman Coulter DU 730 UV/Vis spectrophotometer, Beckman Coulter Inc., San Diego, CA, USA). CAT activity was expressed as U/mg protein/min, one unit (U) decomposing 1 μM H_2O_2 in 1 min (25 °C, pH = 7) [47].

2.14. Peroxidase Activity

Peroxidase (POX, ***EC*** 1.11.1.7) activity was assessed in the samples collected at days 3 and 8. First, samples were ground in 5 mL of TRIS—0.1 M HCl buffer containing 1 mM dithiothreitol and 1 mM EDTA (pH = 7.8) in an ice bath and further centrifuged at 4500 rpm for 15 min (Hettich Mikro 22R centrifuge, Tuttlingen, Germany). A volume of 0.1 mL of supernatant was mixed with 3 mL of buffered *o*-dianisidine and 0.2 mL of H_2O_2 0.05%, incubated at 20 °C for 5 min, followed by the addition of 5 mL of H_2SO_4 50% to block POX activity. The absorbance was read at 540 nm (Beckman Coulter DU 730 UV/Vis spectrophotometer, Beckman Coulter Inc., San Diego, CA, USA). PO activity was expressed as U/mg protein/min, one unit (U) decomposing 1 μM H_2O_2 in 1 min at 20 °C [48].

2.15. Statistical Analysis

All experiments were performed in triplicate, the results being presented as mean ± standard deviation (SD). Data were processed using SPSS software version 18.0 and Tukey HSD test (95% confidence interval). A p value lower than 0.05 was considered to be statistically significant.

3. Results

3.1. Generation and Analysis of Non-Thermal Plasma-Activated Water

Two types of PAW were produced in the present study. The concentrations of reactive species in PAW strongly depend on the electrical parameters of electrical discharge that generates PAW. The electrical parameters (current, voltage, frequency) as well as the physical ones (air and water flow rates) were selected in order to obtain the desired concentrations of reactive species in PAW: a level of NO_3^- (source of nitrogen for wheat sprouts) with a positive impact on sprouting and a level of H_2O_2 (strong oxidizing agent) which does not affect negatively the development of wheat sprouts. In PAW1, generated at 60 Hz, the concentrations of NO_3^-, NO_2^-, and H_2O_2 were 25.0 ± 2.5, 4.0 ± 0.4, and 6.0 ± 0.5 mg/mL, respectively. The higher frequency (100 Hz) used to produce PAW2 resulted in elevated levels of reactive species (35.0 ± 3.5 mg/L NO_3^-, 5.0 ± 0.5 mg/L NO_2^-, and 7.5 ± 0.7 mg/L H_2O_2) (Figure 2). PAW treatment was applied 2 h after preparation when measurements revealed that the concentrations of reactive species (NO_3^-, NO_2^-, H_2O_2) had stabilized. The EEf values used for the generation of H_2O_2, NO_2^-, and NO_3^- at both frequencies are illustrated in Table 2.

Table 2. Energy efficiency (EEf) values used for the generation of H_2O_2, NO_2^-, and NO_3^- in PAW1 and PAW2.

PAW	Frequency (Hz)	EEf (g/kWh)		
		H_2O_2	NO_2^-	NO_3^-
PAW1	60	1.7	1.4	8.6
PAW2	100	1.6	1.2	8.1

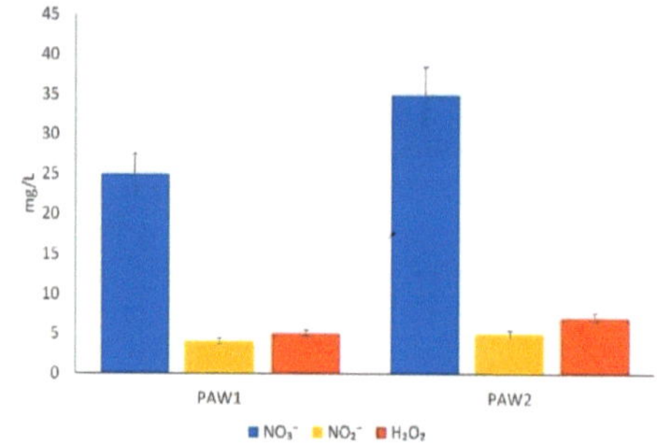

Figure 2. Concentrations of H_2O_2, NO_2^-, and NO_3^- in PAW1 and PAW2.

3.2. Germination Rate

PAW1 and PAW2 enhanced the germination of wheat caryopses by 4.66% and 5.33%, respectively, in comparison with the control (evaluation at day 1, Table 3), with no statistically significant differences between groups ($p > 0.05$).

Table 3. Effect of PAW treatment on the germination of *Triticum aestivum* L. cv. Glosa caryopses.

Treatment	Germination (%)
Control	42.67 ± 1.89
PAW1	47.33 ± 4.99
PAW2	48.00 ± 3.27

3.3. Growth Parameters

PAW treatment positively influenced the biometric characteristics of shoots and roots (Figure 3). At day 3, PAW2 induced the most pronounced increase in the lengths of shoots and roots (13.2 and 17.54 mm, respectively, vs. 1.81 and 8.14 mm, respectively, in the control group). At day 8, PAW1-treated caryopses showed the longest shoots (107.29 vs. 77.68 mm in the control group), whereas PAW2-treated ones had the most developed roots (97.02 vs. 75.05 mm in the control group). The shoot lengths in the PAW1-treated group at day 2 and the PAW2-treated group at days 2 and 3 were significantly different from the control group ($p < 0.05$). With respect to the root lengths, significant differences in comparison with the control were determined for the PAW2-treated group at days 2, 3, and 8 and the PAW1-treated group at day 2 ($p < 0.05$).

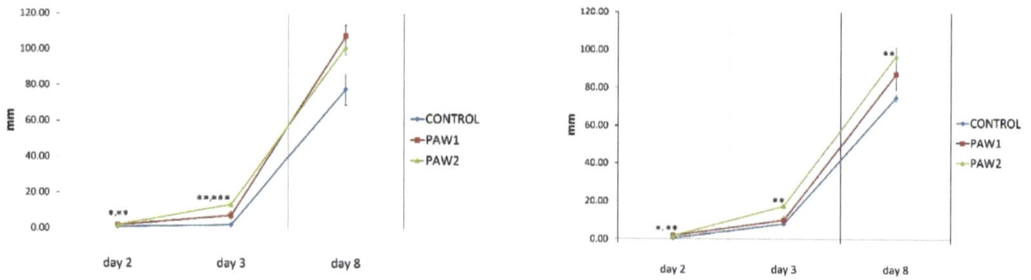

(a) (b)

Figure 3. Effect of PAW treatment on the lengths of shoots (**a**) and roots (**b**) of *Triticum aestivum* L. cv. Glosa sprouts; * significant difference ($p < 0.05$) between control and PAW1; ** significant difference ($p < 0.05$) between control and PAW2; *** significant difference ($p < 0.05$) between PAW1 and PAW2. Each point represents the mean ± SD of three determinations.

3.4. Total Protein Content

Total protein content increased under PAW treatment (Figure 4). PAW1 and PAW2 caused important increase in the protein content of wheat sprouts at days 1 (~43 and ~31%, respectively) and 3 (~19 and ~18%, respectively). In the shoots at day 8, the increase in the protein content was smaller (~12% in the PAW1-treated group and ~17% in the PAW2-treated group). However, significant differences in comparison with the control were determined only for the PAW1-treated group at day 1 and the PAW2-treated group at day 8 ($p < 0.05$). There were no statistically significant differences between the PAW1- and PAW2-treated groups at all four collection days ($p > 0.05$).

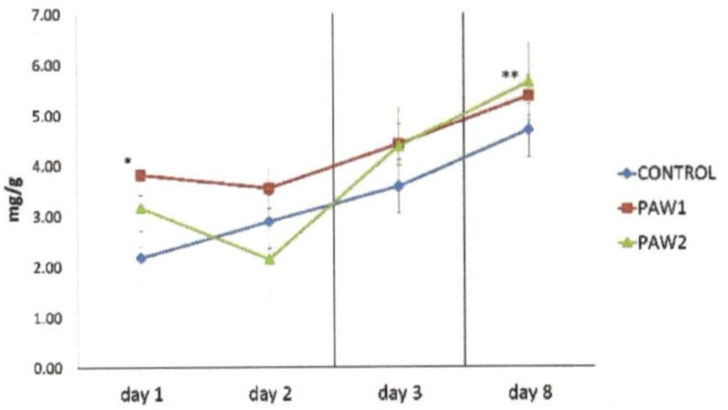

Figure 4. Effect of PAW treatment on the total protein content in *Triticum aestivum* L. cv. Glosa sprouts (days 1, 2, and 3) and shoots (day 8); * significant difference ($p < 0.05$) between control and PAW1; ** significant difference ($p < 0.05$) between control and PAW2. Each point represents the mean ± SD of three determinations.

3.5. Photosynthetic Pigments Content

Both chlorophylls and carotenoids in the shoots at day 8 increased under PAW treatment (Figure 5). PAW2 induced a more pronounced increase in chlorophyll a, chlorophyll b, total chlorophylls, and carotenoid contents in comparison with PAW1 (1.32-, 1.8-, 1.45-, and 2.18-fold, respectively vs. 1.24-, 1.13-, 1.21-, and 1.98-fold, respectively). The contents of chlorophyll a and total chlorophylls in PAW1- and PAW2-treated groups were significantly different from the control group ($p < 0.05$). With respect to the chlorophyll b content, a significant increase in comparison with the control was determined only for the PAW2-treated group ($p < 0.05$). Differences in chlorophyll b and total chlorophyll contents between the PAW1- and PAW2-treated groups were found to be statistically significant ($p < 0.05$). There were no statistically significant differences between the control and treated groups, on one hand, and between the PAW1- and PAW2-treated groups, on the other hand, with respect to the carotenoid content ($p > 0.05$).

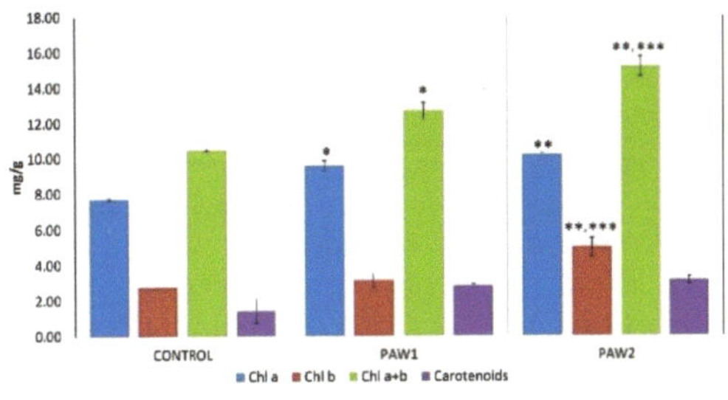

Figure 5. Effect of PAW treatment on the photosynthetic pigments content in *Triticum aestivum* L. cv. Glosa shoots at day 8; * significant difference ($p < 0.05$) between control and PAW1; ** significant difference ($p < 0.05$) between control and PAW2; *** significant difference ($p < 0.05$) between PAW1 and PAW2. Each point represents the mean ± SD of three determinations.

3.6. Total Phenolic Content

Our results revealed that PAW treatment enhanced the content in free and bound phenolic compounds (Figure 6). At day 3, both PAW1 and PAW2 induced a significant increase in the free phenolic contents in wheat sprouts in comparison with the control (48.05 vs. 38.49 mg% and 51.69 vs. 38.49 mg%, respectively) ($p < 0.05$). In addition, PAW1 enhanced the free phenolics in the sprouts at day 1 (39.02 vs. 33.27 mg% in the control group, $p < 0.05$) and shoots at day 8 (121.19 vs. 108.49 mg% in the control group, $p > 0.05$). PAW1 and PAW2 treatment also resulted in elevated bound phenolic contents in the sprouts at day 3 (125.94 and 119.52 mg% vs. 83.81 mg% in the control group) and shoots at day 8 (165.79 and 164.70 mg% vs. 139.41 mg% in the control group), with no significant differences between each treated group and control group, and also between PAW1- and PAW2-treated groups ($p > 0.05$).

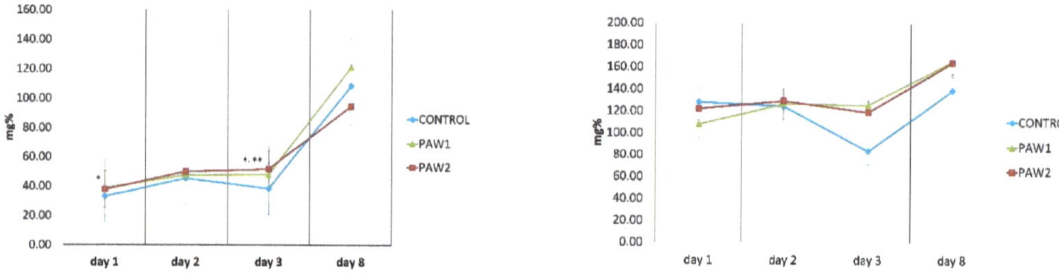

Figure 6. Effect of PAW treatment on the free (**a**) and bound (**b**) phenolic contents in *Triticum aestivum* L. cv. Glosa sprouts (days 1, 2, and 3) and shoots (day 8); * significant difference ($p < 0.05$) between control and PAW1; ** significant difference ($p < 0.05$) between control and PAW2. Each point represents the mean ± SD of three determinations.

3.7. Antioxidant Activity

PAW treatment positively influenced the antioxidant capacity of free phenolic fractions, except those isolated from PAW1-treated sprouts (day 1) and PAW2-treated shoots (day 8). A substantial increase in the antioxidant capacity was noticed for the free phenolic fractions obtained from PAW2-treated sprouts at days 2 (17.01 vs. 10.74% for the control group, $p > 0.05$) and 3 (18.32 vs. 10.15% for the control group, $p < 0.05$). At day 8, the antioxidant activity of the free phenolic fraction of PAW1-treated shoots was significantly higher in comparison with the one of PAW2-treated shoots (92.09 vs. 79.94%, $p < 0.05$). A burst in antioxidant capacity was noted for the bound phenolic fraction obtained from the shoots at day 8 under treatment with PAW1 (91.19 vs. 57.49% in the control group, $p < 0.05$) (Figure 7).

3.8. Antioxidant Enzymes Activity

PAW1 and PAW2 substantially stimulated SOD and CAT in the sprouts at day 3 and shoots at day 8, the differences between each treated group and control group and also between PAW1- and PAW2-treated groups being significant ($p < 0.05$). The increase in activity in both enzymes was more prominent at day 3, with a 143.72 and 91.53% increase in SOD activity under PAW1 and PAW2 exposure, respectively, and 50.21 and 36.34% increase in CAT activity under treatment with PAW1 and PAW2, respectively. On the contrary, PAW1 and PAW2 considerably elevated POX activity only on day 3 (by 207.32 and 244.72%, respectively) and reduced it at day 8 (by 39.02 and 18.98%, respectively) ($p < 0.05$ between each treated group and control group and between PAW1- and PAW2-treated groups) (Figure 8).

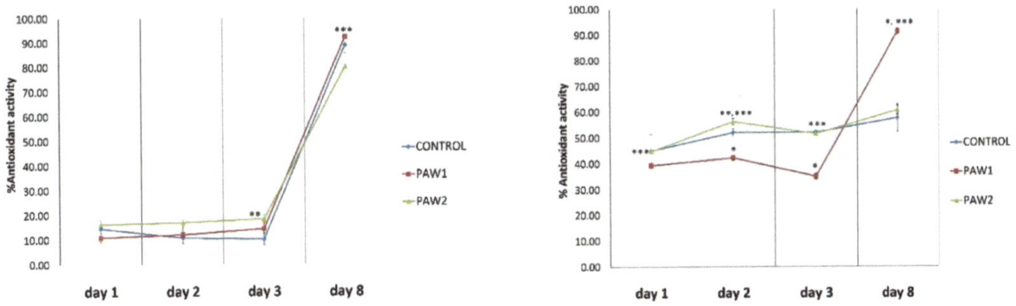

(a) (b)

Figure 7. Antioxidant activity of the free (**a**) and bound (**b**) phenolic fractions isolated from *Triticum aestivum* L. cv. Glosa sprouts (days 1, 2, and 3) and shoots (day 8); * significant difference ($p < 0.05$) between control and PAW1; ** significant difference ($p < 0.05$) between control and PAW2; *** significant difference ($p < 0.05$) between PAW1 and PAW2. Each point represents the mean ± SD of three determinations.

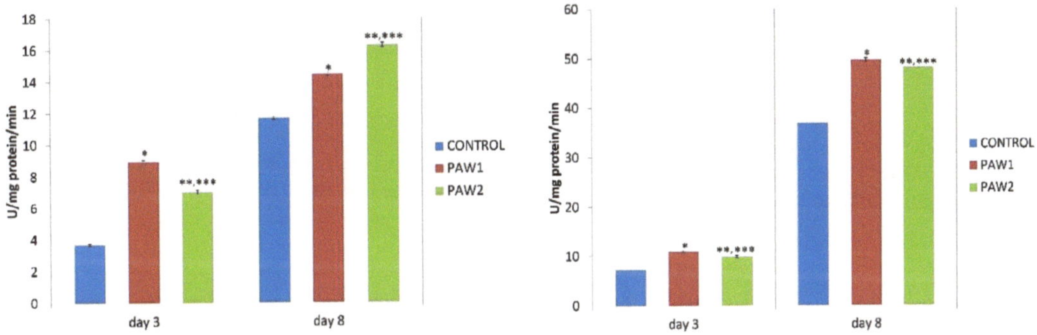

(a) (b)

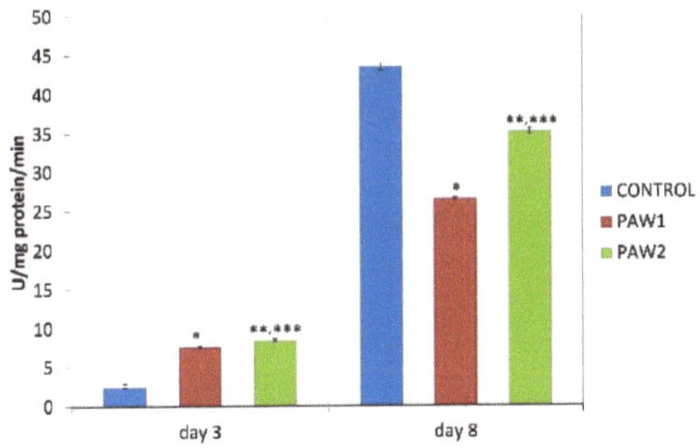

(c)

Figure 8. Activity of antioxidant enzymes SOD (**a**), CAT (**b**), and POX (**c**) in *Triticum aestivum* L. cv. Glosa sprouts (day 3) and shoots (day 8); * significant difference ($p < 0.05$) between control and PAW1; ** significant difference ($p < 0.05$) between control and PAW2; *** significant difference ($p < 0.05$) between PAW1 and PAW2. Each point represents the mean ± SD of three determinations.

4. Discussion

Non-thermal PAW has gained increasing attention in recent years as several studies showed that it significantly impacts germination, development of seedlings as well as accumulation of nutritional and bioactive compounds in plants [29,49–58]. In the present study, we investigated the impact of two types of PAW, generated by exposing distilled water to high-voltage electric discharge, on the caryopses of *Triticum aestivum* L. cv. Glosa, widely cultivated in Romania due to its productivity and resistance to various stress factors.

The influence of PAW1 and PAW2 on the germination of wheat caryopses was evaluated after 24 h as previous studies showed that the increase in germination is more pronounced after 24 h than after several days [29]. PAW1 and PAW2 slightly enhanced the germination of wheat caryopses (by 4.66% and 5.33%, respectively). The positive influence of PAW on germination is ascribed to its reactive oxygen and nitrogen species, especially H_2O_2 and NO. H_2O_2 decreases abscisic acid (hormone responsible for seed dormancy) while increasing gibberellic acid (hormone promoting germination) [48]. In addition, H_2O_2 is decomposed by CAT to water and O_2; O_2 enhances mitochondrial respiration and metabolic activities which results in a positive impact on germination. H_2O_2 oxidizes and inactivates inhibitors of germination present in pericarp [29]. On the other hand, nitrogenous compounds also enhance seed germination. NO_3^- and NO_2^- are a source of NO; the latter was reported to stimulate light-induced germination mediated by phytochrome A, downregulate abscisic acid signaling, and increase the expression of aquaporin genes, thus enhancing water absorption and sprout development [49]. Other studies also found a positive impact of PAW on germination. Kučerová et al. reported that PAW (produced by transient spark discharge) significantly influenced the germination of *Triticum aestivum* L. cv. IS Gordius caryopses depending on the type of water used to generate PAW (tap or deionized water), water activation time, and chemical composition of PAW. PAW obtained from deionized water enhanced germination by 26% whereas PAW produced from tap water, containing higher content of NO_3^- (~0.5 mM), induced 103% increase in germination [29]. The discrepancies between our results and the ones reported by Kučerová et al. could be attributed, in large part, to the cultivar (Glosa in our study, IS Gordius in Kučerová's study) and different physicochemical characteristics of PAW used in both studies. An increase in the germination rate was also reported by Wang et al. who investigated the effects of PAW generated by atmosphere pressure Ar–O_2 plasma jet (1–5 min activation time) on *Triticum aestivum* L. (Jimai 23) caryopses. After 7 days, PAW activated for 3 min ($\sim$2 mg/L NO_3^-, $\sim$0.5 mg/L NO_2^-) induced 100% germination vs. ~98.5% germination rate in the control group (distilled water) [50]. Zhang et al. reported a significant increase in the germination rate of lentil (*Lens culinaris* Medik.) seeds (to almost 80% at day 14, the control showing a germination rate lower than 40%) when treated with PAW ($\sim$600 μM NO_3^-, 160 μM H_2O_2); PAW was generated from tap water in an atmospheric pressure plasma jet (He as carrier gas) [51]. In another study, Fan et al. prepared PAW by exposing distilled water to non-thermal plasma for 15, 30, 60, and 90 s in an atmospheric pressure plasma jet device. The concentrations of reactive oxygen and nitrogen species increased with exposure time, reaching 118.39 mg/L NO_3^-, 62.05 mg/L NO_2^-, and 0.90 mg/L H_2O_2 90 s after activation. After 30 h of treatment, only PAW produced with the shortest discharge time (15 s) increased the germination rate of the mung bean (*Vigna radiata* (L.) R. Wilczek, Zhonglv 5) seeds to 93.01% vs. 84.41% in the control (distilled water). PAW produced with a longer discharge time (30, 60, and 90 s) decreased the germination rate, most likely due to the toxic effects induced by the high levels of the reactive species [52]. Another previous report showed significant enhancement in the germination rate of radish (*Raphanus sativus* L.) seeds treated with two types of PAW containing 8.90 mg/L NO_3^- and 0.11 mg/L H_2O_2 and 16.22 mg/L NO_3^- and 0.10 mg/L H_2O_2, respectively. The germination rate was determined daily for 3 days after sowing. After 24 h, the germination rate of the treated seeds was significantly higher in comparison to control (60 and 100% vs. 40%) [56]. Lo Porto et al. observed an enhancement in the germination of soybean (*Glycine max* (L.) Merr) seeds exposed to two types of PAW containing various amounts of NO_3^- and H_2O_2 (20 µM NO_3^-, 20 µM H_2O_2 and 170 µM NO_3^-, 100 µM H_2O_2), with significant differences between samples at day 2 [53]. PAW also increased the germination rate of black gram (*Vigna mungo* (L.) Hepper) seeds (10–15% increase after 24 h submersion in PAW, followed by 12 h treatment) [57] and rye seeds (by up to 50% at 72 h) [54]. On the contrary, Stoleru et al. found no significant increase in the germination rate of lettuce (*Lactuca sativa* L.) seeds exposed to PAW (containing 1.5 mg/L NO_3^- and 0.5 mg/L H_2O_2 or 3.0 mg/L NO_3^- and 1.65 mg/L H_2O_2) at 3 days after sowing [58].

In our study, the influence of PAW1 and PAW2 on the sprout growth was assessed at days 2, 3, and 8. PAW2 caused significant increases in the lengths of roots at days 2, 3, and 8 (185%, 115.48%, and 29.27%, respectively) and shoots at days 2 and 3 (58.66% and 629.28%, respectively). The growth promoting effects of PAW are mainly triggered by H_2O_2 via stimulation of sucrose phosphate synthase and modulation of NO and Ca^{2+} signaling pathways [59]. NO_3^- and NO_2^- also play an important role in plant growth and development. For example, NO_2^- upregulates genes encoding expansins (proteins inducing cell wall loosening and extension) and auxins (growth hormones) [60]. Kučerová et al. reported a weak effect of PAW (water activation time 0.5–1 min/mL) on the length of wheat seedlings (increase by only 6–7% after 6 days of cultivation). A longer water activation time (1.5–3 min/mL) produced PAW with higher reactive species content which caused a reduction in the seedling length [29]. Similar results were reported by Fan et al. and Sajib et al. PAW obtained with 15 s discharge time had positive effects on the growth of mung bean sprouts (8.09% increase in stem length) whereas PAW produced with longer exposure times (30, 60, and 90 s) negatively impacted the seedling growth [52]. After 7 days of cultivation, PAW, generated by 3 min high voltage discharge in deionized water, caused a pronounced elongation of black gram roots and shoots (to 5.46 vs. 4.10 cm in control and 7.67 vs. 4.83 cm in control, respectively), whereas PAW, produced by longer plasma treatment of deionized water (6–15 min), decreased the lengths of roots and shoots [57]. Other studies reported a positive impact of PAW treatment on seedling growth. PAW (1.5 mg/L NO_3^-, 0.5 mg/L H_2O_2) increased the lettuce average radicle length by 5.58% (53 vs. 50.2 mm in control) and the average hypocotyl length by 34.85% (8.9 vs. 6.6 mm in control) at day 8 after sowing [58]. PAW (20 µM NO_3^-, 20 µM H_2O_2) stimulated the growth of soybean stems, especially at days 10 and 20 [53]. PAW also induced the elongation of the roots and coleoptiles of rye seeds and radish seedlings [54,56]. In case of radish seedlings, a direct relationship between NO_3^- content in PAW and seedling length was observed [56].

Proteins are important components of plant enzymes and key players in plant growth [29]. In our study, PAW1 (25 mg/L NO_3^-) induced a significant increase in the protein content (76.04%) in the sprouted caryopses at day 1. At day 8, PAW2 (35 mg/L NO_3^-) was more efficient causing 20.51% enhancement in the protein content of shoots. The increase in protein content is due to NO_3^- and NO_2^- present in PAW; both species are important sources of nitrogen which is required for protein synthesis [29,61]. Kučerová et al. also observed an increase in the soluble protein content in the roots and above-ground parts of wheat plants after 4 weeks of cultivation in perlite substrate and irrigation with PAW produced from tap water (water activation time 0.5 and 2 min/mL). The increase in soluble protein content was higher in the roots in comparison with the above-ground parts (~43% and ~69%, respectively, vs. ~19% and ~14%, respectively) [29]. After 7 days of cultivation, Sajib et al. reported an increase in the total soluble protein content in the roots (19.18%) and leaves (33.28%) of black gram plants developed from the seeds treated with PAW (produced by 6 min exposure of deionized water to high voltage discharge). Longer discharge times (9–15 min) resulted in the reduction in soluble protein content [57].

Chlorophylls and carotenoids are involved in photosynthesis and other processes (metabolic reactions, protection against oxidative stress) [62]. In our study, both PAW1 and PAW2 enhanced the chlorophyll content in wheat shoots at day 8. PAW 2 was more potent increasing chlorophyll a, chlorophyll b, and total chlorophyll by 32.17, 80.07, and 44.93%, respectively. Moreover, PAW1 and PAW2 induced 2- and 2.2-fold increase in the carotenoid content, respectively, but with no significant differences between groups. H_2O_2 and NO_3^-

are responsible, to a great extent, for the boost in photosynthetic pigments. H_2O_2 increases the stomatal opening, thus enhancing CO_2 uptake, photosynthetic rate, and accumulation of photosynthetic pigments [58]. In case of NO_3^-, the effect is attributed to up-regulation of the genes encoding δ-amino levulinic acid dehydratase, enzyme involved in the biosynthesis of chlorophyll [63]. Kučerová et al. determined chlorophylls and carotenoids in the wheat plants after 4 weeks of cultivation (conditions mentioned above) and found that contents increased by 17 and 12%, respectively [29]. Moreover, Maniruzzaman et al. showed that PAW, generated using either Ar (containing mainly H_2O_2) or air (containing mainly NO_3^-), positively influenced the relative chlorophyll content in wheat seedlings grown in a potting mix up to 4 weeks or in a soil-free plant growth system up to 14 days [55]. A previous study conducted by Stoleru et al. showed that the exposure to PAW did not significantly affect the chlorophyll content in the lettuce leaves after transplanting in pots. Only treatment with PAW containing 3.0 mg/L NO_3^- and 1.65 mg/L H_2O_2, in a bigger volume of substrate (3200 cm^3), determined 10.71% increase in the chlorophyll content in lettuce leaves at day 64 after transplanting in pots [58]. Sajib et al. investigated the effects of PAW treatment of black gram seeds on the chlorophyll content of leaves after 7 days of cultivation. Treatment with PAW (produced by 6 min exposure of deionized water to high voltage discharge) enhanced the leaf chlorophyll level by 23.80% [57].

Reactive oxygen and nitrogen species in PAW stimulate the production and accumulation of antioxidant phytochemicals in treated plants [29]. Species in PAW, such as H_2O_2 and NO, are potent inductors of the genes involved in phenolic synthesis such as *pal* gene (encoding L-phenylalanine ammonia-lyase, enzyme which deaminates L-phenylalanine) and *chs* gene (encoding chalcone synthase, enzyme implied in flavonoid synthesis) [64]. In our study, exposure to PAW1 and PAW2 enhanced the phenolic content in wheat sprouts; the increase in the free phenolic content induced by PAW1 at days 1 (17.28%) and 3 (24.84%) and PAW2 at day 3 (34.29%) and also the increase in the bound phenolic content induced by PAW2 at day 3 (80.49%) were significant. In general, the impact of PAW1 and PAW2 on the antioxidant potential of phenolic fractions isolated from sprouts/shoots was negligible except the impact of PAW1 on the antioxidant effect of the bound phenolic fraction at day 8 (58.62% boosting effect). To the best of our knowledge, this is the first study reporting on the impact of PAW on the phenolic content and antioxidant effects of wheat sprouts/shoots. Previous investigations evaluated only the total phenolic content and antioxidant activity of the juice obtained from plantlets treated with PAW. Fourteen days of PAW application (generated by atmosphere pressure Ar–O_2 plasma jet, 1–5 min activation time) improved the total phenolic content (by 10.46%) and antioxidant effects (by 5.51–35.34%) of the wheat (*Triticum aestivum* L., Jimai 23) plantlet juice [50]. Data from the literature concerning the impact of PAW treatment on the phenolic content and antioxidant activity of other sprouts are scarce. Fan et al. reported an increase in the total phenolic and flavonoid contents (by 3.53 and 2.22%, respectively) in the mung bean sprouts exposed to PAW (15 s discharge time) with no important influence on the antioxidant potential of the sprouts. The authors observed that a prolonged discharge time reduced the total phenolic and flavonoid contents, and also the antioxidant activity [52].

Reactive oxygen species in PAW trigger not only the production of metabolites with antioxidant properties, but also the activity of antioxidant enzymes. H_2O_2 was reported to increase the expression and activities of SOD, CAT, and POX in various plants, including *Tritium aestivum* L. [65]. In the present study, the activity of these three enzymes was evaluated at the third and eighth days of germination. According to previous investigations, CAT and SOD activity does not change in the first 48 h of imbibition, but significantly increases with the third day, the increase in activity being accompanied by an intensive radicle protrusion. Concerning POX, its activity is related to early seedling development [66]. PAW1 and PAW2 enhanced the activity of all three enzymes in the sprouts at day 3 and SOD and CAT activity in the shoots at day 8. The increase in antioxidant enzymes activity indicates a defense response against PAW-induced oxidative stress. The reduction in POX activity at day 8 could be related to a lower predominance of the enzyme in the aerial parts in comparison to roots [29]. Kučerová et al. investigated the effects of PAW produced from tap water (water activation time 0.5 and 2 min/mL) on the activity of SOD, CAT, and POX in the above-ground parts of wheat but after longer cultivation time (4 weeks) on a perlite substrate and found a reduction in the activity of antioxidant enzymes ranging from 8 to 46% [29]. Puač et al. reported a decrease in CAT activity in *Paulownia tomentosa* Steud. seeds imbibed with PAW immediately and 2 days after exposure to red light pulse (for inducing germination) with a significant increase in activity on the third day; PAW was obtained from distilled water in an atmospheric pressure plasma jet (5, 10, and 20 min treatment) [67]. It seems that the effect of PAW on the activity of antioxidant enzymes depends not only on the level of reactive species in PAW, but also on the germination phase.

Data from the literature clearly show huge differences in plant responses to PAW exposure, which are dependent on the procedure used to generate PAW, physicochemical characteristics of PAW, treatment protocol, and plant species or even cultivar. Therefore, PAW treatment should be optimized for each application [29,52,56].

To the best of our knowledge, this is the first study reporting on PAW impact on wheat sprouts during the short sprouting time (1–3 days) and wheat young shoots (8 days old). Previous investigations evaluated the effects of other types of PAW on wheat caryopses with reference to the growth parameters at different developmental stages [29,50,55], content in various compounds and activity of antioxidant enzymes in leaves after 4 week cultivation [29] and juice of plantlets after 14 days of cultivation [50], and antioxidant activity of the latter.

5. Conclusions

Our results indicate positive effects of PAW1 (25 mg/L NO_3^-, 4 mg/L NO_2^-, and 6 mg/L H_2O_2) and PAW2 (35 mg/L NO_3^-, 5 mg/L NO_2^-, and 7.5 mg/L H_2O_2) on *Triticum aestivum* L. cv. Glosa sprouts. Overall, exposure to PAW2 had a better impact on sprouts with respect to growth parameters, free phenolic content, and antioxidant activity of the free phenolic fraction (day 3), accumulation of proteins and photosynthetic pigments (day 8). The activity of antioxidant enzymes increased under exposure to PAW1 and PAW2. To conclude, PAW treatment increases the quality of *Triticum aestivum* L. cv. Glosa sprouts. Further investigations will focus on: (i) PAW2 impact on the biosynthesis of other nutritional and bioactive compounds (free amino acids, betaine, vitamins B, vitamin E) and (ii) optimization of PAW characteristics and treatment protocol for the most favorable effect on the accumulation of biomass and health-promoting compounds in *Triticum aestivum* L. cv. Glosa sprouts.

Author Contributions: Conceptualization, R.B. and A.M. (Anca Miron); methodology, A.M. (Alexandru Mandici), R.B., D.A., C.R., D.C.T. and A.M. (Anca Miron); software, R.B. and D.A.; validation, A.M. (Alexandru Mandici), D.A. and O.B; formal analysis, A.M. (Alexandru Mandici) and D.A.; investigation, A.M. (Alexandru Mandici), D.A., O.B., C.R., D.E.C. and T.G.A.; resources, R.B. and A.M. (Anca Miron); data curation, A.M. (Alexandru Mandici) and D.A.; writing—original draft preparation, A.M. (Alexandru Mandici), R.B., D.A., C.R. and A.M. (Anca Miron); writing—review and editing, R.B., D.C.T., A.C.A. and A.M. (Anca Miron); visualization, A.M. (Alexandru Mandici), D.A. and O.B.; supervision, R.B. and A.M. (Anca Miron); project administration, R.B. and A.M. (Anca Miron); funding acquisition, R.B. All authors have read and agreed to the published version of the manuscript.

Funding: This research was funded by a grant from the Romanian Ministry of Education and Research, CNCS-UEFISCDI, project PN-III-P2-2.1-PED-2019-0556 (contract 361PED/2020) within PNCDI III.

Institutional Review Board Statement: Not applicable.

Informed Consent Statement: Not applicable.

Data Availability Statement: The data presented in this study are available upon request from the corresponding authors.

Acknowledgments: The authors would like to acknowledge Territorial Inspectorate for Seed and Propagating Material Quality (Iasi, Romania) for providing wheat caryopses (*Triticum aestivum* L. cv. Glosa).

Conflicts of Interest: The authors declare no conflict of interest.

References

1. Neuhouser, M.L. The importance of healthy dietary patterns in chronic disease prevention. *Nutr. Res.* **2019**, *70*, 3–6. [CrossRef] [PubMed]
2. Slavin, J. Whole grains and human health. *Nutr. Res. Rev.* **2004**, *17*, 99–110. [CrossRef]
3. Stevenson, L.; Phillips, F.; O'Sullivan, K.; Walton, J. Wheat Bran: Its Composition and Benefits to Health, a European Perspective. *Int. J. Food Sci. Nutr.* **2012**, *63*, 1001–1013. [CrossRef]
4. Štěrbová, L.; Bradová, J.; Sedláček, T.; Holasová, M.; Fiedlerová, V.; Dvořáček, V.; Smrčková, P. Influence of technological processing of wheat grain on digestibility and resistant starch content. *Starch* **2016**, *68*, 593–602. [CrossRef]
5. Russell, W.R.; Hoyles, L.; Flint, H.J.; Dumas, M.-E. Colonic bacterial metabolites and human health. *Curr. Opin. Microbiol.* **2013**, *16*, 246–254. [CrossRef] [PubMed]
6. Ward, J.L.; Poutanen, K.; Gebruers, K.; Piironen, V.; Lampi, A.-M.; Nyström, L.; Andersson, A.A.M.; Åman, P.; Boros, D.; Rakszegi, M.; et al. The HEALTHGRAIN cereal diversity screen: Concept, results and prospects. *J. Agric. Food Chem.* **2008**, *56*, 9699–9709. [CrossRef]
7. Gonzalez-Thuillier, I.; Salt, L.; Chope, G.; Penson, S.; Skeggs, P.; Tosi, P.; Powers, S.J.; Ward, J.L.; Wilde, P.; Shewry, P.R.; et al. Distribution of Lipids in the Grain of Wheat (cv. Hereward) Determined by Lipidomic Analysis of Milling and Pearling Fractions. *J. Agric. Food Chem.* **2015**, *63*, 10705–10716. [CrossRef]
8. Prinsen, P.; Gutiérrez, A.; Faulds, C.B.; Del Río, J.C. Comprehensive Study of Valuable Lipophilic Phytochemicals in Wheat Bran. *J. Agric. Food Chem.* **2014**, *62*, 1664–1673. [CrossRef]
9. Shewry, P.R.; Hey, S.J. The contribution of wheat to human diet and health. *Food Energy Secur.* **2015**, *4*, 178–202. [CrossRef] [PubMed]
10. Lampi, A.-M.; Nurmi, T.; Ollilainen, V.; Piironen, V. Tocopherols and Tocotrienols in Wheat Genotypes in the HEALTHGRAIN Diversity Screen. *Agric. Food Chem.* **2008**, *56*, 9716–9721. [CrossRef] [PubMed]
11. Rizvi, S.; Raza, S.T.; Ahmed, F.; Ahmad, A.; Abbas, S.; Mahdi, F. The role of vitamin E in human health and some diseases. *Sultan Qaboos Univ. Med. J.* **2014**, *14*, e157–e165.
12. Vasanthi, H.R.; Parameswari, R.P.; Das, D.K. Multifaceted role of tocotrienols in cardioprotection supports their structure: Function relation. *Genes Nutr.* **2012**, *7*, 19–28. [CrossRef]
13. Hussain, A.; Larsson, H.; Johansson, E. Carotenoid Extraction from Locally and Organically Produced Cereals Using Saponification Method. *Processes* **2021**, *9*, 783. [CrossRef]
14. Trono, D. Carotenoids in Cereal Food Crops: Composition and Retention throught Grain Storage and Food Processing. *Plants* **2019**, *8*, 551. [CrossRef]
15. Gan, R.-Y.; Lui, W.-Y.; Wu, K.; Chan, C.-L.; Dai, S.-H.; Sui, Z.-Q.; Corke, H. Bioactive compounds and bioactivities of germinated edible seeds and sprouts: An updated review. *Trends Food Sci. Technol.* **2017**, *59*, 1–14. [CrossRef]
16. Likes, R.; Madl, R.L.; Zeisel, S.H.; Craig, S.A.S. The betaine and choline content of a whole wheat flour compared to other mill streams. *J. Cereal Sci.* **2007**, *46*, 93–95. [CrossRef] [PubMed]
17. Tinelli, C.; Di Pino, A.; Ficulle, E.; Marcelli, S.; Feligioni, M. Hyperhomocysteinemia as a Risk Factor and Potential Nutraceutical Target for Certain Pathologies. *Front. Nutr.* **2019**, *6*, 49. [CrossRef]
18. Calinoiu, L.F.; Vodnar, D.C. Whole Grains and Phenolic Acids: A Review in Bioactivity, Functionality, Health Benefits and Bioavailability. *Nutrients* **2018**, *10*, 1615. [CrossRef]
19. Mpofu, A.; Sapirstein, H.D.; Beta, T. Genotype and Environmental Variation in Phenolic Content, Phenolic Acid Composition, and Antioxidant Activity of Hard Spring Wheat. *J. Agric. Food Chem.* **2006**, *54*, 1265–1270. [CrossRef] [PubMed]
20. Li, L.; Shewry, P.R.; Ward, J.L. Phenolic Acids in Wheat Varieties in the HEALTHGRAIN Diversity Screen. *J. Agric. Food Chem.* **2008**, *56*, 9732–9739. [CrossRef]
21. Okarter, N.; Liu, C.-S.; Sorrells, M.E.; Liu, R.H. Phytochemical Content and Antioxidant Activity of Six Diverse Varieties of Whole Wheat. *Food Chem.* **2010**, *119*, 249–257. [CrossRef]
22. Tian, W.; Li, Y. Phenolic Acid Composition and Antioxidant Activity of Hard Red Winter Wheat Varieties. *J. Food Biochem.* **2018**, *42*, e12682. [CrossRef]
23. Van Hung, P.; Hatcher, D.W.; Barker, W. Phenolic acid composition of sprouted wheats by ultra-performance liquid chromatography (UPLC) and their antioxidant activities. *Food Chem.* **2011**, *126*, 1896–1901. [CrossRef]
24. Ding, J.; Feng, H. Controlled germination for enhancing the nutritional value of sprouted grains. In *Sprouted Grains: Nutritional Value, Production, and Applications*; Feng, H., Nemzer, B., DeVries, J.W., Eds.; Woodhead Publishing: Duxford, UK, 2019; pp. 91–112. [CrossRef]
25. Van Hung, P.; Maeda, T.; Morita, N. Improvement of Nutritional Composition and Antioxidant Capacity of High-Amylose Wheat during Germination. *J. Food Sci. Technol.* **2015**, *52*, 6756–6762. [CrossRef] [PubMed]

26. Ceccaroni, D.; Alfeo, V.; Bravi, E.; Sileoni, V.; Perreti, G.; Marconi, O. Effect of the time and temperature of germination on the phenolic compounds of *Triticum aestivum* L. and *Panicum miliaceum* L. *LWT Food Sci. Technol.* **2020**, *127*, 109396. [CrossRef]
27. Perez, S.M.; Biondi, E.; Laurita, R.; Proto, M.; Sarti, F.; Gherardi, M.; Bertaccini, A.; Colombo, V. Plasma activated water as resistance inducer against bacterial leaf spot of tomato. *PLoS ONE* **2019**, *14*, e0217788. [CrossRef]
28. Yodpitaka, S.; Mahatheeranonta, S.; Boonyawand, D.; Sookwonga, P.; Roytrakule, S.; Norkaew, O. Cold plasma treatment to improve germination and enhance the bioactive phytochemical content of germinated brown rice. *Food Chem.* **2019**, *289*, 328–339. [CrossRef]
29. Kučerová, K.; Henselová, M.; Slováková, L.; Hensel, K. Effects of plasma activated water on wheat: Germination, growth parameters, photosynthetic pigments, soluble protein content, and antioxidant enzymes activity. *Plasma Process. Polym.* **2019**, *16*, e1800131. [CrossRef]
30. Yang, F.; Basu, T.K.; Ooraikul, B. Studies on germination conditions and antioxidant contents of wheat grain. *Int. J. Food Sci. Nutr.* **2001**, *52*, 319–330. [CrossRef]
31. Hsieh, K.C.; Wandell, R.J.; Bresch, S.; Locke, B.R. Analysis of hydroxyl radical formation in a gas-liquid electrical discharge plasma reactor utilizing liquid and gaseous radical scavengers. *Plasma Process. Polym.* **2017**, *14*, e1600171. [CrossRef]
32. Wandell, R.J.; Wang, H.; Tachibana, K.; Makled, B.; Locke, B.R. Nanosecond pulsed plasma discharge over a flowing water film: Characterization of hydrodynamics, electrical, and plasma properties and their effect on hydrogen peroxide generation. *Plasma Process. Polym.* **2018**, *15*, e1800008. [CrossRef]
33. Burlica, R.; Shih, K.Y.; Locke, B.R. Formation of H_2 and H_2O_2 in a water-spray gliding arc nonthermal plasma reactor. *Ind. Engin. Chem. Res.* **2010**, *49*, 6342–6349. [CrossRef]
34. Cretu, D.E.; Astanei, D.; Burlica, R.; Beniuga, O.; Tesoi, D. The Influence of NTP Reactor Geometry on H_2O_2 Generation in Water. In Proceedings of the 2020 International Conference and Exposition on Electrical and Power Engineering (EPE), Iasi, Romania, 22–23 October 2020; pp. 632–635. [CrossRef]
35. Colorimetric Test Kit VISOCOLOR ECO Nitrate. Available online: https://www.mn-net.com/media/pdf/a5/a5/8f/Instruction-931041-931241-Colorimetric-test-kit-VISOCOLOR-ECO-Nitrate.pdf (accessed on 15 October 2022).
36. Colorimetric Test Kit VISOCOLOR ECO Nitrite. Available online: https://www.mn-net.com/media/pdf/ee/4f/f2/Instruction-931044-931244-Colorimetric-test-kit-VISOCOLOR-ECO-Nitrite.pdf (accessed on 15 October 2022).
37. Mughal, I.; Shah, Y.; Tahir, S.; Haider, W.; Fayyaz, M.; Yasmin, T.; Ilyas, M.; Farrakh, S. Protein quantification and enzyme activity estimation of Pakistani wheat landraces. *PLoS ONE* **2020**, *15*, e0239375. [CrossRef] [PubMed]
38. Lichtenthaler, H.K. Chlorophylls and carotenoids: Pigments of photosynthetic membranes. *Methods Enzymol.* **1987**, *148*, 350–382. [CrossRef]
39. Vochita, G.; Oprica, L.; Gherghel, D.; Mihai, C.-T.; Boukherroub, R.; Lobiuc, A. Graphene oxide effects in early ontogenetic stages of *Triticum aestivum* L. seedlings. *Ecotoxicol. Environ. Safety* **2019**, *181*, 345–352. [CrossRef]
40. Stagnari, F.; Glieni, A.; D'Egidio, S.; Falcinelli, B.; Pagnani, G.; Pace, R. Effects of sprouting and salt stress on polyphenol composition and antiradical activity of einkorn, emmer and durum wheat. *Ital. J. Agron.* **2017**, *12*, 293–301. [CrossRef]
41. Singleton, V.L.; Rossi, J.A. Colorimetry of total phenolics with phosphomolybdic-phosphotungstic acid reagents. *Am. J. Enol. Vitic.* **1965**, *16*, 144–158.
42. Vasincu, A.; Paulsen, B.S.; Diallo, D.; Vasincu, I.; Aprotosoaie, A.C.; Bild, V.; Charalambous, C.; Constantinou, A.I.; Miron, A.; Gavrilescu, C.M. *Veronia kotschyana* Roots: Therapeutic Potential via Antioxidant Activity. *Molecules* **2014**, *19*, 19114–19136. [CrossRef]
43. Naidoo, C.M.M.; Naidoo, Y.; Dewir, Y.H.; Singh, M.; Daniels, A.N.; El-Ramady, H. In Vitro Investigation of the Antioxidant and Cytotoxic Potential of *Tabernaemontana ventricosa* Hochst. ex A. DC. Leaf, Stem, and Latex Extracts. *Horticulturae* **2022**, *8*, 91. [CrossRef]
44. Unal, M.S.; Gundesli, M.A.; Ercisli, S.; Kupe, M.; Assouguem, A.; Ullah, R.; Almeer, R.; Najda, A. Cultivar Differences on Nutraceuticals of Grape Juices and Seeds. *Horticulturae* **2022**, *8*, 267. [CrossRef]
45. Winterbourn, C.; Hawkins, R.; Brian, M.; Carrel, R. The estimation of red cell superoxide dismutase activity. *J. Lab.Clin. Med.* **1975**, *85*, 337–341.
46. Artenie, V.; Ungureanu, E.; Negură, A.M. *Metode de Investigare a Metabolismului Glucidic şi Lipidic*; PIM: Iaşi, Romania, 2008; p. 182.
47. Beers, R.; Sizer, I. A spectrophotometric method for measuring the breakdown of hydrogen peroxide by catalase. *J. Biol. Chem.* **1952**, *195*, 133–140. [CrossRef]
48. Su, L.; Lan, Q.; Pritchard, H.W.; Xue, H.; Wang, X. Reactive oxygen species induced by cold stratification promote germination of *Hedysarum scoparium* seeds. *Plant Physiol. Biochem.* **2016**, *109*, 406–415. [CrossRef] [PubMed]
49. Šírová, J.; Sedlářová, M.; Piterková, J.; Luhová, L.; Petřivalský, M. The role of nitric oxide in the germination of plant seeds and pollen. *Plant Sci.* **2011**, *181*, 560–572. [CrossRef] [PubMed]
50. Wang, J.; Cheng, J.-H.; Sun, D.-W. Enhancement of Wheat Germination, Seedling Growth and Nutritional Proprieties of Wheat Plantlet Juice by Plasma Activated Water. *J. Plant Growth Regul.* **2022**. [CrossRef] [PubMed]
51. Zhang, S.; Rousseau, A.; Dufour, T. Promoting lentil germination and stem growth by plasma activated tap water, demineralized water and liquid fertilizer. *RSC Adv.* **2017**, *7*, 31244. [CrossRef]
52. Fan, L.; Liu, X.; Ma, Y.; Xiang, Q. Effects of plasma-activated water treatment on seed germination and growth of mung bean sprouts. *J. Taibah Univ. Sci.* **2020**, *14*, 823–830. [CrossRef]

53. Lo Porto, C.; Ziuzina, D.; Los, A.; Boehm, D.; Palumbo, F.; Favia, P.; Tiwari, B.; Bourke, P.; Cullen, P.J. Plasma activated water and airborne ultrasound treatments for enhanced germination and growth of soybean. *Innov. Food Sci. Emerg. Technol.* **2018**, *49*, 13–19. [CrossRef]
54. Naumova, I.K.; Maksimov, A.I.; Khlyustova, A.V. Stimulation of the Germinability of Seeds and Germ Growth under Treatment with Plasma-Activated Water. *Surf. Eng. Appl. Elecvtrochem.* **2011**, *47*, 263–265. [CrossRef]
55. Maniruzzaman, M.; Sinclair, A.J.; Cahill, D.M.; Wang, X.; Dai, X.J. Nitrate and Hydrogen Peroxide Generated in Water by Electrical Discharges Stimulate Wheat Seedling Growth. *Plasma Chem. Plasma Process.* **2017**, *37*, 1393–1404. [CrossRef]
56. Sivachandiran, L.; Khacef, A. Enhanced seed germination and plant growth by atmospheric pressure cold air plasma: Combined effect of seed and water treatment. *RSC Adv.* **2017**, *7*, 1822. [CrossRef]
57. Sajib, S.A.; Billah, M.; Mahmud, S.; Mlah, M.; Hossain, F.; Omar, F.B.; Roy, N.C.; Hoque, K.M.F.; Talukder, M.R.; Kabir, A.H.; et al. Plasma activated water: The next generation eco-friendly stimulant for enhancing plant seed germination, vigor and increased enzyme activity, a study on black gram (*Vigna mungo* L.). *Plasma Chem. Plasma Process.* **2020**, *40*, 119–143. [CrossRef]
58. Stoleru, V.; Burlică, R.; Mihalache, G.; Dîrlău, I.D.; Pădureanu, S.; Teliban, G.-C.; Astanei, D.; Cojocaru, A.; Beniugă, O.; Patraș, A.J. Plant growth promotion effect of plasma activated water on *Lactuca sativa* L. cultivated in two different volumes of substrate. *Nat. Sci. Rep.* **2020**, *10*, 20920. [CrossRef]
59. Nurnaeimah, N.; Mat, N.; Suryati Mohd, S.; Badaluddin, N.A.; Yusoff, N.; Sajili, M.H.; Mahmud, K.; Mohd Adnan, A.F.; Khandaker, M.M. The Effects of Hydrogen Peroxide on Plant Growth, Mineral Accumulation, as Well as Biological and Chemical Properties of *Ficus deltoidea*. *Agronomy* **2020**, *10*, 599. [CrossRef]
60. Wang, X.; Li, Y.; Fang, G.; Zhao, Q.; Zeng, Q.; Li, X.; Gong, H.; Li, Y. Nitrite Promotes the Growth and Decreases the Lignin Content of *indica* Rice Calli: A Comprehensive Transcriptome Analysis of Nitrite-Responsive Genes during In Vitro Culture of Rice. *PLoS ONE* **2014**, *9*, e95105. [CrossRef] [PubMed]
61. Punith, N.; Harsha, R.; Lakshminarayana, R.; Hemanth, M.; Anand, M.S.; Dasappa, S. Plasma Activated Water Generation and its Application in Agriculture. *Adv. Matter. Lett.* **2019**, *10*, 700–704. [CrossRef]
62. Pérez-Gálvez, A.; Viera, I.; Roca, M. Carotenoids and Chlorophylls as Antioxidants. *Antioxidants* **2020**, *9*, 505. [CrossRef]
63. Agnihotri, A.; Seth, C.S. Exogenously applied nitrate improves the photosynthetic performance and nitrogen metabolism in tomato (*Solanum lycopersicum* L. cv Pusa Rohini) under arsenic (V) toxicity. *Physiol. Mol. Biol. Plants* **2016**, *22*, 341–349. [CrossRef]
64. Zhang, R.-Q.; Zhu, H.-H.; Zhao, H.-Q.; Yao, Q. Arbuscular mycorrhizal fungal inoculation increases phenolic synthesis in clover roots via hydrogen peroxide, salicylic acid and nitric oxide signaling pathways. *J. Plant Physiol.* **2013**, *170*, 74–79. [CrossRef]
65. Niu, L.; Liao, W. Hydrogen Peroxide Signaling in Plant Development and Abiotic Responses: Crosstalk with Nitric Oxide and Calcium. *Front. Plant. Sci.* **2016**, *7*, 230. [CrossRef]
66. Bogdanović, J.; Radotić, K.; Mitrović, A. Changes in activities of antioxidant enzymes during *Chenopodium murale* seed germination. *Biol. Plant.* **2008**, *52*, 396–400. [CrossRef]
67. Puač, N.; Škoro, N.; Spasić, K.; Živković, S.; Miltunović, M.; Malović, G.; Petrović, Z.L.J. Activity of catalase enzyme in *Paulownia tomentosa* seeds during the process of germination after treatments with low pressure plasma and plasma activated water. *Plasma Process. Polym.* **2018**, *15*, e1700082. [CrossRef]

horticulturae

MDPI

Article

Volatile Profile of Garden Rose (*Rosa hybrida*) Hydrosol and Evaluation of Its Biological Activity In Vitro

Biljana Božanić Tanjga [1], Biljana Lončar [2,*], Milica Aćimović [3], Biljana Kiprovski [3], Olja Šovljanski [2], Ana Tomić [2], Vanja Travičić [2], Mirjana Cvetković [4], Vidak Raičević [5] and Tijana Zeremski [3]

1 Pheno Geno Roses Ltd., 23326 Ostojićevo, Serbia
2 Faculty of Technology Novi Sad, University of Novi Sad, 21000 Novi Sad, Serbia
3 Institute of Field and Vegetable Crops, 21000 Novi Sad, Serbia
4 Institute of Chemistry, Technology and Metallurgy, University of Belgrade, 11000 Belgrade, Serbia
5 Department of Chemistry, Biochemistry and Environmental Protection, Faculty of Science, University of Novi Sad, 21000 Novi Sad, Serbia
* Correspondence: cbiljana@uns.ac.rs

Abstract: Garden rose, *Rosa hybrida*, is primarily used for decoration and has a wide range of growing area, contrary to *R. damascena* that has a limited area of distribution (Turkey and Bulgaria), yet it is extensively used for commercial production of valuable and expensive rose oil. Since the content of essential oil in rose petals is low (0.03–0.04%), its production is quite limited; however, during this process, a significant amount of rose hydrosol is obtained as a secondary product. The aim of this research was to determine the chemical composition of garden rose hydrosols and to evaluate their biological properties. Obtained results show that *R. hybrida* hydrosol containing phenylethyl alcohol, nerol, linalool, and geraniol may be used as an alternative for *R. damascena* hydrosol. However, the total phenolic content was quite low (4.96 μg GAE/mL), which is related to a low level of observed antioxidant activity based on different antioxidant activity assays. Furthermore, *R. hybrida* hydrosol did not exhibit antimicrobial activity against several gram-positive and gram-negative bacteria, as well as yeast and fungi. Anti-inflammatory activity was also low, while no antihyperglycemic activity was detected. With these results in mind, no potential is evident for the therapeutic application of rose hydrosol beyond that found in complimentary medicine such as aromatherapy.

Keywords: hydrolate; rose water; antioxidant activity; antimicrobial activity; antihyperglycemic activity; anti-inflammatory activity

Citation: Tanjga, B.B.; Lončar, B.; Aćimović, M.; Kiprovski, B.; Šovljanski, O.; Tomić, A.; Travičić, V.; Cvetković, M.; Raičević, V.; Zeremski, T. Volatile Profile of Garden Rose (*Rosa hybrida*) Hydrosol and Evaluation of Its Biological Activity In Vitro. *Horticulturae* **2022**, *8*, 895. https://doi.org/10.3390/horticulturae8100895

Academic Editor: Ivan Salamon

Received: 15 September 2022
Accepted: 27 September 2022
Published: 29 September 2022

1. Introduction

Roses have been valued as ornamental and fragrance-bearing plants since ancient times. Medieval civilizations in the Eastern Mediterranean region cultivated, grew, and boiled roses with water to produce aromatic water, used because of its apparent refreshing effects on the mind and spirit [1,2]. This was a harbinger of the modern rose hydrosol. According to Persian folk medicine, Damask rose (*Rosa damascena* Mill.) aromatic water was used for mental refreshing, the improvement of cognitive abilities, its strengthening or stimulating effect on the nerves, prevention or reduction of fatigue, and as a sedative [3,4]. Today, rose hydrosols or rose aromatic water have great commercial importance [5]. They are mainly used for flavoring food (Turkish delight, lokkum), tonic beverage (Jollab), and other drinks, ice creams, yogurt, cake, rice pudding, jam, marmalade, cake, and syrups; it is also sprinkled on many meat dishes [6–13]. It has been traditionally used for skin care and healing-related diseases such as erythema, itchiness, and swelling in the form of soaps, cosmetics, toiletries, and perfumes [6,14].

Rose oil is one of the most valuable and expensive flavor and fragrance products, as well as an aroma-therapeutic agent. Taking into account the increased demand for rose oil on the international market, the climate-related limitations on cultivation area required for optimum-yield production (Bulgaria, Turkey, India), as well as the fact that it flowers only once per year (May–June), there are attempts to extend the cultivation of essential-oil-bearing roses into different agroclimatic regions, to find other cultivars suitable for essential oil production, or to improve growing management and promote processing [3,7,15]. Apart from *R. damascena*, which is the superior oil-bearing rose globally (with 0.03–0.04% of essential oil content), *R. gallica*, *R. centifolia*, *R. moschata*, *R. rugosa*, *R. bourbonica*, and *R. alba* are also widely used for commercial production of rose oil and rose water, as well as other products such as rose absolute and concrete (Morocco, France, Egypt, China, etc.) [7,16–18]. The hydrosols are produced as by-products during the process of essential oil distillation in quite significant quantities [19]. In some countries (Iran, Tunisia), rose hydrosol is traditionally used in religious ceremonies and as a therapeutic product [6,20,21]. The literature review of the volatile compounds of rose hydrosols, which addresses the commercial samples of *R. damascena*, *R. alba*, *R. brunonii*, *R. canina*, *R. centifolia* and *R. rugose*, is not enough to analyze all difference between these samples. However, there is no data about the volatile composition of the one of the most contemporary roses, *R. hybrida*.

Hybrid roses (*R. hybrida*) as a group are one of the major contemporary roses cultivated as garden plants nowadays. Owing to their complex hybridization history, numerous varieties have appeared, with different flower colors and shapes, as well as various scent notes [22,23]. Due to the mentioned world-wide cultivation and its use in different fields of daily life of people, this study aims to gather more insight into the volatiles of nursery-produced rose hybrids and their potential applications. The main goal of this study was to investigate the quality of the by-product during essential oil distillation of *R. hybrida* flowers grown in Serbia. Furthermore, even though the quality of rose essential oil is regulated by international standards (ISO 9842:2003), there are no boundary values set for rose hydrosols. Therefore, another aim of this investigation was to collect literature data about rose hydrosol quality and compare it with the one obtained from *R. hybrida*.

2. Materials and Methods

2.1. Plant Material

Rosa hybrida cv Mileva™ Frayla® has large soft pink petals arranged in very double flowers with a sweet and strong fragrance and long-term period flowering (from May to October). Voucher specimens were confirmed by Milica Rat, PhD, and deposited at the BUNS Herbarium, University of Novi Sad, under number 2-0693. The plantation was established in 2018 on Pheno Geno Roses Serbia fields; the harvest was performed by hand in June 2021, during the full-flowering stage.

2.2. Hydrosol Extraction

Immediately after harvest, fresh flowers were subjected to hydrodistillation. A description of the procedure for rose hydrodistillation is as follows: approximately 50 kg of fresh rose flowers were placed in a distilling vessel (0.8 m^3), and water was added to soak the petals (approximately half of vessel volume) and heated to boiling. Steam was supplied through a manifold pipe into the bottom of the vessel from a high-pressure boiler and routed upwards through a plumbing system to the vessel, with plant material being extracted. The steam, water vapor, and entrained volatiles exited the tank near the top via a 10 cm diameter pipe and were carried to a water-cooled condenser that was mounted vertically, which acted as a pipe heat exchanger (the distillate flows through a pipe system and is immersed into a cooling fluid (water) in with the re-circulation flow rate of 2.5 m^3/h). After 3 h, the obtained essential oil was decanted, and hydrosol was collected and stored in plastic containers at ambient temperature. Furthermore, liquid–liquid extraction of volatile compounds from *R. hybrida* hydrosol was conducted by dichloromethane, using the Likens–Nickerson apparatus for 2 h.

2.3. *Gas Chromatography–Mass Spectrometry*

A gas chromatograph equipped with flame ionization (FID), mass selective (MSD) detectors, and non-polar capillary column HP-5MS was used for the analysis of volatile compounds [24]. The injection volume was 1 μL, and the split ratio was 100:1 for all samples, under GC operating conditions: injector temperature 250 °C, MS source temperature 230 °C, and interface temperature 315 °C. The mass spectra were obtained in electron ionization mode at electron energy 70 eV, with a mass scan range of m/z 40–600. Identification of the compounds was conducted according to their linear retention indices (RI_{exp}) and comparison with mass spectral libraries (RI_{lit}). The relative abundance of each detected compound was calculated from GC/FID chromatograms as a percentage area of each peak.

2.4. *Determination of pH*

The *R. hybrida* hydrosol pH value was measured by WTW Digital pH-meter (±0.005 pH).

2.5. *Total Phenolics Content*

Determination of total phenolics content of the rose hydrosol was performed by the Folin–Ciocalteu method (slightly modified [25]). The reaction medium consisted of 5 mL H_2O, 0.5 mL 33% Folin–Ciocalteu reagent, and 0.5 mL of hydrosol (water was added in the blank sample). After 3–6 min, 1 mL of 20% Na_2CO_3 was added and incubation at room temperature maintained for 60 min. The content of total phenolics was calculated from a calibration curve formed from absorbances of different concentrations of gallic acid (0.1–11 μg/mL) measured at λ = 756 nm. Results were expressed as μg of gallic acid equivalents in ml of hydrosol (μg GAE/mL).

2.6. *Antioxidant Tests*

Rose hydrosol antioxidant activity was determined by the method based on the difference in the activity of removing 1,1-diphenyl-2-picrylhydrazyl radical (DPPH radical) between the blank and the working sample [26]. Briefly, hydrosol (0.2 mL) was added to 0.5 mL of 0.04% DPPH ethanol solution and 2.3 mL of methanol, while 0.2 mL of distilled water was added to the blank instead of the sample and, after 30 min, the absorbance was read at λ = 517 nm. DPPH radical scavenging activity is expressed as % relative to the blank.

The superoxide anion ($O_2^{\bullet}-$) scavenging test (NBT-test) was based on a riboflavin-light-NBT system [27]. Briefly, the reaction mixture contained 0.5 mL of phosphate buffer (50 mM, pH 7.6), 0.3 mL riboflavin (50 mM), 0.25 mL PMS (20 mM), and 0.1 mL *p*-Nitro-Blue tetrazolium chloride (NBT, 0.5 mM), prior to the addition of 1 mL of hydrosol. The reaction was started by illuminating the reaction mixture with different concentrations using LED tube lights. After 5–15 min of incubation, the absorbance was measured at 560 nm. Superoxide anion radical scavenging activity (NBT test) was expressed as % inhibition or % relative to the blank.

Hydroxyl radical ($^{\bullet}OH$) scavenging activity of hydrosol was assayed by the method of the deoxyribose degradation assay [28]. Briefly, the reaction mixture contained 2.7 mL phosphate buffer (0.1 M, pH 7), 0.1 mL 2-deoxyribose (0.05 M), 0.1 mL ferric chloride (10 mM), 0.1 mL H_2O_2 (0.015%), and hydrosol (0.1 mL). Correction tubes contained phosphate buffer instead of 2-deoxyribose, while control tubes contained distilled water instead of sample. After incubation on 37 °C, for 60 min, 0.2 mL 0.1 M EDTA and 2 mL TBA reagent solution (20% TCA and 0.5% TBA) were added to all tubes. The mixture was heated for 15 min on a boiling water bath, cooled, and measured at 532 nm. The scavenging activity of the hydroxyl radical (OH test) was expressed as % inhibition or % relative to the blank.

2.7. *Antimicrobial Activity*

A standard screening method (disc diffusion method) with nine microorganisms (*Bacillus cereus* ATCC 11778, *Staphylococcus aureus* ATCC 25923, *Enterococcus faecalis* ATCC 19433, *Escherichia coli* ATCC 25922, *Pseudomonas aeruginosa* ATCC 27853, *Salmonella* Typhimurium

ATCC 13311, *Saccharomyces cerevisiae* ATCC 9763, *Candida albicans* ATCC 10231, and *Aspergillus brasiliensis* ATCC 16404) was used for evaluation of antimicrobial properties of rose hydrosol. The preparation of cultures and setup of the used disk diffusion method has been described in detail previously [29]. The obtained results after the appropriate period of incubation were expressed as the diameter of halo zones around disks and interpreted as follows: diameter of halo zone lower than 22 mm—resistant; diameter of halo zone between 22 and 26 mm—intermediary effect and diameter of halo zone above 26 mm—sensitive microorganism.

2.8. Anti-Inflammatory Assay

The in vitro anti-inflammatory activity of the rose hydrosol was tested by protein denaturation bioassay using egg albumin in 96-well micro plates [30]. Briefly, 5 mL of the reaction mixture (0.2 mL of egg albumin + 2.8 mL of phosphate buffered saline + 2 mL of hydrosol in varying concentrations—100, 200, 300, 400, 500 μg/mL) was incubated at 37 °C for 15 min, heated at 70 °C for 5 min, and then cooled. The absorbance was measured at 660 nm, while acetylsalicylic acid (at the same concentrations as hydrosol) was used as a reference drug and used for the determination of absorbance. The results were computed as the inhibitory concentration of extract (mg/mL) that inhibits 50% of protein denaturation (IC_{50}).

2.9. Antihyperglycemic Assay

The in vitro antihyperglycemic activity of the rose hydrosol was tested by α-glucosidase inhibitory potential in 96-well micro plates [30]. Briefly, 0.1 mL of the mixture (2 mmol/L 4-nitrophenyl-α-D-glucopyranoside in 10 mmol/L potassium phosphate buffer + 0.02 mL of the samples, diluted in buffer + 0.1 mL of the enzyme solution) was incubated at 37 °C for 10 min. The absorbance was measured at 405 nm. The increase in absorbance was compared with that of the control (buffer instead of sample solution) to calculate the inhibitory activity. The sample concentration providing 50% inhibition of α-glucosidase enzyme activity (IC50α-GIP) was calculated from the graph of α-GIP (%) against extract concentration.

2.10. Statistical Analysis

Statistical analysis of volatile compounds in different samples of rose hydrosols from literature and their visualization with an unrooted tree diagram was conducted using cluster analysis (R"APE" package—Analysis of Phylogenetics and Evolution), while calculation of the distance matrix was conducted using the Euclidean method [31].

3. Results

3.1. Chemical Composition of Rose hydrosol

Volatile Composition of *R. hybrida* Hydrosol

The *R. hybrida* hydrosol is a colorless clear liquid, with a pleasant rose scent. There were 44 volatile compounds detected in the hydrosol by using GC–MS (Table 1), among which the dominant were phenylethyl alcohol (23.5%), nerol (17.2%), linalool (13.2%), and geraniol (8.3%). A representative gas chromatogram of *R. hybrida* hydrosol is presented in Figure 1. Fragrance is determined by phenylethyl alcohol, which possesses a warm odor, mild-bland rose floral impression, shaded with slight honey-green tones [32], as well as nerol and geraniol, which have floral and fruity odors [33], while linalool odor is described as floral, citric, fresh, and sweet [34]. All these compounds are highly valuable, and are some of the most widely used in cosmetic formulations and in the fragrance industry, from personal care, household, and laundry products to food and tobacco [35].

Table 1. Volatile compounds from *R. hybrida* hydrosol.

No.	R.T.	Compound	Class	RI_{exp}	%
1	4.136	(Z)-3-Hexenol	O	854	0.5
2	4.189	NI-1	-	857	0.2
3	4.334	*n*-Hexanol	O	866	1.1
4	7.347	6-Methyl-5-hepten-2-one	O	987	4.3
5	7.499	Dehydro-1,8-cineole	O	991	1.3
6	7.569	6-Methyl-5-hepten-2-ol	O	993	0.4
7	8.873	1,8-Cineole	OMT	1031	0.4
8	10.427	*cis*-Linalool oxide (furanoid)	OMT	1074	0.2
9	11.052	*trans*-Linalool oxide (furanoid)	OMT	1091	0.1
10	11.561	Linalool	OMT	1106	13.2
11	11.773	*cis*-Thujone	OMT	1108	0.2
12	11.957	*cis*-Rose oxide	OMT	1113	0.2
13	12.264	Phenylethyl alcohol	O	1124	32.5
14	12.621	*trans*-Rose oxide	OMT	1126	0.1
15	13.020	*cis*-*p*-Mentha-2,8-dien-1-ol	OMT	1139	0.1
16	13.391	Camphor	OMT	1146	0.2
17	13.815	Nerol oxide	OMT	1157	0.1
18	14.407	*p*-Mentha-1,5-dien-8-ol	OMT	1171	1.5
19	14.834	Terpinen-4-ol	OMT	1180	0.3
20	15.455	α-Terpineol	OMT	1194	4.2
21	15.487	NI-2	-	1195	1.8
22	16.018	Isopiperitenol	OMT	1205	0.1
23	17.146	Nerol	OMT	1231	17.2
24	17.647	Neral	OMT	1243	2.0
25	17.791	Carvone	OMT	1246	0.2
26	18.023	NI-3	-	1250	0.1
27	18.312	Geraniol	OMT	1258	8.3
28	18.821	Orcinol dimethyl ether	O	1270	0.6
29	18.977	Geranial	OMT	1273	3.2
30	19.635	NI-4	-	1287	0.2
31	20.135	Thymol	OMT	1302	0.7
32	20.575	Carvacrol	OMT	1311	0.1
33	22.910	Eugenol	O	1362	0.6
34	23.379	(E)-3,7-Dimethyl-2,6-octadienoic acid	O	1372	1.5
35	24.943	Methyl eugenol	O	1408	0.6
36	26.315	Dihydro-β-ionone	O	1441	tr
37	26.834	Dihydro-β-ionol	O	1453	0.8
38	32.677	Viridiflorol	OST	1596	0.1
39	32.744	Cubeban-11-ol	OST	1598	0.1
40	33.108	NI-5	-	1608	0.1
41	33.343	Humulene epoxide II	OST	1613	0.2

Table 1. *Cont.*

No.	R.T.	Compound	Class	RI_{exp}	%
42	34.249	γ-Eudesmol	OST	1636	0.2
43	34.981	β-Eudesmol	OST	1655	0.1
44	35.121	α-Eudesmol	OST	1658	0.1
		Oxygenated Monoterpenes (OMT)			52.6
		Oxygenated Sesquiterpenes (OST)			0.8
		Other (O)			44.2
		Not identified (NI)			2.4
		Total			100.0

R.T. (min)—retention time; RI_{exp}—retention index determined experimentally by C8–C36 *n*-alkanes on a HP-5MS non-polar capillary column; tr—trace (less than 0.05%).

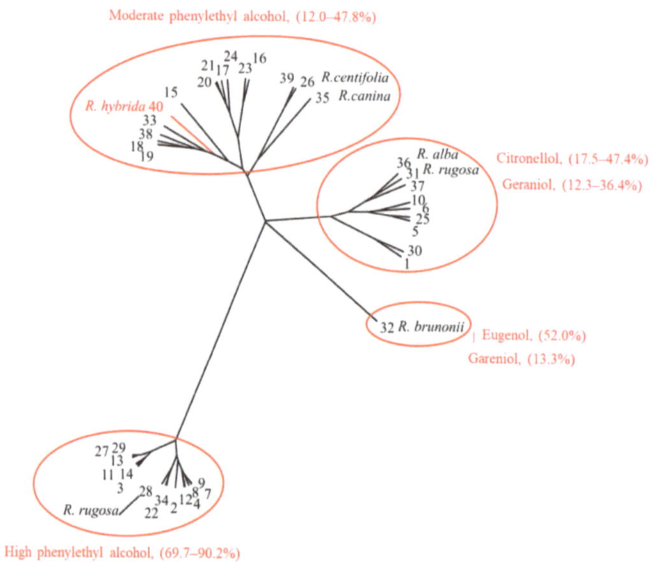

Figure 1. The unrooted cluster tree for different *Rosa* sp. hydrosols samples (according to Table 2).

Table 2. Volatile compounds (represented in percentages, %) of different rose hydrosols according to literature.

No.	Reference	Phenyl Ethyl Alcohol	Citronellol	Geraniol	Eugenol	Nerol	Dibuthyl Phthalate *	Linalool	Curzerene	Methyl Eugenol	Nonadecane
1	[6]	1.7	47.4	22.6	1.5	0.0	0.0	5.3	0.0	1.9	10.8
2	[36]	69.7	7.2	7.0	0.4	4.2	0.0	2.9	0.0	0.4	0.9
3	[36]	81.6	1.8	0.9	0.7	0.2	0.0	3.3	0.0	0.8	0.6
4	[36]	73.9	2.3	1.2	0.8	0.6	0.0	1.5	0.0	0.9	1.2
5	[36]	21.5	22.7	12.3	0.3	11.6	0.0	2.4	0.0	0.1	0.5

Table 2. *Cont.*

No.	Reference	Phenyl Ethyl Alcohol	Citronellol	Geraniol	Eugenol	Nerol	Dibuthyl Phthalate *	Linalool	Curzerene	Methyl Eugenol	Nonadecane
6	[37]	23.7	29.4	30.7	0.0	16.1	0.0	0.0	0.0	0.0	0.0
7	[38]	76.0	3.5	6.6	0.0	1.5	0.0	0.0	0.0	0.0	1.0
8	[38]	80.7	3.1	4.2	1.6	1.4	0.0	0.3	0.0	0.0	2.0
9	[38]	76.7	3.8	7.9	0.1	1.9	0.0	0.0	0.0	0.0	1.5
10	[39]	17.2	36.7	21.5	5.0	10.7	0.0	1.4	0.0	4.4	0.0
11	[40]	82.3	5.7	0.0	0.0	0.0	0.0	0.6	0.0	0.0	0.0
12	[40]	77.4	4.3	0.0	0.0	0.0	0.0	0.2	0.0	0.0	1.3
13	[40]	89.4	2.6	0.0	0.0	0.0	0.0	0.0	0.0	0.0	0.0
14	[40]	83.2	1.0	0.0	0.0	0.0	0.0	0.0	0.0	0.0	2.4
15	[21]	12.0	5.9	11.7	17.8	0.0	0.0	9.0	0.0	0.0	2.4
16	[21]	20.8	3.4	4.0	2.5	0.0	18.8	1.5	5.4	0.0	0.0
17	[21]	47.8	5.9	2.3	3.6	0.0	11.4	0.0	3.9	1.0	0.0
18	[21]	33.4	10.3	24.0	3.1	0.0	6.8	4.1	2.0	0.7	0.0
19	[21]	39.5	8.6	12.6	2.3	0.0	8.8	2.0	2.7	0.0	0.0
20	[21]	33.8	5.6	0.0	0.0	0.0	10.1	0.5	7.3	1.3	0.0
21	[21]	37.7	8.2	0.0	2.9	0.0	18.0	1.4	3.0	1.1	0.0
22	[21]	73.3	8.4	0.0	3.6	0.0	4.0	1.9	1.2	1.5	0.0
23	[21]	26.8	6.5	0.0	0.0	0.0	9.9	0.7	4.6	0.0	0.0
24	[21]	34.9	2.7	0.0	0.0	0.0	13.9	0.0	11.0	0.0	0.0
25	[41]	25.0	20.9	21.2	2.1	10.8	0.0	1.9	0.0	2.7	0.0
26	[41]	45.6	24.6	11.3	3.1	3.8	0.0	2.0	0.0	2.4	0.0
27	[42]	90.2	4.5	0.6	1.5	0.0	0.0	0.0	0.0	1.0	0.0
28	[42]	78.7	13.5	tr	5.7	0.0	0.0	0.1	0.0	1.0	0.0
29	[42]	87.9	2.4	1.5	1.7	0.0	0.0	2.3	0.0	0.7	0.0
30	[43]	2.0	40.1	16.0	2.3	0.0	0.0	3.2	0.0	0.0	0.0
31	[44]	14.3	17.5	30.6	0.1	7.5	0.0	0.0	0.0	0.1	0.0
32	[45]	9.4	0.6	13.3	52.0	1.2	0.0	1.7	0.0	0.0	2.8
33	[46]	35.6	8.3	27.9	6.2	12.7	0.0	0.0	0.0	1.2	0.0
34	[3]	77.0	12.7	2.5	5.1	0.0	0.0	0.0	0.0	2.8	0.0
35	[3]	46.9	8.3	2.9	28.8	0.0	0.0	1.0	0.0	0.6	0.0
36	[47]	6.0	28.7	36.4	0.1	6.1	0.0	3.6	0.0	0.1	0.3
37	[47]	5.0	28.7	16.4	2.7	10.8	0.0	2.1	0.0	1.8	2.1
38	[48]	36.0	19.2	13.2	6.0	5.8	0.0	8.4	0.0	3.7	0.0
39	[49]	45.4	34.1	12.2	2.2	0.1	0.0	2.1	0.0	1.3	0.0
40	This study	32.5	0.0	8.3	0.6	17.2	0.0	13.2	0.0	0.6	0.0
	AVERAGE	46.3	12.5	9.8	4.2	3.1	2.5	2.0	1.0	0.8	0.7

* contaminants (dibutyl phthalate was probably transferred into the samples from the storage containers).

3.2. Chemical Diversity of Volatile Compounds Rose hydrosols

According to the literature review, there are 17 papers (39 samples) dealing with volatile compounds of rose hydrosols, among which are the commercial samples *R. damascena*, *R. alba*, *R. brunonii*, *R. canina*, *R. centifolia*, and *R. rugosa*. However, this is the first time that the volatile composition of *R. hybrida* is determined. The ten volatile compounds—most represented on average for all these samples are given in Table 2—and their cluster analysis are shown in Figure 1.

According to the unrooted cluster tree, there are four main chemotypes: (1) high phenylethyl alcohol (69.7–90.2%), (2) moderate phenylethyl alcohol (12.0–47.8%), (3) citronellol + geraniol (17.5–47.4% and 12.3–36.4%, respectively), and (4) eugenol + geraniol (52.0% and 13.3%, respectively). In addition, investigations conducted in Latvia with seven cultivars of winter-hardy *R. rugosa* show that the main volatile compounds in hydrosol were phenylethyl alcohol (28.6–79.9%), citronellol (28.0–57.0%), and nerol (up to 39.0%), depending on variety, while linalool was not detected [50].

High phenylethyl alcohol chemotype includes samples of *R. damascena* from India [36,38,40] and Iran [3,21], as well as *R. damascena* and *R. rugosa* from China [42]. The high content of phenylethyl alcohol is mainly linked with residue rose water, which remains after cohobation (repeated distillation of rose hydrosol) [46]. This process is used in commercial production of rose oil, known as *rose otto*, which is a mixture of essential oil and re-extracted volatile oil from hydrosol [51].

The moderate phenylethyl alcohol chemotype includes samples of *R. centifolia* [41] and *R. canina* [3], as well as in the *R. hybrida* sample from Serbia investigated in this study. *R. damascena* hydrosols from Turkey [46,48,49] and Iran [21] also belong to this chemotype.

The citronellol + geraniol chemotype could also be divided into three subgroups: the first subgroup includes samples with high content of citronellol, more than 40% [6,43]. The second subgroup contains citronellol, geraniol. and nerol; it is noted in *R. damascena* and *R. alba* from Bulgaria [47], as well as *R. rugosa* from Poland [44]. The third subgroup is a combination with phenylethyl alcohol, which is noted in four samples [36,37,39,41]. The last chemotype, a combination of eugenol and geraniol, is noted only in the *R. brunonii* hydrosol sample [52].

3.3. pH Value

The *R. hybrida* hydrosol is slightly acidic, pH = 6.07 (Table 3). A study conducted in France with commercial hydrosols of *R. damascena* and *R. centifolia* showed that the pH value ranged from 4.0 to 7.1 [41], while for *R. damascena* hydrosol from Morocco it was 6.91 [53].

Table 3. pH value, total phenolic content, and biological activities of *R. hybrida* hydrosol.

Activity	Method	Result
pH		6.07
total phenolics	Folin–Ciocalteu method	4.96 µg GAE/mL
antioxidant	DPPH	24.66%
	NBT	nd
	•OH scavenging assay	12.07%
antimicrobial	disc diffusion method (*A. brasiliensis*, *B. cereus*, *C. albicans*, *E. faecalis*, *E. coli*, *P. aeruginosa*, *S. cerevisiae*, *S. typhimurium*, *S. aureus*)	nd
anti-inflammatory	protein denaturation bioassay using egg albumin	IC_{50} = 3.28 mg/L
antihyperglycemic	α-glucosidase inhibitory potential	nd

nd—not detected.

3.4. Total Phenolic Compounds

The total phenolic content (TPC) in *R. hybrida* hydrosol was 4.96 μg GAE/mL (Table 3). Similar results (5.2 μg GAE/mL) were obtained from Turkish *R. damascena* hydrosol [37]. Significantly higher values of TPC are noted in *R. damascena*, from 32.52 μg GAE/mL [47] to 57.02 μg GAE/mL [53], while hydrosol of *R. alba* contains 72.72 μg GAE/mL [47].

3.5. Biological Activity of Rose hydrosol

Rose hydrosol contains minute amounts of diluted water-soluble essential oil compounds (below 0.1%). The biological activities of hydrosols are consequently low [39], but organoleptic properties remain pronounced [41]. The pH value, total phenolic content, and biological activities of *R. hybrida* hydrosol investigated in this study are shown in Table 3.

3.6. Antioxidant Activity

According to all tests, the antioxidant activity of *R. hybrida* hydrosol was low or even absent (Table 3). The free radical scavenging activity according to the DPPH test was 24.66%, hydroxyl radical showed 12.07%, while superoxide anion activity was not detected. Similarly, results obtained in Poland by the FRAP (ferric reducing antioxidant power) method revealed that *R. damascena* hydrosol had low antioxidant activity (below 20%), while *R. centifolia* exhibited even lower activity (below 5%) [54]. A study conducted in Morocco showed that the antioxidant capacity of *R. damascena* hydrosol according to the DPPH test was 47.07%, while according to FRAP it was 100.71 AAE μg/mL [53]. Furthermore, *R. damascena* and *R. alba* hydrosols from Bulgaria had higher TPC values (32.52 and 72.72 μg GAE/mL, respectively) and demonstrated low capacity to inhibit lipid peroxidation (approximately 20%) as well as to neutralize hydroxyl radical and superoxide anion (under 40%) [47].

3.7. Antimicrobial Activity

The results obtained in this study indicate that rose hydrosol does not possess antimicrobial properties against the tested panel of microorganisms (Table 3). These results are consistent with the findings of other authors [37,55]. There is only one report which claims that rose water has a great potential of reducing the number of *Candida albicans* and methicillin-resistant *Staphylococcus aureus* (MRSA), followed by attenuation of neutrophil stimulation using modified testing methods. One of the newer research efforts on the *Rose damascena* hydrosol was also related to its in vivo antibacterial performance after hand-rubbing [49]; the results obtained in the mentioned study show that rose hydrosol has no impact on skin flora, confirming the results presented in this study. The low activity of rose hydrolate can be related to chemical composition, i.e., the low percentages of well-known antimicrobial plant biomolecules. However, the antimicrobial potency of rose essential oil and different extracts has been confirmed against a broad spectrum of microorganisms in many studies so far [37,56,57]. Nevertheless, the presence of an antimicrobial effect of essential oils and extracts does not guarantee that rose hydrosol has to have such an effect, due to qualitatively and quantitatively different chemical compositions.

3.8. Anti-Inflammatory Activity

The anti-inflammatory activity of *R. hybrida* hydrosol, according to protein denaturation bioassay using egg albumin, was found to have an IC50 value of 3.28 mg/L (Table 3). Rose water from Bulgaria (containing phenylethyl alcohol 43.7%, citronellol 18.2%, and geraniol 14.8%) suppressed neutrophil activation induced by TNF-α, with an IC50 of 3–15% [14].

3.9. Antihyperglycemic Activity

In vitro evidence of *R. hybrida* hydrosol antihyperglycemic activity according to the α-glucosidase inhibitory method was not found (Table 3). However, a methanolic extract of *R. damascena* was previously found to exhibit antihyperglycemic properties in diabetic and normal rats, and it is useful in insulin resistant therapy because of the increased insulin sensitivity in animals [57].

4. Conclusions

Rosa hybrida cv Mileva™ Frayla® hydrosol contains phenylethyl alcohol, nerol, linalool, and geraniol as the main volatile compounds. According to the unrooted cluster tree, it could be classified as a moderate phenylethyl alcohol chemotype, together with many other commercial samples of rose water from Iran and Turkey. Because of its pleasant odor, rose hydrosol is widely used in food flavoring and as fragrance in perfumery and the cosmetic industries. However, antioxidant, antimicrobial, anti-inflammatory, and antihyperglycemic effects were found to be either low or absent, which is attributable to the low content of volatile and phenolic compounds detected in the hydrosol.

Author Contributions: Conceptualization, M.A.; methodology, B.K., M.C. and V.R.; software, B.L.; validation, B.K., M.C., T.Z. and V.T.; formal analysis, O.Š., A.T. and V.T.; investigation, M.A., T.Z., B.K. and O.Š.; resources, B.B.T.; data curation, B.L.; writing—original draft preparation, M.A. and B.B.T.; writing—review and editing, V.R. and B.L.; visualization, T.Z.; supervision, M.A.; project administration, M.A.; funding acquisition, B.L., M.A., B.K. and T.Z. All authors have read and agreed to the published version of the manuscript.

Funding: This research was funded by the Ministry of Education, Science and Technological Development of the Republic of Serbia, grant numbers 451-03-68/2022-14/200032, 451-03-68/2022-14/200134, 451-03-9/2022-14/200026, and 451-03-68/2022-14/200125.

Data Availability Statement: Not applicable.

Conflicts of Interest: The authors declare no conflict of interest.

References

1. Gochev, V.; Wlcek, K.; Buchbauer, G.; Stoyanova, A.; Dobreva, A.; Schmidt, E.; Jirovetz, L. Comparative evaluation of antimicrobial activity and composition of rose oils from various geographic origins, in particular Bulgarian rose oil. *Nat. Prod. Commun.* **2008**, *3*, 1063–1068. [CrossRef]
2. Husnu, K.; Baser, C. Rose mentioned in the works of scientists of the medieval east and implications in modern science. *Nat. Prod. Commun.* **2017**, *12*, 1327–1330. [CrossRef]
3. Hamedi, A.; Pasdaran, A.; Zabarjad, Z.; Moen, M. A survey on chemical constituents and indications of aromatic waters soft drinks (hydrosols) used in Persian nutrition culture and folk medicine for neurological disorders and mental health. *J. Evid. Based Complement. Altern. Med.* **2017**, *22*, 744–752. [CrossRef] [PubMed]
4. Akram, M.; Raiz, M.; Munir, N.; Akhter, N.; Zafar, S.; Jabeen, F.; Shariati, M.A.; Akhtar, N.; Riaz, Z.; Altaf, S.H.; et al. Chemical constituents, experimental and clinical pharmacology of *Rosa damascena*: A literature review. *J. Pharm. Pharmacol.* **2020**, *72*, 161–174. [CrossRef] [PubMed]
5. Canabay, H.S. Effectiveness of liquid-liquid extraction, solid phase extraction, and headspace technique for determination of some volatile water-soluble compounds of rose aromatic water. *Int. J. Anal. Chem.* **2017**, *2017*, 4870671. [CrossRef]
6. Eikani, M.H.; Golmohammad, F.; Rowshanzamir, S.; Mirza, M. Recovery of water-soluble constituents of rose oil using simultaneous distillation-extraction. *Flavour Fragr. J.* **2005**, *20*, 555–558. [CrossRef]
7. Pal, P.K. Evaluation, genetic diversity, recent development of distillation method, challenges and opportunities of *Rosa damascena*: A review. *J. Essent. Oil-Bear. Plants* **2013**, *16*, 1–10. [CrossRef]
8. Batu, A.; Arslan, A. Biochemical and sensory evaluations of Turkish delight (lokum) enriched with black grape and sour cherry syrups. *Turk. J. Agric. For.* **2014**, *38*, 561–569. [CrossRef]
9. Pasalar, M.; Choopani, R.; Mosaddegh, M.; Kamalinejad, M.; Mohagheghzadeh, A.; Fattahi, M.R.; Zarshenas, M.M.; Jafari, P.; Lankarani, K.B. Efficacy of Jollab in the Treatment of Depression in Dyspeptic Patients: A Randomized Double-Blind Controlled Trial. *J. Evid. Based Complement. Altern. Med.* **2015**, *20*, 104–108. [CrossRef]
10. Aggarwal, P.; Kaur, S. Technology development for the preparation, concentration and utilization of rose extract in different valuable products and by products with retention of color and flavor. *J. Pharm. Innov.* **2017**, *6*, 189–193.
11. Tomi, K.; Sakaguchi, E.; Ueda, S.; Matsumura, Y.; Hayashi, T. Physiological and psychological effects of rose 'wishing' flowers and their hydrosols on the human autonomic nervous system and mood state. *J. Hortic.* **2017**, *86*, 105–112. [CrossRef]
12. Lavilla, I.L.; Gaytos, C.E.G.; Doblon–Merilles, R.; Lacaba, T.V.G.D.M. Sensory evaluation of rose petal (Rosaceae) ice cream. *Int. J. Eng. Res. Dev.* **2019**, *2*, 317–353.
13. Verma, A.; Srivastava, R.; Sonar, P.K.; Yadav, R. Traditional, phytochemical, and biological aspects of *Rosa alba* L.: A systematic review. *Future J. Pharm. Sci.* **2020**, *6*, 114. [CrossRef]

14. Maruyama, N.; Tansho-Nagakawa, S.; Miyazaki, C.; Shimomura, K.; Ono, Y.; Abe, S. Inhibition of neutrophil adhesion and antimicrobial activity by diluted hydrosol prepared from *Rosa damascena*. *Biol. Pharm. Bull.* **2017**, *40*, 161–168. [CrossRef]
15. Kazaz, S.; Erbas, S.; Baydar, H. The effects of storage temperature and duration on essential oil content and composition oil rose (*Rosa damascena* Mill.). *Turk. J. Field Crop.* **2003**, *14*, 89–96.
16. Agaoglu, Y.S. Rose oil industry and the production of oil rose (*Rosa damascena* Mill.) in Turkey. *Biotechnol. Biotechnol. Equip.* **2000**, *14*, 8–15. [CrossRef]
17. Najem, W.E.; Beyrouthy, M.; Wakim, L.H.; Neema, C.; Ouaini, N. Essential oil composition of *Rosa damascena* Mill. Rom different localities in Lebanon. *Acta Bot. Gallica* **2011**, *158*, 365–373. [CrossRef]
18. Ozel, M.Z.; Gogus, F.; Lewis, A.C. Comparison of direct thermal desorption with water distillation and superheated water extraction for the analysis of volatile components of *Rosa damascena* Mill. Using GC×GC-TOF/MS. *Anal. Chim. Acta* **2006**, *566*, 172–177. [CrossRef]
19. Aćimović, M.; Tešević, V.; Smiljanić, K.; Cvetković, M.; Stanković, J.; Kiprovski, B.; Sikora, V. Hydrolates—by-products of essential oil distillation: Chemical composition, biological activity and potential uses. *Adv. Technol.* **2020**, *9*, 54–70. [CrossRef]
20. Ayadi, A.J.; Ayed, N.; Karamous, T.; Bessiere, J.M.; Talou, T. Tunisian aromatic waters profile. *J. Essent. Oil-Bear. Plants* **2004**, *7*, 136–145. [CrossRef]
21. Moein, M.; Zarshenas, M.M.; Delnavaz, S. Chemical composition analysis of rose water samples from Iran. *Pharm. Biol.* **2014**, *52*, 1358–1361. [CrossRef] [PubMed]
22. Nakamura, S. Scent and component analysis of the hybrid tea rose. *Perfum. Flavor* **1987**, *12*, 43–46.
23. Ryu, J.; Lyu, J.I.; Kim, D.G.; Kim, J.M.; Jo, Y.D.; Kang, S.Y.; Kim, J.B.; Ahn, J.W.; Kim, S.H. Comparative analysis of volatile compounds of gamma-irradiated mutants of rose (*Rosa hybrida*). *Plants* **2020**, *9*, 1221. [CrossRef] [PubMed]
24. Aćimović, M.; Stanković Jeremić, J.; Todosijević, M.; Kiprovski, B.; Vidović, S.; Vladić, J.; Pezo, L. Comparative Study of the Essential Oil and Hydrosol Composition of Sweet Wormwood (*Artemisia annua* L.) from Serbia. *Chem. Biodivers.* **2022**, *19*, e202100954. [CrossRef] [PubMed]
25. Mikulic-Petkovsek, M.; Slatnar, A.; Stampar, F.; Veberic, R. HPLC-MSn identification and quantification of flavonol glycosides in 28 wild and cultivated berry species. *Food Chem.* **2012**, *135*, 2138–2146. [CrossRef] [PubMed]
26. Panda, S.K. Assay Guided Comparison for Enzymatic and Non-Enzymatic Antioxidant Activities with Special Reference to Medicinal Plants. In *Biochemistry, Genetics and Molecular Biology*; El-Missiry, M.A., Ed.; InTech: Rijeka, Croatia, 2012. [CrossRef]
27. Cheng, C.W.; Chen, L.Y.; Chou, C.W.; Liang, J.Y. Investigations of riboflavin photolysis via coloured light in the nitro blue tetrazolium assay for superoxide dismutase activity. *J. Photochem. Photobiol. B Biol.* **2015**, *148*, 262–267. [CrossRef]
28. Li, X. Solvent effects and improvements in the deoxyribose degradation assay for hydroxyl radical-scavenging. *Food Chem.* **2013**, *141*, 2083–2088. [CrossRef]
29. Micić, D.; Đurović, S.; Riabov, P.; Tomić, A.; Šovljanski, O.; Filip, S.; Tosti, T.; Dojčinović, B.; Božović, R.; Jovanović, D.; et al. Rosemary Essential Oils as a Promising Source of Bioactive Compounds: Chemical Composition, Thermal Properties, Biological Activity, and Gastronomical Perspectives. *Foods* **2021**, *10*, 2734. [CrossRef]
30. Aćimović, M.; Šovljanski, O.; Šeregelj, V.; Pezo, L.; Zheljazkov, V.D.; Ljujić, J.; Tomić, A.; Ćetković, G.; Čanadanović-Brunet, J.; Miljković, A.; et al. Chemical Composition, Antioxidant, and Antimicrobial Activity of Dracocephalum moldavica L. Essential Oil and Hydrolate. *Plants* **2022**, *11*, 941. [CrossRef]
31. Aćimović, M.; Pezo, L.; Zeremski, T.; Lončar, B.; Marjanović Jeromela, A.; Stanković Jeremić, J.; Cvetković, M.; Sikora, V.; Ignjatov, M. Weather conditions influence on hyssop essential oil quality. *Processes* **2021**, *9*, 1152. [CrossRef]
32. Akbar, N.; Raghav, C.S. Phenyl Ethyl Alcohol—A Valuable Aroma Chemical. *J. Essent. Oil-Bear. Plants* **2006**, *9*, 304–308. [CrossRef]
33. Shibusawa, N.; Nohara, I.; Ohsawa, R. Interspecific variation of scent characteristics in the Cyclamen genus and the utility of the variation. *Hortic. Sci.* **2018**, *45*, 193–204. [CrossRef]
34. Elsharif, S.A.; Banerjee, A.; Buettner, A. Structure-odor relationships of linalool, linalyl acetate and their corresponding oxygenated derivatives. *Front. Chem.* **2015**, *3*, 57. [CrossRef] [PubMed]
35. Kamatou, G.P.P.; Viljoen, A.M. Linalool—A review of a biologically active compound of commercial importance. *Nat. Prod. Commun.* **2008**, *3*, 1183–1192. [CrossRef]
36. Agarwal, S.G.; Gupta, A.; Kapahi, B.K.; Baleshwar; Thappa, R.K.; Suri, O.P. Chemical composition of rose water volatiles. *J. Essent. Oil Res.* **2005**, *17*, 265–267. [CrossRef]
37. Ulusoy, S.; Bosgelmez-Tinaz, G.; Secilmis-Canabay, H. Tocopherol, carotene, phenolic contents and antibacterial properties of rose essential oil, hydrosol and absolute. *Curr. Microbiol.* **2009**, *59*, 554–558. [CrossRef]
38. Verma, R.S.; Padalia, R.C.; Chauhan, A.; Singh, A.; Yadav, A.K. Volatile constituents of essential oil and rose water of damask rose (*Rosa damascena* Mill.) cultivars from North Indian hills. *Nat. Prod. Res.* **2011**, *25*, 1577–1584. [CrossRef]
39. Baydar, H.; Kuleasan, H.; Kara, N.; Secilmis-Canbay, H.; Kineci, S. The effect of pasteurization, ultraviolet radiation and chemical preservatives on microbial spoilage and scent composition of rose water. *J. Essent. Oil-Bear. Plants* **2013**, *16*, 151–160. [CrossRef]
40. Lohani, H.; Andola, H.C.; Chauhan, N.K.; Gwari, G.; Bhandari, U. Volatile constituents of rose water of damask rose (*Rosa damascena* Mill.) from Uttarakhand Hymalayas. *Med. Plants—Int. J. Phytomed. Relat. Ind.* **2013**, *5*, 102. [CrossRef]
41. Labadie, C.; Ginies, C.; Guinebretiere, M.H.; Renard, C.M.G.C.; Cerutti, C.; Carlin, F. Hydrosols of orange blossom (*Citrus aurantium*), and rose flower (*Rosa damascena* and *Rosa centifolia*) support the growth of a heterogeneous spoilage microbiota. *Food Res. Int.* **2015**, *76*, 576–586. [CrossRef]

42. Lei, G.; Wang, L.; Liu, X.; Zhang, A. Fast quantification of phenylethyl alcohol in rose water and chemical profiles of rose water and oil of *Rosa damascena* and *Rosa rugosa* from southeast China. *J. Liq. Chromatogr. Relat. Technol.* **2015**, *38*, 823–832. [CrossRef]
43. Koksal, N.; Saribas, R.; Kafkas, E.; Aslancan, H.; Sadighazadi, S. Determination of volatile compounds of the first rose oil and the first rose water by HS-SPME/GC/MS techniques. *Afr. J. Tradit. Complement. Altern. Med.* **2015**, *12*, 145–150. [CrossRef]
44. Maciag, A.; Kalemba, D. Composition of rugosa rose (*Rosa rugosa Thunb.*) hydrolate according to the time of distillation. *Phytochem. Lett.* **2015**, *11*, 373–377. [CrossRef]
45. Verma, R.S.; Padalia, R.C.; Chauhan, A. Chemical composition of essential oil and rose-water extract of Hymalayan musk rose (*Rosa brunonii Lindl.*) from Kumaon region of Western Himalaya. *J. Essent. Oil Res.* **2016**, *28*, 332–338. [CrossRef]
46. Erbas, S.; Baydar, H. Variation in scent compounds of oil-bearing rose (*Rosa damascena* Mill.) produced by headspace solid phase microextraction, hydrodistillation and solvent extraction. *Rec. Nat. Prod.* **2016**, *10*, 555–565.
47. Georgieva, A.; Dobreva, A.; Tzvetanova, E.; Alexandrova, A.; Mileva, M. Comparative study of phytochemical profiles and antioxidant properties of hydrosols from Bulgarian *Rosa alba* L. and *Rosa damascena* Mill. *J. Essent. Oil-Bear. Plants* **2019**, *22*, 1362–1371. [CrossRef]
48. Demirbolat, I.; Ekinci, C.; Nuhoglu, F.; Kartal, M.; Yildiz, P.; Gecer, M.O. Effects of oraly consumed *Rosa damascena* Mill. Hydrosol on hematology, clinical chemistry, lens enzymatic activity, and lens pathology in streptozotocin-induced diabetic rats. *Molecules* **2019**, *24*, 4069. [CrossRef]
49. Bayhan, G.I.; Gumus, T.; Alan, B.; Savas, I.K.; Cam, S.A.; Sahin, E.A.; Arslan, S.O. Influence of *Rosa damascena* hydrosol on skin flora (contact culture) after hand-rubbing. *GMS Hyg. Infect. Control* **2020**, *15*, 32974119. [CrossRef]
50. Sparinska, A.; Rostoks, N. Volatile organic compounds of hybrid rugosa roses in Latvia. *Proc. Latv. Acad. Sci.* **2015**, *69*, 57–61. [CrossRef]
51. Kovacheva, N.; Rusanov, K.; Atanassov, I. Industrial cultivation of oil bearing rose and rose oil production in Bulgaria during 21st century, directions and challenges. *Biotechnol. Biotechnol. Equip.* **2010**, *24*, 1793–1798. [CrossRef]
52. El Bouny, H.; Ouahzizi, B.; Sellam, K.; Alem, C. Reconsidering the hydrosols of four aromatic plants from the Toudgha region (South East of Morocco). *J. Anal. Sci. Technol.* **2021**, *3*, 126–132. [CrossRef]
53. Kalemba-Drozdz, M.; Cierniak, A. Antioxidant and genprotective properties of extracts from edible flowers. *J. Food Nutr. Res.* **2019**, *58*, 42–50.
54. Aliasghari, A.; Khorasgani, M.R.; Khoroushi, M. The effect of vinegar, rosewater and ethanolic extract green tea against oral Streptococci, an in vitro study. *J. Probiotics Health* **2017**, *5*, 4. [CrossRef]
55. Bayoub, K.; Baibai, T.; Mountassif, D.; Retmane, A.; Soukri, A. Antibacterial activities of the crude ethanol extracts of medicinal plants against Listeria monocytogenes and some other pathogenic strains. *Afr. J. Biotechnol.* **2010**, *9*, 4251–4258.
56. Mileva, M.; Ilieva, Y.; Jovtchev, G.; Gateva, S.; Zaharieva, M.M.; Georgieva, A.; Dimitrova, L.; Dobreva, A.; Angelova, T.; Vilhelmova-Ilieva, N.; et al. Rose Flowers—A Delicate Perfume or a Natural Healer? *Biomolecules* **2021**, *11*, 127. [CrossRef] [PubMed]
57. Mohammadi, A.; Fallah, H.; Gholamhosseinian, A. Antihypergycemic effect of *Rosa damascena* in mediated by PPAR.γ gene expression in animal model of insulin resistance. *Iran. J. Pharm. Sci.* **2017**, *16*, 1080–1088.

horticulturae

MDPI

Article

Seasonal Change in Phytochemical Composition and Biological Activities of *Carissa macrocarpa* (Eckl.) A. DC. Leaf Extract

Reshika Ramasar [1], Yougasphree Naidoo [1], Yaser Hassan Dewir [2,*] and Antar Nasr El-Banna [3]

1 School of Life Sciences, University of KwaZulu-Natal, Westville Campus, Private Bag X54001, Durban 4000, South Africa
2 Plant Production Department, College of Food & Agriculture Sciences, King Saud University, Riyadh 11451, Saudi Arabia
3 Federal Research Centre for Cultivated Plants, Julius Kühn-Institut (JKI), Messeweg 11–12, 38104 Braunschweig, Germany
* Correspondence: ydewir@ksu.edu.sa

Abstract: Genus *Carissa* represents several species that are reported to be of great phyto-medicinal and ethnopharmacological value. However, *Carissa macrocarpa* is relatively understudied. Furthermore, environmental conditions such as seasonal changes are known to affect the phytochemical composition of medicinal plants. Therefore, this study aimed to investigate the phytochemical composition and biological activity of the leaf extracts of *C. macrocarpa* in the summer and winter seasons. The phytochemical screening of *C. macrocarpa* leaves showed positive results for a variety of phytochemicals, such as alkaloids, tannins, phenols, naphthoquinones, flavonoids, saponins, steroids, proteins, carbohydrates, mucilage, gum and resin. The methanolic extract was evaluated for its antibacterial activity against *Escherichia coli* and *Staphylococcus aureus* using the agar well diffusion method. The winter leaf extract was distinguished for its potential antibacterial activity against both bacterial strains with inhibition zones (mm) of 8.17 ± 1.04 and 6.83 ± 0.58 at 10 mg/mL. The antioxidant activity of the leaf extracts was evaluated using the 2, 2′-diphenyl-1-picrylhydrazyl (DPPH) assay. The percentage scavenging activities of the different extracts were significantly greater than that of the control. Furthermore, at 15, 30, 60, 120 and 240 μg/mL, the percentage scavenging activities of the winter methanol leaf extract were 74.65, 78.31, 85.45, 90.02 and 95.68%, and those of the summer one were 71.66, 73.57, 84.05, 88.22 and 96.28%, respectively, indicating that the methanol leaf extract had greater percentage scavenging activity in winter than in summer. In winter, the IC_{50} value of the methanol leaf extract (0.67 μg/mL) was lower than that of ascorbic acid (8.26 μg/mL). It is concluded that winter is the optimal season to harvest leaves of *C. macrocarpa* for medicinal use. To the best of our knowledge, this is the first report that relates the phytochemical composition and medicinal properties of *C. macrocarpa* to changes in seasons. The results obtained are promising, and this species should be further explored to decipher its pharmacological worth.

Keywords: antibacterial; antioxidant; DPPH; *Escherichia coli*; phytochemical; seasonal change; *Staphylococcus aureus*

Citation: Ramasar, R.; Naidoo, Y.; Dewir, Y.H.; El-Banna, A.N. Seasonal Change in Phytochemical Composition and Biological Activities of *Carissa macrocarpa* (Eckl.) A. DC. Leaf Extract. *Horticulturae* **2022**, *8*, 780. https://doi.org/10.3390/horticulturae8090780

Academic Editors: Ivan Salamon and Milica Aćimović

Received: 23 July 2022
Accepted: 12 August 2022
Published: 28 August 2022

1. Introduction

In many developing countries, plants of medicinal value are being used by billions of people. This is due to the continual inadequate supply of modern, commercial medicines; thus, people prefer to use medicinal plants as they are of low cost and effective [1,2]. Approximately 80% of black South Africans utilize traditional medicine to fulfil their primary health care needs [3]. There are two types of metabolites produced within plants, namely, primary and secondary metabolites [4,5]. Secondary metabolites such as phenolics, flavonoids, alkaloids, lignans and terpenoids are abundant in medicinal plants. Secondary metabolites, also known as phytochemicals, are bioactive compounds that are required for a plant's survival (under environmental stress) and its protection against insects and predators [6]. Since ancient times, secondary metabolites have been widely utilized in the treatment of different illnesses and disorders, and they are still quite popular today [7]. However, there are many environmental factors that may affect the phytochemical composition of medicinal plants. Among the many factors, there are soil type, change in season, salinity, light, altitude and humidity [8–10]. Determining the effect of seasonal change on plant phytochemical composition provides fundamental information on the best time/season to harvest individual plant species (with maximum phytochemical composition) [11,12].

Over the years, the growth and spread of bacterial infections have been controlled; however, many are now emerging due to mutations and drug resistance. Thus, research on and development of new or improved antibacterial medications possibly derived from plants are proposed [13]. Phytochemicals such as phenols, flavonoids, saponins, alkaloids, tannins and steroids exhibit antibacterial activity; hence, they have been of great value in the pharmaceutical industry [14]. Furthermore, phytochemicals are also known to have antioxidant activity [15]. Antioxidants are substances that can lower or prevent oxidative stress caused by reactive oxygen species. This is because antioxidants can neutralize or scavenge reactive oxygen species via hydrogen donation, single-electron transfer and the ability to chelate transition metals [16,17]. Reactive oxygen species (ROS) such as hydroxyl (OH^-), superoxide (O_2^-), nitric oxide (NO), peroxyl (RO_2^-) and lipid peroxyl (LOO^-) are free radicals that are naturally produced in the human body as products of cellular metabolism. When they are produced in abundance, they can cause major damage to cellular structures (namely, nucleic acids, proteins and lipids). This can lead to the development of chronic diseases such as diabetes mellitus, cancer, hypertension, cardiac disorders and neurodegenerative diseases [18–21]. Therefore, the discovery of natural antioxidants from plants must be emphasized. Furthermore, the use of medicinal plants to develop drugs is of great interest as they are more readily available, affordable and effective and have fewer harmful side-effects than synthetic drugs [6,22].

The Apocynaceae family is believed to be one of the largest families in the plant kingdom [23]. The majority of the species belonging to this family are rich in phytochemicals, especially alkaloids, which are known to have great medicinal value [23]. Members of this family are distributed mainly in tropical, subtropical and temperate regions of the world [24]. *Carissa macrocarpa* (Eckl.) A. DC. syn. *C. grandiflora* (E.Mey.) A. DC. [25] is an ornamental shrub that is commonly known as the Natal plum. This species belongs to the Apocynaceae family and is indigenous to South Africa, where it is used in traditional medicine [26,27]. *Carissa* trees are perfect plants for hedges due to their numerous large, y-shaped thorns. *Carissa macrocarpa* has white–yellow scented flowers that are star-shaped and have red fruits containing considerable amounts of vitamins, carbon, calcium and magnesium [15,26,28]. The fruit of *C. macrocarpa* is renowned for its several medicinal benefits. They are used to treat anaemia and increase haemoglobin and possess immune-boosting properties [28,29]. On the other hand, the leaves are used by Zulu people to treat diarrhoea in cattle, and they are also used to treat coughs and venereal diseases in humans [29–31].

Phytochemical studies on *C. macrocarpa* plant parts revealed the presence of alkaloids, flavonoids, saponins, triterpenoids, steroids, quinones, tannins, carbohydrates and phenols [27,28,32–34]. Studies on *C. macrocarpa* extracts that have previously been conducted reported pharmacological properties, such as antioxidant and antibacterial activities [28,33,35,36]. However, the impact of seasonal variations on the production of phytochemicals within this species has not been evaluated. Seasonal change plays a vital role in the production of phytochemicals and the overall medicinal potential of a plant. Hence, this study aimed to screen phytochemical composition using qualitative phytochemical tests and to evaluate the in vitro antibacterial and antioxidant properties of *C. macrocarpa* leaves in summer and winter.

2. Materials and Methods

2.1. Plant-Material Collection

Fresh leaves of *C. macrocarpa* were harvested from the University of KwaZulu-Natal Westville campus (29.817° S 30.940° E), Durban, South Africa. A voucher specimen was deposited at Bews Herbarium, University of KwaZulu-Natal, Pietermaritzburg campus with accession number 92,175 (voucher number 04).

2.2. Preparation of Plant Material for Extraction

The fresh leaves were collected in summer (December 2020) and winter (July 2021) and dried in an oven at 35 °C for 72 h, until all water molecules that were present completely evaporated. Thereafter, the dried leaves were ground to powder form using a Waring blender (Kenwood Ltd., Havant, UK).

2.3. Extraction of C. Macrocarpa Leaves

Carissa macrocarpa powdered leaves (10 g) (summer) were placed into a round-bottom flask. Thereafter, 100 mL of hexane was placed into the flask, and the latter was then attached to a reflux apparatus. The solvent was heated under reflux for three 3 h sessions to obtain the crude leaf extract. This process was then performed utilizing the same leaf material for chloroform and methanol, respectively. Phytochemicals were extracted using various solvents of different polarities, from non-polar (hexane and chloroform) to more polar solvents (methanol). This was to ensure that a broad polarity range of phytocompounds could be extracted. The resultant crude leaf extracts (hexane, chloroform and methanol) were filtered through Whatman No.1 filter paper (Merck, Darmstadt, Germany). This procedure was conducted again for the extraction of winter leaves.

2.4. Evaporation of Extracts

The extracted material was left in a dark room at room temperature until the solvent completely evaporated. Once solvent evaporated, the crude extract remained at the bottom of the jar. Thereafter, the percentage yields for the 3 different extracts were calculated using the formula below:

$$Extract\ yield\ (\%) = \frac{weight\ of\ dried\ extract\ (g)}{weight\ of\ leaf\ material\ (g)} \times 100$$

2.5. Phytochemical Screening

The phytochemical screening of the different leaf extracts obtained from *C. macrocarpa* was conducted utilizing standard qualitative protocols adapted from other studies. Sufficient amounts of the dried extracts (hexane, chloroform and methanol) were dissolved in their respective solvents. These preparations were used for phytochemical testing as detailed below.

2.5.1. Detection of Alkaloids

Mayer's test [37]: Two drops of Mayer's reagent (potassium mercuric iodide (K_2HgI_4) solution) are added along the side of the test tube to 1 mL of each crude extract. A creamy-white (turbid) precipitate confirms the presence of alkaloids.

Wagner's test [37]: In a test tube containing 1 mL of crude extract, two drops of Wagner's reagent (aqueous iodine in potassium iodide solution) are added along the side of the test tube. The formation of a reddish brown or yellow precipitate confirms the presence of alkaloids.

2.5.2. Detection of Tannins

Ferric chloride test [38]: In a test tube containing 1 mL of crude extract, 1 mL of 10% of iron (III) chloride ($FeCl_3$) solution is added and mixed. A blue-black or brownish green colour confirms the presence of tannins.

2.5.3. Detection of Phenols

Lead acetate test [39]: In a test tube containing 1 mL of crude extract, 1 mL of 10% of lead acetate solution is added. A white precipitate confirms the presence of phenolic compounds.

2.5.4. Detection of Quinones

Gelatine test [40]: In a test tube containing 1 mL of crude extract, three drops of 1% gelatine solution are added. The formation of white precipitate confirms the presence of tannins, specifically naphthoquinone.

2.5.5. Detection of Flavonoids

Alkaline-reagent test (alkaline hydrolysis) [41]: In a test tube containing 2 mL of crude extract, 2 mL of 5% sodium hydroxide (NaOH) is added. A yellow-coloured solution that becomes decolorized upon the insertion of 1 mL of 50% H_2SO_4 confirms the presence of flavonoids.

Acid-hydrolysis test [42]: In a test tube containing 1 mL of crude extract, 1 mL of concentrated H_2SO_4 was added. A yellow-coloured solution that is intense confirms the presence of flavones and flavonols.

2.5.6. Detection of Saponins and Steroids

Saponins

Foam test [43]: In a test tube containing 5 mL of crude extract, 20 mL of distilled water is added. The test tube is then vigorously shaken for 15 min. The appearance of a froth layer is indicative of the presence of saponins (triterpene glycosides). The observed results are recorded as negative if no froth forms, positive for the presence of 1.2 cm high froth, strongly positive for froth formation greater than 2 cm in height and weakly positive for froth formation less than 1 cm in height.

Olive oil test [40]: In a test tube containing 2 mL of crude extract, two drops of olive oil are added. The test tube is vigorously shaken for 5 min. Thereafter, a soluble emulsion confirms the presence of saponins.

Steroids

Salkowski's test [40]: In a test tube containing 1.5 mL of crude extract, 1 mL of Chloroform is added and thoroughly mixed. Then, with care, 1.5 mL of concentrated H_2SO_4 is inserted along the side of the test tube. A reddish-brown colour at the interface confirms the presence of a steroid ring.

Liebermann–Burchard test [44]: In a test tube containing 1 mL of crude extract, 1 mL of chloroform is added. The test tube contents are thoroughly mixed. Thereafter, 1 mL of acetic acid is inserted into the test tube. The resultant solution is cooled in ice for approximately 10 min. Then, 1 mL of concentrated sulfuric acid is inserted along the side of the test tube. After a few minutes, the appearance of a reddish-brown ring at the interface confirms the presence of steroids.

2.5.7. Detection of Proteins

Biuret test [42]: In a test tube containing 1 mL of crude extract, 1 mL of 10% NaOH is added. The contents of the test tube are thoroughly mixed, and the addition of 95% ethanol follows. Thereafter, 1 mL of 0.5% copper sulphate is inserted along the side of the test tube. The formation of a purplish-violet or pink-violet colour confirms the presence of a protein.

2.5.8. Detection of Carbohydrates

Molisch's test [38]: In a test tube containing 3 mL of crude extract, 2 mL distilled water is added. The solution is then filtered. Three drops of Molisch's reagent are added to the filtrate. Thereafter, 1 mL of concentrated H_2SO_4 is carefully inserted along the side of the test tube to form a layer without shaking. The resultant solution is allowed to stand for two minutes. The appearance of a purplish ring at the interface is taken as indicative of the presence of carbohydrates.

2.5.9. Detection of Mucilage and Gum

Precipitation test [40]: In a test tube containing 1.5 mL of crude extract, 2 mL of distilled water is added. To this diluted solution, 2 mL of absolute ethanol is added with continuous stirring. The formation of a white or cloudy precipitate confirms the presence of gums or mucilage.

Ruthenium-red test [40]: In a test tube containing 2 mL of crude extract, two drops of ruthenium red are added. The colour change of the solution into a pink-coloured solution confirms the presence of mucilage.

2.5.10. Detection of Resins

Acetone test [42]: In a test tube containing 1 mL of crude extract, 1 mL of acetone is added. The contents of the test tube are thoroughly mixed. Thereafter, 2 mL of distilled water is inserted into this mixture. A turbid solution confirms the presence of resins.

2.6. Antibacterial Activity

In this investigation, two biological strains were used: Gram-negative *Escherichia coli* (ATCC 25922) and Gram-positive *Staphylococcus aureus* (ATCC 43300). These bacterial strains were provided by Professor Johnson Lin, Department of Microbiology, University of KwaZulu-Natal, and maintained in 75% glycerol at −80 °C. The methanol crude extracts from the summer and winter leaves of *C. macrocarpa* were dissolved in 10% DMSO at the different concentrations of 10, 5.2, 2.5, 1.25 and 0.625 mg/mL. The agar well diffusion procedure described by Perez et al. [45] was used to conduct the antibacterial assay (in vitro) on the *C. macrocarpa* leaf extract (summer and winter), with some modifications. To 10 mL of nutrient broth (Merck, Darmstadt, Germany), a loopful of each bacterial strain stock was inserted. In a test-tube shaker, the cultures were grown overnight for 24 h at 36 ± 1 °C. Thereafter, the bacterial cultures (inoculum) were diluted further with nutrient broth (sterile) to an optical density (OD) of 0.08–0.1 to yield a final concentration of about 1×10^8–1×10^9 bacteria cells/mL. Utilizing Mueller–Hinton agar (MHA), agar plates were prepared. Agar was poured into Petri dishes and allowed to set (solidify) at room temperature. Thereafter, sterile cotton swabs were used to swab the cultures of bacteria onto the agar plates. A six-millimetre-diameter sterile cork borer was used to punch wells in the agar plates. Subsequently, 100 µL of each of the prepared concentrations of the methanol leaf extracts (various concentration) was pipetted into the wells. The plates were then incubated at 36 °C to encourage the growth of bacterial colonies. After 24 h of incubation, the plates were assessed for antibacterial activity by measuring the diameter of the bacterial clearance zone of inhibition (mm). The negative control was 10% DMSO, and the positive controls were 10 µg/mL Gentamicin (for Gram-negative bacteria) and Streptomycin (for Gram-positive bacteria). The tests were carried out in three replicates, and the results were expressed as means ± standard deviations.

2.7. DDPH Scavenging Activity

Summer and winter leaf extracts were used. Stock solutions of 1 mg/mL in methanol of leaf extracts (hexane, chloroform and methanol) were prepared. From these, 15, 30, 60, 120 and 240 µg/mL were prepared for in vitro antioxidant screening. The procedure described by Braca et al. [46] was used to establish the antioxidant activity, such as the 2, 2′-diphenyl-1-picrylhydrazyl (DPPH)-radical-scavenging activity of the crude extracts of leaves of *C. macrocarpa*. A total volume of 100 µL of each crude extract at standard concentrations (15, 30, 60, 120 and 240 µg/mL) was mixed with 50 µL of 0.3 mM DPPH solution prepared in methanol. For 30 min, the microplate was incubated in the dark at room temperature (24 °C). Subsequently, the absorbance was measured at 517 nm (Synergy

HTX Multi-mode reader; BioTek Instruments Inc., Winooski, VT, USA). Ascorbic acid was utilized as the standard drug-positive control.

The potential of the crude extracts to scavenge DPPH radicals were calculated using the following equation:

$$\text{DPPH scavenging activity } (\%) = \left(\frac{\text{Abs control} - \text{Abs sample}}{\text{Abs control}}\right) * 100$$

where Abs control is the absorbance of DPPH and methanol and Abs sample is the absorbance of DPPH radical + sample (sample or standard).

The analysis was performed in three replicates, and all data were reported as means ± standard deviations. The IC50 was used to evaluate the antioxidant activity (in vitro). The IC_{50} is the concentration of the antioxidant agent that inhibits 50% of the oxidant. The percentage inhibition (scavenging activity) was plotted against the various concentrations (logarithmic scale). The inhibition curves in this graph were used to obtain the IC_{50}. The IC_{50} values were calculated.

2.8. Statistical Analysis

The results obtained from antioxidant screening were subjected to statistical analyses using SPSS 28 for windows, IBM Corporation, New York, NY, USA. The percentage scavenging values were subjected to Tukey's honestly significant difference multiple-range post hoc test, and the *p*-value was obtained. The significant difference was established at $p < 0.05$.

3. Results and Discussion

3.1. Yield of Extract and Screening of Phytochemicals

The methanol extract of *C. macrocarpa* leaves obtained the highest percentage yield, and a similar observation was reported by Abbas et al. [33]. The hexane extract had an intermediate percentage yield, followed by the chloroform extract with the lowest metabolite yield (Table 1). The winter extracts (methanol, hexane and chloroform) had higher percentage metabolite yields than the summer extracts (Table 1). In winter, the maximum metabolite yield was obtained from the crude methanol extract (24.3%), while the lowest yield obtained was from the crude chloroform extract (6.5%) (Table 1). In summer, the highest metabolite yield was also obtained from the crude methanol extract (15%), while the lowest yield was obtained from chloroform (4.8%) (Table 1). The results indicate that there were more polar compounds in the leaves of *C. macrocarpa*. In addition to that, methanol was more effective in extracting phytochemicals from *C. macrocarpa* leaves than other extraction solvents (Table 1). This is because methanol is a polar solvent and easily interacts with and dissolves polar phytocompounds. Chloroform is a non-polar solvent and would be effective for extracting non-polar compounds such as oils [47].

Table 1. Yield percentage of the dried extracts of summer and winter leaves of *C. macrocarpa*.

Solvent	Yield (Summer)		Yield (Winter)	
	(g)	(%)	(g)	(%)
Hexane	0.82	8.2	0.98	9.8
Chloroform	0.48	4.8	0.65	6.5
Methanol	1.5	15	2.43	24.3

In this study, secondary metabolites were present in both the summer and winter leaves of *C. macrocarpa*; however, the presence varied amongst the different extracts (hexane, chloroform and methanol) (Table 2). The results revealed the presence of ten phytochemicals in the methanol leaf extract, which included alkaloids, tannins, phenols, naphthoquinones, flavonoids, saponins, steroids, proteins, carbohydrates, mucilage and gums, and the absence of resin. On the other hand, hexane extracted nine phytochemicals from the leaves, which included alkaloids, tannins, phenols, naphthoquinones, flavonoids, steroids, proteins, mucilage, gums and resin, but not saponins or carbohydrates. The chloroform extract of the leaves tested positive for eight phytochemicals, which included alkaloids, tannins, phenols, saponins, proteins, carbohydrates, mucilage, gums and resin, but tested negative for naphthoquinones, flavonoids and steroids (Table 2). The methanol extract of the leaves tested positive for the majority of the phytochemicals, followed by the hexane extract and then the chloroform extract (Table 2). This could be due to the methanol extract having the highest metabolite yield, while chloroform had the lowest (Table 1).

Table 2. Qualitative preliminary screening of phytochemicals of *C. macrocarpa* leaf extracts for summer and winter seasons.

Phytochemical Constituent	Type of Test	Winter			Summer		
		H	C	M	H	C	M
Alkaloids	Wagner's	-	-	+	-	++	++
	Meyer's	++	+	-	++	+	-
Tannins	Ferric chloride	+	+	++	+	+	++
Phenols	Lead acetate	+	+	++	++	++	++
Tannins (naphthoquinone)	Gelatine	++	-	+	+	-	+
Flavonoids	Alkaline-reagent test	+	-	+	+	-	+
	Acid-hydrolysis test	+	-	++	+	-	++
Saponins	Foam test	-	-	+	-	-	+
	Olive oil test	-	++	++	-	+	++
Steroids (terpenoids)	Salkowski's test	+	-	+	+	-	+
	Lieberman–Bouchard test	+	-	+	+	-	+
Proteins	Biuret test	+	+	+	+	+	+
Carbohydrates	Molisch's test	-	++	++	-	+	+
Mucilage + gums	Precipitation test	+	+	-	+	+	-
	Ruthenium-red test	-	+	+	-	+	+
Resin	Acetone test	+	++	-	+	++	-

- Absent, + Present, ++ Intense positive, H = Hexane, C = Chloroform, M = Methanol.

Several of the above-mentioned phytochemicals found in *C. macrocarpa* were reported to be present in other *Carissa* species, such as *C. opaca, C. spinarum, C. carandas* and *C. edulis* [34]. A study conducted by Khalil et al. [27] revealed the presence of a variety of phytochemicals in the leaves of *C. macrocarpa*, namely, saponins, flavonoids, triterpenoids, steroids, tannins, anthraquinones and carbohydrates, and the absence of cardiac glycosides and alkaloids. In addition, the hexane leaf extract did not show any presence of phytochemicals. A similar study conducted by Abbas et al. [33] revealed the presence of alkaloids, steroids and terpenoids in the leaves of *C. grandiflora* and the absence of flavonoids and tannins. All around the world, carbohydrates are probably the most familiar organic substance [48]. This phytochemical is used in pharmacy for the preparation of tablets (sucrose and lactose), in anti-diarrhoea drugs (pectin), antacids, diuretic drugs (mannitol and sorbitol), etc. [48]. When a plant is injured due to unfavourable conditions such as drought, the cell walls of the plant break down, which causes the formation of gums [49]. Mucilage is a common product of plant metabolism and is formed within the cells of plants [49]. Hence, gums are considered as pathological products, and mucilage is considered as a physiological product of plants [49]. These phytocompounds are used in medicine for their anti-inflammatory and anti-irritant activity in cough suppression. Therefore, the presence of mucilage and gums in the leaves of *C. macrocarpa* (Table 2) may substantiate its use in traditional medicine to treat coughs [29–31]. Furthermore, mucilage and gums are used in several ways in the pharmaceutical industry. Due to their ability to easily dissolve in water, they are used as tablet binders, to coat capsules and as disintegrants (agents added to tablet formations to help to break down the tablet into small particles) [50].

Terpenoids have been found to be useful in the prevention and treatment of several diseases, including cancer. This phytochemical is well known for possessing antimicrobial, antifungal, antiparasitic, anti-inflammatory, and anti-allergenic properties. It also aids in the regulation of immune systems [51,52]. Furthermore, steroids can calm airway inflammation in asthma patients [53] and are effective in reducing cholesterol [54]. Flavonoids and tannins have been reported to possess anti-inflammatory and antimicrobial properties [55,56]; therefore, the leaves of C. *macrocarpa* could be used for the treatment of wounds due to the presence of flavonoids and tannins. For centuries, alkaloids have been employed for medicinal uses. The most common biological properties of this phytochemical are its toxicity against cells of foreign organisms and anti-asthmatic, anti-inflammatory and anti-anaphylactic properties [57,58]. Furthermore, saponins possess several pharmacological properties (antibacterial, antiviral) and are effective in slowing down or preventing inflammation [59,60]. From this study, it was found that C. *macrocarpa* leaf extracts exhibited several medicinal properties due to the presence of many different phytochemicals.

3.2. Antibacterial Activity

The antibacterial activity of the methanol leaf extract of C. *macrocarpa* was evaluated using five different concentrations (0.625, 1.25, 2.5, 5 and 10 mg/mL) against Gram-positive bacteria (*Staphylococcus aureus*) and Gram-negative bacteria (*Escherichia coli*). In Table 3, it is evident that the zone of inhibition increased with the increase in the concentration of the extracts. The methanol extract of the leaves inhibited the growth of the two bacterial strains (*S. aureus* and *E. coli*) but exhibited modest antibacterial activity compared with the control antibiotics. This may have been due to the small quantities of crude extract used. Future studies should increase the concentration of crude extracts used, which may result in a more favourable positive inhibitory effect. The leaf extracts at the different concentrations showed a greater inhibitory effect against the Gram-positive bacteria (*S. aureus*) than against the Gram-negative bacteria (*E. coli*) (Table 3). In summer, the zones of inhibition of the methanol leaf extract at the concentrations of 10, 5 and 2.5 mg/mL were 7.75 ± 1.77, 7.25 ± 0.35 and 6.75 ± 0.35 for *S. aureus* and 6.75 ± 0.35, 6.75 ± 0.35 and 6.5 ± 0.35 for *E. coli*, respectively. In winter, the zones of inhibition of the methanol leaf extract at the concentrations of 10, 5 and 2.5 mg/mL were 8.17 ± 1.04, 8.17 ± 1.04 and 7.25 ± 0.35 for *S. aureus* and 6.83 ± 0.58, 6.83 ± 0.58 and 6.75 ± 0.00 for *E. coli*, respectively. This is because the impermeable cell wall of Gram-negative bacteria makes them more resistant to certain antibiotics and antibacterial compounds than Gram-positive bacteria [61,62]. The results also revealed that the methanol extracts of C. *macrocarpa* leaves collected in winter showed greater antibacterial activity than those collected in summer (Table 3). Since *E. coli* and *S. aureus* are known to infect the digestive system [63], the findings from this study suggest that the methanol extract of C. *macrocarpa* leaves could be used to treat the infections caused by these two bacteria.

Table 3. Means and standard deviations of the zones of inhibition (mm) of the methanol extract of C. *macrocarpa* leaves (summer and winter) at different concentrations (mg/mL).

Bacterial Strain	Concentrations (mg/mL)					Positive Control
	10	5	2.5	1.25	0.625	
			Summer			
S. aureus	7.75 ± 1.77	7.25 ± 0.35	6.75 ± 0.35	7.0 ± 0.00	7.0 ± 0.00	10 ± 1.41
E. coli	6.75 ± 0.35	6.75 ± 0.35	6.5 ± 0.35	6.5± 0.00	6.5 ± 0.00	11 ± 1.41
			Winter			
S. aureus	8.17 ± 1.04	8.17 ± 1.04	7.25 ± 0.35	7.0 ± 0.00	7.0 ± 0.00	10 ± 1.32
E. coli	6.83 ± 0.58	6.83 ± 0.58	6.75 ± 0.00	6.5± 0.00	6.5 ± 0.00	11 ± 1.00

Studies conducted by Moodley et al. [28] and Abbas et al. [33] reported the antibacterial activities of C. *macrocarpa* and C. *grandiflora*. Moodley et al. [28] reported that the pentacyclic triterpenoids extracted from C. *macrocarpa* leaves and fruits showed antibacterial activity against *S. aureus*, *E. faecium*, *S. saprophyticus*, *E. coli*, *K. pneumonia* and *P. aeruginosa*. Abbas et al. [33] reported that the root, stem and leaf extracts of C. *grandiflora* showed antibacterial activity against *S. aureus* and *S. epidermidis*. The above findings imply that C. *macrocarpa* can be utilized for treatment against various bacterial strains. Moreover, phytochemicals such as phenols, saponins, flavonoids, tannins and steroids are renowned to be biologically active and thus partially responsible for the antimicrobial activities of plants [34]. Thus, the presence of these phytochemicals in the leaf extracts of *C. macrocarpa* (Table 2) may substantiate the plant's use in traditional medicine to treat venereal diseases caused by bacteria [29].

3.3. Antioxidant Activity

The DDPH-radical-scavenging assay was used to determine the antioxidant activity of the crude hexane, chloroform and methanol extracts of the leaves of *C. macrocarpa* (Figure 1).

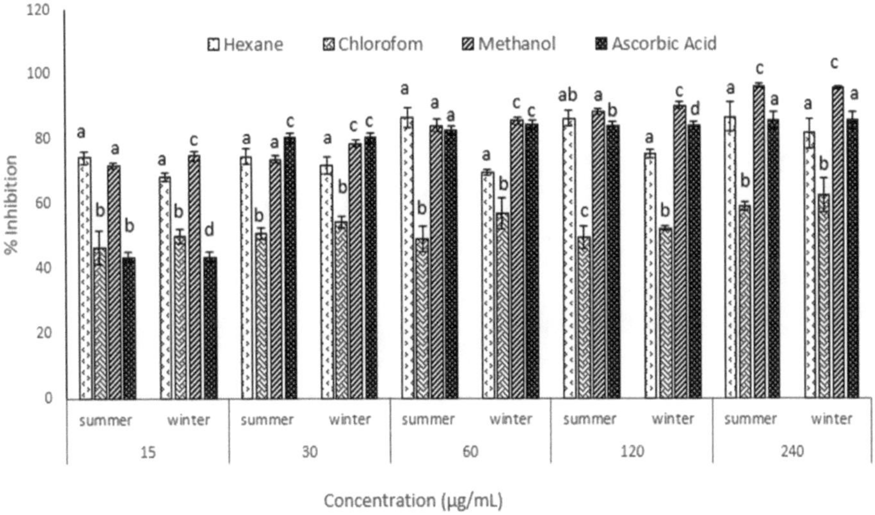

Figure 1. In vitro antioxidant activity (% scavenging) at different concentrations (μg/mL) of crude extracts from summer and winter leaves of *C. macrocarpa*. Different letters means significant at $p < 0.05$.

For both seasons, the methanol extract of the leaves exhibited the most significant radical-scavenging activity ($p < 0.05$) at 240 μg/mL, followed by the hexane extract and then the chloroform extract (Figure 1). For each season, the percentage scavenging activity of the DPPH radical increased with the increase in the concentration of the crude methanol leaf extract (Figure 1), which may imply an increased ability to donate hydrogen ions [64]. For the methanol leaf extract, there was a significant difference in the percentage scavenging activities between summer and winter ($p < 0.05$) at most concentrations. In summer, the percentage scavenging activities of the methanol leaf extract were 71.66, 73.57, 84.05 and 88.22% at 15, 30, 60 and 120 μg/mL, respectively. In winter, the percentage scavenging activities of the methanol leaf extract were 74.65, 78.31, 86.15 and 90.02% at 15, 30, 60 and 120 μg/mL, respectively (Figure 1). Overall, the methanol leaf extract showed a greater percentage of scavenging activity in winter. In summer, the percentage scavenging activities of the hexane, chloroform and methanol extracts at 15 μg/mL were 74.18, 46.64 and 71.66%, respectively, and that of ascorbic was 43.47%. In winter, the percentage scavenging activities of the hexane, chloroform and methanol extracts at 15 μg/mL were 68.27, 49.87 and 74.65%, respectively, and that of ascorbic was 43.47% (Figure 1). The percentage scavenging activities of the different extracts were significantly greater than that of the positive control ($p < 0.05$), which indicates that the extracts of *C. macrocarpa* leaves have appreciable antioxidant activity.

The results revealed that in summer, the IC_{50} values (concentration required for 50% inhibition) of the hexane extract (0.15 μg/mL) and methanol extract (1.72 μg/mL) were lower than that of ascorbic acid (8.26 μg/mL) (Table 4). In winter, the IC_{50} values of the hexane extract (0.29 μg/mL) and methanol extract (0.67 μg/mL) were also lower than that of ascorbic acid (8.26 μg/mL), whereas the chloroform extract had a greater IC_{50} value than ascorbic acid (8.26 μg/mL) for summer (44.76 μg/mL) and winter (13.06 μg/mL) (Table 4). A high percentage scavenging activity together with a low IC_{50} is indicative of strong antioxidant activity. Therefore, hexane and methanol extracts of *C. macrocarpa* leaves showed strong antioxidant activity. However, the methanol extract of winter leaves showed stronger antioxidant activity than that of summer leaves. While the chloroform extract of winter leaves showed moderate antioxidant activity, that of summer leaves showed low antioxidant activity. The result from this study corresponded to those of a study conducted by [35], whereby the DPPH assay revealed appreciable antioxidant activity in the leaves of *C. macrocarpa*. A similar study conducted by Abbas et al. [33] revealed that the methanol extracts of *C. grandiflora* leaves exhibited the highest free-radical-scavenging activity (antioxidant activity), while the hexane fractions exhibited the lowest one. Many phytochemicals possess antioxidant activity, protect our cells against damage caused by oxidative stress and lower the risk of developing certain types of cancer. From previous studies, it is known that flavonoids and phenols exhibit antioxidant properties [51,65]. Flavonoids help to manage diabetes caused by oxidative stress [51]. Phenols were found to possess free-radical-scavenging capabilities and act as food antioxidants [65]. The presence of phenolic compounds in the leaf extracts of *C. macrocarpa* (Table 2) may have enhanced the scavenging activity of the hexane and methanol extracts [66], thus causing the significantly low IC_{50} values. In addition, the collective effect of resins, flavonoids, triterpenes and phenols that were present in the leaf extracts (Table 2) may have also positively influenced the percentage inhibition and the IC_{50} values observed [67–71].

Table 4. IC_{50} values showing the antioxidant activity of the various solvent extracts from summer and winter leaves of *C. macrocarpa*.

Sample	IC_{50} (μg/mL)	
	Summer	Winter
Hexane	0.15	0.29
Chloroform	44.76	13.06
Methanol	1.72	0.67
Ascorbic acid	8.26	8.26

When plants are stressed, they produce more phytochemicals to withstand the unfavourable/harmful conditions [72]. Previous studies were performed on plants under stress conditions, and it was revealed that greater quantities of flavonoids, anthocyanins and mucilage were produced by these plants [73]. Therefore, the observed differences in metabolite yield (%) and antibacterial and antioxidant activities between summer and winter could be due to the differences in the environmental conditions experienced in different seasons. From previous studies, it was suggested that environmental temperatures associated with seasonal change play a vital role in phytochemical composition and biological activity and that it is more pronounced in cold weather [72]. The results from this study indicated that winter provided favourable stimuli (such as low temperatures) to bring about improved phytochemical production and biological activity in the leaves of *C. macrocarpa*.

4. Conclusions

Carissa macrocarpa leaves are found to contain many bioactive metabolites from various extracts, which implies that the leaves have several medicinal applications. The results from this study suggest that *C. macrocarpa* extracts possess antibacterial and antioxidant activity; thus, these results are expected to open the possibility of deriving clinically effective drugs from this plant species. It is also found that the winter leaf extracts have potentially higher antibacterial activity and more potent antioxidant activity than the summer leaf extracts, thus inferring that winter is the best season to harvest *C. macrocarpa* leaves for medicinal use. To the best of our knowledge, this is first report to relate the phytochemical composition and medicinal properties of *C. macrocarpa* to changes in seasons. It is recommended that further quantitative phytochemical analyses are conducted on this plant in summer and winter.

This would improve the understanding of how seasonal changes influence phytochemical composition. This would also elucidate which bioactive compounds are responsible for the potent antioxidant activity in the leaves of *C. macrocarpa*. The findings from this study add to the existing body of knowledge on South African ethnobotany and aid in better understanding the use and contribution of *C. macrocarpa* leaves in treating various illnesses.

Author Contributions: Conceptualization, R.R. and Y.N.; methodology, R.R. and Y.N.; formal analysis, R.R., Y.N. and Y.H.D.; investigation and data curation, R.R., Y.N. and Y.H.D.; validation, Y.H.D. and A.N.E.-B.; visualization, A.N.E.-B.; writing—original draft preparation, R.R. and Y.N.; writing—review and editing, Y.N., Y.H.D. and A.N.E.-B., supervision, Y.N. and Y.H.D. All authors have read and agreed to the published version of the manuscript.

Funding: The authors acknowledge Researchers Supporting Project number (RSP-2021/375), King Saud University, Riyadh, Saudi Arabia.

Institutional Review Board Statement: Not applicable.

Informed Consent Statement: Not applicable.

Data Availability Statement: All data are presented within the article.

Acknowledgments: The authors acknowledge Researchers Supporting Project number (RSP-2021/375), King Saud University, Riyadh, Saudi Arabia. Funding by National Research Foundation (NRF-South Africa) is greatly appreciated.

Conflicts of Interest: The authors declare no conflict of interest.

References

1. Shanley, P.; Luz, L. The Impacts of Forest Degradation on Medicinal Plant Use and Implications for Health Care in Eastern Amazonia. *BioScience* **2003**, *53*, 573. [CrossRef]
2. Sheldon, J.W.; Balick, M.J.; Laird, S.A.; Milne, G.M. Medicinal plants: Can utilization and conservation coexist? *Adv. Econ. Bot.* **1997**, *12*, 1–104.
3. Mbatha, N.; Street, R.A.; Ngcobo, M.; Gqaleni, N. Sick certificates issued by South African traditional health practitioners: Current legislation, challenges and the way forward: Issues in medicine. *South Afr. Med. J.* **2012**, *102*, 129–131. [CrossRef] [PubMed]
4. Svoboda, K.P.; Svoboda, T.G.; Syred, A.D. *Secretory Structures of Aromatic and Medicinal Plants: A Review and Atlas of Micrographs*; Microscopix Publications: Hoddesdon, UK, 2000.
5. Figueiredo, A.C.; Barroso, J.G.; Pedro, L.G.; Scheffer, J.J. Factors affecting secondary metabolite production in plants: Volatile components and essential oils. *Flavour Fragr. J.* **2008**, *23*, 213–226. [CrossRef]
6. Yadav RN, S.; Agarwala, M. Phytochemical analysis of some medicinal plants. *J. Phytol.* **2011**, *3*, 10–14.
7. Wakeel, A.; Jan, S.A.; Ullah, I.; Shinwari, Z.K.; Xu, M. Solvent polarity mediates phytochemical yield and antioxidant capacity of *Isatis tinctoria*. *PeerJ* **2019**, *7*, e7857. [CrossRef] [PubMed]
8. Nchabeleng, L.; Mudau, F.N.; Mariga, I.K. Effects of chemical composition of wild bush tea (*Athrixia phylicoides* DC.) growing at locations differing in altitude, climate and edaphic factors. *J. Med. Plants Res.* **2012**, *6*, 1662–1666. [CrossRef]
9. Jayanthy, A.; Prakash, K.U.; Remashree, A.B. Seasonal and geographical variations in cellular characters and chemical contents in *Desmodium gangeticum* (L.) DC.—An ayurvedic medicinal plant. *Int. J. Herb. Med.* **2013**, *1*, 34–37.
10. Odjegba, V.J.; Alokolaro, A.A. Simulated drought and salinity modulate the production of phytochemicals in *Acalypha wilkesiana*. *J. Plant Stud.* **2013**, *2*, 105. [CrossRef]
11. Kale, V.S.; Shitole, M.G. Variable rates of primary and secondary metabolites during different seasons and physiological stages in *Datura metel* L. *J. Maharashtra Agric. Univ.* **2010**, *35*, 371–374.
12. Gololo, S.S.; Shai, L.J.; Agyei, N.M.; Mogale, M.A. Effect of seasonal changes on the quantity of phytochemicals in the leaves of three medicinal plants from Limpopo province, South Africa. *J. Pharmacogn. Phytother.* **2016**, *8*, 168–172.
13. Valgas, C.; Souza SM, D.; Smânia, E.F.; Smânia, A., Jr. Screening methods to determine antibacterial activity of natural products. *Braz. J. Microbiol.* **2007**, *38*, 369–380. [CrossRef]
14. Babu, H.R.; Savithramma, N. Screening of secondary metabolites of underutilized species of *Cyperaceae*. *Int. J. Pharm. Sci. Rev. Res.* **2014**, *24*, 182–187.
15. Lim, T.K. *Carissa macrocarpa*. In *Edible Medicinal and Non-Medicinal Plants*; Springer: Dordrecht, The Netherlands, 2012; pp. 237–239.
16. Halliwell, B.; Gutteridge, J.M. *Free Radicals in Biology and Medicine*; Oxford University Press: New York, NY, USA, 2015.
17. Erkan, N. Antioxidant activity and phenolic compounds of fractions from *Portulaca oleracea* L. *Food Chem.* **2012**, *133*, 775–781. [CrossRef]

18. Valko, M.; Leibfritz, D.; Moncol, J.; Cronin, M.T.; Mazur, M.; Telser, J. Free radicals and antioxidants in normal physiological functions and human disease. *Int. J. Biochem. Cell Biol.* **2007**, *39*, 44–84. [CrossRef]
19. Mayne, S.T. Antioxidant nutrients and chronic disease: Use of biomarkers of exposure and oxidative stress status in epidemiologic research. *J. Nutr.* **2003**, *133*, 933S–940S. [CrossRef]
20. Rakesh, S.U.; Patil, P.R.; Mane, S.R. Use of natural antioxidants to scavenge free radicals: A major cause of diseases. *Int. J. PharmTech Res.* **2010**, *2*, 1074–1081.
21. Kapadiya, D.B.; Dabhi, B.K.; Aparnathi, K.D. Spices and herbs as a source of natural antioxidants for food. *Int. J. Curr. Microbiol. Appl. Sci.* **2016**, *5*, 280–288. [CrossRef]
22. Tadhani, M.B.; Patel, V.H.; Subhash, R. In vitro antioxidant activities of *Stevia rebaudiana* leaves and callus. *J. Food Compos. Anal.* **2007**, *20*, 323–329. [CrossRef]
23. Bhadane, B.S.; Patil, M.P.; Maheshwari, V.L.; Patil, R.H. Ethnopharmacology, phytochemistry, and biotechnological advances of family Apocynaceae: A review. *Phytother. Res.* **2018**, *32*, 1181–1210. [CrossRef]
24. El-Taher, A.M.; EL-Gendy, A.G.; Lila, M.I. Morphological and anatomical studies on some taxa of family Apocynaceae. *Al-Azhar J. Agric. Res.* **2019**, *44*, 136–147. [CrossRef]
25. The Plant List. Carissa. 2022. Available online: http://www.theplantlist.org/tpl1.1/record/kew-34192 (accessed on 8 August 2022).
26. Moodley, R.; Koorbanally, N.; Jonnalagadda, S.B. Elemental composition and fatty acid profile of the edible fruits of Amatungula (*Carissa macrocarpa*) and impact of soil quality on chemical characteristics. *Anal. Chim. Acta* **2012**, *730*, 33–41. [CrossRef] [PubMed]
27. Khalil, H.E.; Aljeshi, Y.M.; Saleh, F.A. Authentication of *Carissa macrocarpa* cultivated in Saudi Arabia; botanical, phytochemical and genetic study. *J. Pharm. Sci. Res.* **2015**, *7*, 497.
28. Moodley, R.; Chenia, H.; Jonnalagadda, S.B.; Koorbanally, N. Antibacterial and anti-adhesion activity of the pentacyclic triterpenoids isolated from the leaves and edible fruits of *Carissa macrocarpa*. *J. Med. Plants Res.* **2011**, *5*, 4851–4858.
29. Alsudani AA AA, Z.; Altameme HJ, M. A Taxonomical study of *Carissa macrocarpa* (Eckl.) A. Dc (Apocynaceae) in Iraq. *Rev. Int. Geogr. Educ.* **2021**, *11*, 1328–1341.
30. National Research Council. *Lost Crops of Africa: Volume III: Fruits*; National Academies Press: Washington, DC, USA, 2008; Volume 3, pp. 262–269.
31. Ibrahim, H.; Abdurahman, E.M.; Shok, M.; Ilyas, N.; Bolaji, R. Preliminary phytochemical and antimicrobial studies of the leaves of *Carissa edulis* Vahl. In *Standardization and Utilization of Herbal Medicines: Challenges of the 21st Century, Proceedings of the 1st International Workshop on Herbal Medicinal Products, Ibadan, Nigeria, 22–24 November 1998*; University of Ibadan, Department of Pharmacognosy: Ibadan, Nigeria, 1999; pp. 217–222.
32. Khalil, H.E.; Mohamed, M.E.; Morsy, M.A.; Kandeel, M. Flavonoid and phenolic compounds from *Carissa macrocarpa*: Molecular docking and cytotoxicity studies. *Pharmacogn. Mag.* **2018**, *14*, 304. [CrossRef]
33. Abbas, M.; Rasool, N.; Riaz, M.; Zubair, M.; Abbas, M.; Noor Ul, H.; Hayat, N. GC-MS profiling, antioxidant, and antimicrobial studies of various parts of *Carissa grandiflora*. *Bulg. Chem. Commun.* **2014**, *46*, 831–839.
34. Dhatwalia, J.; Kumari, A.; Verma, R.; Upadhyay, N.; Guleria, I.; Lal, S.; Thakur, S.; Gudeta, K.; Kumar, V.; Chao, J.C.C.; et al. Phytochemistry, pharmacology, and nutraceutical profile of *Carissa* species: An updated review. *Molecules* **2021**, *26*, 7010. [CrossRef]
35. Souilem, F.; Dias, M.I.; Barros, L.; Calhelha, R.C.; Alves, M.J.; Harzallah-Skhiri, F.; Ferreira, I.C. Phenolic profile and bioactive properties of *Carissa macrocarpa* (Eckl.) A. DC.: An in vitro comparative study between leaves, stems, and flowers. *Molecules* **2019**, *24*, 1696. [CrossRef]
36. Souilem, F.; El Ayeb, A.; Djlassi, B.; Ayari, O.; Chiboub, W.; Arbi, F.; Ascizzi, R.; Flamini, G.; Harzallah-Skhiri, F. Chemical composition and activity of essential oils of *Carissa macrocarpa* (Eckl.) A. DC. Cultivated in Tunisia and its anatomical features. *Chem. Biodivers.* **2018**, *15*, e1800188. [CrossRef]
37. Surendra, T.V.; Roopan, S.M.; Arasu, M.V.; Al-Dhabi, N.A.; Sridharan, M. Phenolic compounds in drumstick peel for the evaluation of antibacterial, hemolytic and photocatalytic activities. *J. Photochem. Photobiol. B Biol.* **2016**, *161*, 463–471. [CrossRef] [PubMed]
38. Sofowara, A. *Medicinal Plants and Traditional Medicine in Africa*; Spectrum Books Ltd.: Ibadan, Nigeria, 1993.
39. Tamilselvi, N.; Krishnamoorthy, P.; Dhamotharan, R.; Arumugam, P.; Sagadevan, E. Analysis of total phenols, total tannins and screening of phytocomponents in *Indigofera aspalathoides* (Shivanar Vembu) Vahl EX DC. *J. Chem. Pharm. Res.* **2012**, *4*, 3259–3262.
40. Trease, G.E.; Evans, W.C. *Pharmacognosy*; Bahiv Tinal: London, UK, 1985.
41. Khandelwal, K. *Practical Pharmacognosy*; Pragati Books Pvt. Ltd.: Nagpur, India, 2008.
42. Harborne, A.J. *Phytochemical Methods a Guide to Modern Techniques of Plant Analysis*; Springer Science & Business Media: Dordrecht, The Netherlands, 1998.
43. Bargah, R.K. Preliminary test of phytochemical screening of crude ethanolic and aqueous extract of *Moringa pterygosperma* Gaertn. *J. Pharmacogn. Phytochem.* **2015**, *4*, 7–9.
44. Kumar, R.S.; Balasubramanian, P.; Govindaraj, P.; Krishnaveni, T. Preliminary studies on phytochemicals and antimicrobial activity of solvent extracts of *Coriandrum sativum* L. roots (Coriander). *J. Pharmacogn. Phytochem.* **2014**, *2*, 74–78.
45. Perez, C. Antibiotic assay by agar-well diffusion method. *Acta Biol. Med. Exp.* **1990**, *15*, 113–115.
46. Braca, A.; Sortino, C.; Politi, M.; Morelli, I.; Mendez, J. Antioxidant activity of flavonoids from *Licania licaniaeflora*. *J. Ethnopharmacol.* **2002**, *79*, 379–381. [CrossRef]

47. Doss, A.; Mubarack, H.M.; Dhanabalan, R. Antibacterial activity of tannins from the leaves of *Solanum trilobatum* Linn. *Indian J. Sci. Technol.* **2009**, *2*, 41–43. [CrossRef]
48. Ruther, J.; Meiners, T.; Steidle, J.L.M. Allelochemical Reactions Involving Heterotrophic Microorganisms. *Biogeochem. Inland Waters* **2010**, *22*, 436.
49. Geetha, B.; Gowda, K.P.; Kulkarni, G.T.; Badami, S. Microwave assisted fast extraction of mucilages and pectins. *Indian J. Pharm. Educ. Res.* **2009**, *43*, 260–265.
50. Sungthongjeen, S.; Pitaksuteepong, T.; Somsiri, A.; Sriamornsak, P. Studies on pectins as potential hydrogel matrices for controlled-release drug delivery. *Drug Dev. Ind. Pharm.* **1999**, *25*, 1271–1276. [CrossRef]
51. Rahal, A.; Kumar, A.; Singh, V.; Yadav, B.; Tiwari, R.; Chakraborty, S.; Dhama, K. Oxidative stress, prooxidants, and antioxidants: The interplay. *BioMed Res. Int.* **2014**, *2014*, 761264. [CrossRef] [PubMed]
52. Rabi, T.; Bishayee, A. Terpenoids and breast cancer chemoprevention. *Breast Cancer Res. Treat.* **2009**, *115*, 223–239. [CrossRef] [PubMed]
53. Krishnaiah, D.; Devi, T.; Bono, A.; Sarbatly, R. Studies on phytochemical constituents of six Malaysian medicinal plants. *J. Med. Plants Res.* **2009**, *3*, 67–72.
54. Shah, B.A.; Qazi, G.N.; Taneja, S.C. Boswellic acids: A group of medicinally important compounds. *Nat. Prod. Rep.* **2009**, *26*, 72–89. [CrossRef]
55. Peter, H.; Pavel, T.; Marie, S. Flavonoids-potent and versatile biologically active compounds interacting with cytochrome P450. *Chem. Biol. Interact.* **2002**, *139*, 1–21.
56. Killedar, S.G.; More, H.N. Estimation of tannins in different parts of *Memecylon umbellatum* Burm. *J. Pharm. Res.* **2010**, *3*, 554–556.
57. Ganguly, T.; Sainis, K.B. Inhibition of cellular immune responses by *Tylophora indica* in experimental models. *Phytomedicine* **2001**, *8*, 348–355. [CrossRef]
58. Staerk, D.; Lykkeberg, A.K.; Christensen, J.; Budnik, B.A.; Abe, F.; Jaroszewski, J.W. In Vitro Cytotoxic Activity of Phenanthroindolizidine Alkaloids from *Cynanchum vincetoxicum* and *Tylophora tanakae* against Drug-Sensitive and Multidrug-Resistant Cancer Cells. *J. Nat. Prod.* **2002**, *65*, 1299–1302. [CrossRef]
59. Just, M.J.; Recio, M.C.; Giner, R.M.; Cuéllar, M.J.; Máñez, S.; Bilia, A.R.; Ríos, J.L. Anti-inflammatory activity of unusual lupane saponins from *Bupleurum fruticescens*. *Planta Med.* **1998**, *64*, 404–407. [CrossRef]
60. Estrada, A.; Katselis, G.S.; Laarveld, B.; Barl, B. Isolation and evaluation of immunological adjuvant activities of saponins from *Polygala senega* L. *Comp. Immunol. Microbiol. Infect. Dis.* **2000**, *23*, 27–43. [CrossRef]
61. Nikaido, H.; Vaara, M. Molecular basis of bacterial outer membrane permeability. *Microbiol. Rev.* **1985**, *49*, 1–32. [CrossRef] [PubMed]
62. Rabe, T.; Van Staden, J. Antibacterial activity of South African plants used for medicinal purposes. *J. Ethnopharmacol.* **1997**, *56*, 81–87. [CrossRef]
63. Sekirov, I.; Russell, S.L.; Antunes LC, M.; Finlay, B.B. Gut microbiota in health and disease. *Physiol. Rev.* **2010**, *90*, 859–904. [CrossRef]
64. Silva, C.D.; Herdeiro, R.S.; Mathias, C.J.; Panek, A.D.; Silveira, C.S.; Rodrigues, V.P.; Renno, M.N.; Falcao, D.Q.; Cerqueira, D.M.; Minto AB, M.; et al. Evaluation of antioxidant activity of Brazilian plants. *Pharmacol. Res.* **2005**, *52*, 229–233. [CrossRef] [PubMed]
65. Kasote, D.M.; Katyare, S.S.; Hegde, M.V.; Bae, H. Significance of antioxidant potential of plants and its relevance to therapeutic applications. *Int. J. Biol. Sci.* **2015**, *11*, 982. [CrossRef] [PubMed]
66. Srivastava, J.; Kumar, S.; Vankar, P.S. Correlation of antioxidant activity and phytochemical profile in native plants. *Nutr. Food Sci.* **2012**, *42*, 71–79. [CrossRef]
67. Kessler, M.; Ubeaud, G.; Jung, L. Anti-and pro-oxidant activity of rutin and quercetin derivatives. *J. Pharm. Pharmacol.* **2003**, *55*, 131–142. [CrossRef]
68. Talla, E.; Tamfu, A.N.; Gade, I.S.; Yanda, L.; Mbafor, J.T.; Laurent, S.; Elst, L.V.; Popova, M.; Bankova, V. New mono-ether of glycerol and triterpenes with DPPH radical scavenging activity from *Cameroonian propolis*. *Nat. Prod. Res.* **2017**, *31*, 1379–1389. [CrossRef]
69. Miao, J.; Li, X.; Zhao, C.; Gao, X.; Wang, Y.; Gao, W. Active compounds, antioxidant activity and α-glucosidase inhibitory activity of different varieties of *Chaenomeles* fruits. *Food Chem.* **2018**, *248*, 330–339. [CrossRef]
70. Akwu, N.A.; Naidoo, Y.; Singh, M.; Nundkumar, N.; Lin, J. Phytochemical screening, in vitro evaluation of the antimicrobial, antioxidant and cytotoxicity potentials of *Grewia lasiocarpa* E. Mey. ex Harv. *S. Afr. J. Bot.* **2019**, *123*, 180–192. [CrossRef]
71. Assimopoulou, A.N.; Zlatanos, S.N.; Papageorgiou, V.P. Antioxidant activity of natural resins and bioactive triterpenes in oil substrates. *Food Chem.* **2005**, *92*, 721–727. [CrossRef]
72. Iqbal, S.; Bhanger, M.I. Effect of season and production location on antioxidant activity of *Moringa oleifera* leaves grown in Pakistan. *J. Food Compos. Anal.* **2006**, *19*, 544–551. [CrossRef]
73. Kaplan, F.; Kopka, J.; Sung, D.Y.; Zhao, W.; Popp, M.; Porat, R.; Guy, C.L. Transcript and metabolite profiling during cold acclimation of Arabidopsis reveals an intricate relationship of cold-regulated gene expression with modifications in metabolite content. *Plant J.* **2007**, *50*, 967–981. [CrossRef] [PubMed]

horticulturae

MDPI

Article

Antioxidant Capacity of *Salix alba* (Fam. *Salicaceae*) and Influence of Heavy Metal Accumulation

Demush Bajraktari [1], Biljana Bauer [2] and Lulzim Zeneli [3,*]

1 Faculty of Pharmacy, UBT Higher Education Institution, Lagjja Kalabria, p.n. 10000 Prishtine, Kosovo; demush.bajraktari@ubt-uni.net

2 Faculty of Pharmacy, Ss. Cyril and Methodius University of Skopje, Ul. "Majka Tereza" br. 47, 1000 Skopje, North Macedonia; biba@ff.ukim.edu.mk

3 Faculty of Education, University "Fehmi Agani", Str. "Ismail Qemali", n.n. 50000 Gjakova, Kosovo

* Correspondence: lulzim.zeneli@uni-gjk.org

Abstract: In this study, we analyzed and compared the concentrations of selected metals/metalloids and the antioxidant response of *Salix alba* L. (white willow) bark in the highly polluted area around the Kosovo A and B thermal power plants. The antioxidant capacity of *Salix alba* bark was evaluated in terms of the total phenolics, flavonoids, chlorophylls, and carotenoids, while the metal content in the soil and willow bark was analyzed by ICP-OES (inductively coupled plasma optical emission spectroscopy). For total antioxidant level assessment, FRAP, DPPH, and CUPRAC assays were conducted. The mean concentrations of selected elements in soil and willow dry mass range from 15,698.4 mg kg^{-1} dry mass (soil) to 371.1 mg kg^{-1} (willow bark) for Al; 37.676 mg kg^{-1} (soil) to <2 ppb (willow bark) for As; 14.8 mg kg^{-1} (soil) to 0.62 mg kg^{-1} (willow bark) for Cd; 24.2 mg kg^{-1} (soil) to 1.2 mg kg^{-1} (willow bark) for Cr; 58.8 mg kg^{-1} (soil) to 9.1 mg kg^{-1} (willow bark) for Cu; 16,975.68 mg kg^{-1} (soil) to 385.4 mg kg^{-1} (willow bark) for Fe; 95.0 mg kg^{-1} (soil) to 7.9 mg kg^{-1} (willow bark) for Ni; 185.2 mg kg^{-1} (soil) to <1 ppb (willow bark) for Pb; and 226.7 mg kg^{-1} (soil) to 87.7 mg kg^{-1} (willow bark) for Zn. Additionally, the *Salix alba* bark samples presented mean values of 12,191.6 mg kg^{-1} for Ca, 1306.0 mg kg^{-1} for Mg, and 123.1363 mg kg^{-1} for Mn. The mean phenolic content was 39.292 mg GAE g^{-1} DW, being 28.222 mg CE g^{-1} DW for flavonoids, 38.099 mg g^{-1} FW for CHLa, 49.240 mg g^{-1} FW for CHLb, and 94.976 mg g^{-1} FW for CAR. The results of this study indicate that the bark of *Salix alba* contains significant amounts of phenolic compounds, and strong positive and moderate negative correlations are revealed between total phenolic compounds and iron, and total phenolics and nickel and manganese, respectively.

Keywords: white willow; soil; antioxidant response; heavy metals

Citation: Bajraktari, D.; Bauer, B.; Zeneli, L. Antioxidant Capacity of *Salix alba* (Fam. *Salicaceae*) and Influence of Heavy Metal Accumulation. *Horticulturae* **2022**, *8*, 642. https://doi.org/10.3390/horticulturae8070642

Academic Editors: Ivan Salamon and Milica Aćimović

Received: 8 June 2022
Accepted: 12 July 2022
Published: 15 July 2022

Publisher's Note: MDPI stays neutral with regard to jurisdictional claims in published maps and institutional affiliations.

1. Introduction

From ancient times, members of the genus *Salix* (Family *Salicaceae*) have been used in traditional medicine for the treatment of various health conditions, especially muscular and rheumatic disorders [1]. The analgesic, antipyretic, anti-inflammatory, anti-rheumatic, antiseptic, and astringent properties of willow are well-known, while species of the genus *Salix* have also been used for their diuretic, febrifuge, hypnotic, and sedative effects [2]. Considering these effects, willow bark has been used in the preparation of pharmaceutical formulations to treat many different types of pain, as well as to relieve sore throats and fevers associated with infections of the upper respiratory system and flu-like symptoms [3]. White willow—otherwise known as salicin willow—has been used for thousands of years due to its impact on health [4]. According to the European Pharmacopoeia (2005), willow bark is defined as the dry bark (whole or fragmented) of young or current-year branches of different species of the genus *Salix*. At present, the use of willow bark, both in traditional medicine and in the pharmaceutical industry, is supported by a considerable number of publications that provide data regarding its effects on health [2].

Willow (*Salix* spp.) is a highly diverse genera containing more than 450 species distributed worldwide, from tropical to cold-temperate climates. *Salix alba* L. is one of the most widely distributed and popular willows. It appears from sea level up to 2400 m and generally grows near water on the banks of rivers and lakes, usually surviving for only 20–30 years [5]. *Salix alba* is a native plant in Europe, Central Asia, and North Africa, while in the Americas it was introduced in the late 17th and early 18th centuries. Today, except for in Australia and Antarctica, *Salix alba* is found in all other continents, occurring in pure stands or in a mixture with other willows and poplars [6].

White willow tolerates a broad range of soil types. From a botanical point of view, it is fast-growing and is one of the tallest willow species, with heights up to 30 m and diameters of approximately 1 m in its native habitat [7]. It is characterized by thin, horizontally extended branches that are easily broken, with deeply fissured bark colored dark grey to brown/yellow brown with corky ridges. Its root system is highly developed, with a deep central root and numerous lateral roots.

Heavy metals can be found in surface water, soil, and plants [8]. They are non-biodegradable, have long biological half-lives and have the potential to accumulate. Some of these metals play important roles in the normal development of biological processes in living organisms. Even though plants take nutrients from water, soil, and air, they are not completely selective for basic components, and may take up elements such as metals, which are unsafe even in low amounts [9]. Different plant species in particular have specific tolerance levels for heavy metals/metalloids, such that some plants have adapted to grow in soils rich in heavy metals/metalloids. Studies have confirmed critical differences in the norm of absorption and distribution of these elements and shown crucial changes in the concentration of heavy metals in various parts of the plant. The age of the plant, soil concentration, climate, family, and genetic structure change the ability of a plant to absorb and synthesize chemical content [10]. A few metals are commonly immobilized in the roots within the tree (e.g., lead, chromium, and copper), while cadmium, nickel, and zinc are more easily transported to the shoots; however, the soil properties in different areas, different climatic conditions, and different sampling times during individual studies must be taken into account [11]. As metals compete and replace the essential elements in plants, thus disrupting their normal function, they can be toxic, consequently damaging biodiversity and making the environment unsuitable for plant growth. The damage observed in the presence of toxic concentrations of heavy metals can be attributed to various causes, which usually interact and may include direct and indirect harmful effects (e.g., through oxidation-induced stress) [12]. Toxic levels of heavy metals can cause cell membrane injury and the breakdown of biomolecules and cell organelles due to increased reactive oxygen species (ROS) production in plants [13].

Considerable studies in recent years have established the important connection between air, water, soil, and plant pollution and toxic element accumulation. The zone around the Kosovo thermal power plants (KTPPs), energy producers, is the greatest concern of the Republic of Kosovo in the context of environmental pollution [14]. From a pollution point of view, the most important activities are those in coal (lignite) mines and the production of electric power. Fly and bottom ash from the KTPPs are the main components of this environmental pollution, and these industrial components are rich in elements with high toxic potential, such as Pb, Cd, As, and Ni [15]. A survey conducted during 2005, regarding fly ash release from the KTPPs, showed that EU standards were exceeded in very high percentages, approximately 400–500 times [15,16].

Plants are subjected to a variety of stressors in the natural world, all of which contribute to the over-production of reactive oxygen species [14]. Heavy metals, as abiotic stress factors, also initiate various processes in plants, some of which may even be harmful to their anatomical, morphological, physiological, and biochemical function, causing the production of reactive oxygen species and oxidative stress in plants. Plants have evolutionarily developed mechanisms to defend themselves against the harmful impact of heavy metals, involving the activation of antioxidant systems to neutralize reactive oxygen. Several definitions of chemical, biological, and biomedical antioxidants are known. Generally, any substance which, when existing in low concentrations compared to those of an oxidizable substrate, substantially delays or inhibits the oxidation of this substrate [17], or "any substance that delays, prevents or removes oxidative damage to a target molecule" [18], is defined as an antioxidant. Plants are generally known for their antioxidant properties, as they are rich in substances with antioxidant activity, commonly referred to as polyphenols [19]. A plant's antioxidant composition facilitates radical scavenging, supporting the conversion of radicals into less-reactive species [20]. Antioxidants usually act as reducing agents and neutralize free radicals by oxidizing themselves; examples include thiols, ascorbic acid, and polyphenols [21].

Antioxidants have become an essential part of our lives and, as a result of their action in the body to neutralize or break up ROS, it has been shown that different parts of plants are rich sources of a range biologically active agents—usually phenolics—and phytochemicals, which are plant constituents that possess different biological potentials, such as antioxidant and antimicrobial activities [22–25]. To determine the antioxidant capacity, many different methods based on free radical scavenging have been developed in recent decades; in some cases, these mechanisms have not yet been clearly differentiated. An antioxidant capacity test may involve electron-based transfer (ET)- or hydrogen atom transfer (HAT)-based analyses [26]. The ferric-reducing/antioxidant power (FRAP) method (referring to ferric-reducing ability) measures the potential of antioxidants to reduce ferric iron, and has been presented as a new method for the evaluation of antioxidant capacity [27]. Cupric-reducing antioxidant capacity (CUPRAC), which involves measuring the reduction of Cu (II) to Cu (I) by antioxidants, is an uncomplicated method for extensively relevant analysis of antioxidant capacity, and is beneficial for a broad class of polyphenols, including phenolic acids, flavonoids, carotenoids, hydroxycinnamic acids, thiols, synthetic antioxidants, and vitamins C and E [28,29]. The DPPH (2,2-diphenyl-1-picrylhydrazyl) method measures the ability of antioxidants to reduce 2,2-diphenyl-1-picrylhydrazyl (DPPH), an organic radical, through spectrophotometry [30]. The DPPH assay is the oldest non-direct assay for antioxidant activity determination, based on the potential of the 2,2-diphenyl-1-picrilhydrazyl free radical to interact with phenol hydrogen donors [31].

Considering the importance of *Salix alba* in traditional and modern medicine, in this study, we aim to analyze and evaluate the influence of selected heavy metals in the antioxidant response of *Salix alba* bark from the highly polluted area around the Kosovo thermal power plants. The antioxidant capacity of *Salix alba* bark was evaluated in terms of total phenolics, flavonoids, chlorophylls, and carotenoids, while the metal content in soil and willow bark was analyzed by ICP-OES. For total antioxidant level assessment, FRAP, DPPH, and CUPRAC assays were carried out.

2. Materials and Methods

2.1. Sampling Points and Material

Thermal power plants Kosovo A and Kosovo B, as a part of the local heavy industry and major contributors to environmental pollution in the area, are located about 6.5 km northwest of capital of Kosovo, Pristina, near the municipality of Obiliq. Ninety water, soil, and plant samples (thirty of each) were collected along 30 km of the Sitnica river, which passes near the Kosovo thermal power plants, prior to the determination of heavy metal contents and other parameters. Fifteen samples each of water, soil, and *Salix alba* bark were taken from a starting point near Lipjan municipality, about 17 km from the source of pollution, up to nearby the KTPPs, and fifteen additional samples of each were collected, beginning near the thermal power plants, up to the ending point where the river exits the municipality of Vushtrri, approximately 15 km from the source of pollution. There was an average distance of 1 km between sampling points, the locations of which are shown in Figure 1.

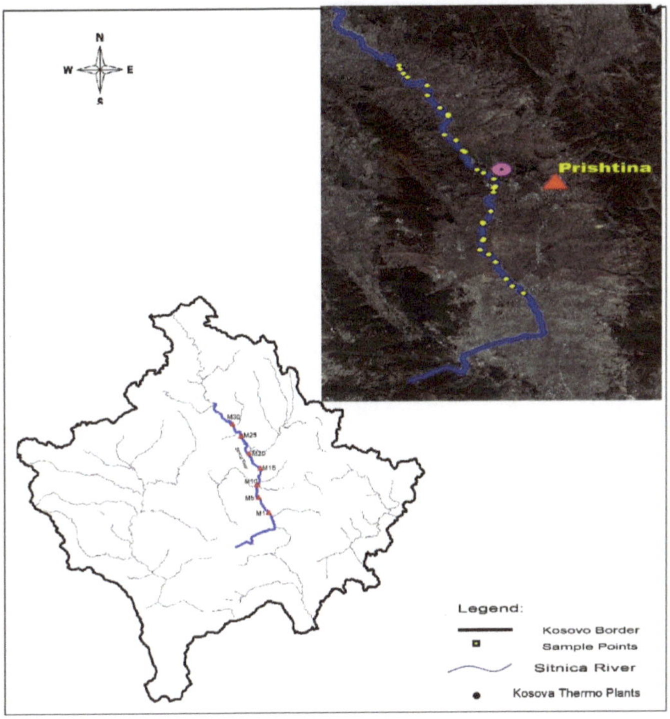

Figure 1. The flow of the river Sitnica in Kosovo and the sampling points.

2.2. *Plant Material*

Bark samples (200–250 g) were taken from *Salix alba* located along the river (Figure 2), collected at a height of 1.5–2 m above the ground and packed in paper bags. To eliminate any surface particles or impurities, the willow bark samples were washed with deionized water and then dried in an oven for 24 h at 65 °C. Then, 6 mL concentrated nitric acid and 2 mL concentrated hydrogen peroxide were added into a microwave digestion vessel containing 500 mg of minced willow bark samples, which were digested in the acid solution and carefully shaken. The vessels were sealed and heated in a microwave and, after cooling to room temperature (20 °C), the solution was filtered and diluted to 50 mL in a volumetric flask with deionized water.

2.3. *Soil Material*

At a depth of 30–40 cm around the willow tree, 2000 g of soil samples were collected and packed in plastic bags. Air-dried and milled samples (2.5 g) were digested with 10 mL aqua regia (concentrated nitric acid and hydrochloric acid in a 1:3 molar ratio) and carefully shaken. The vessels were closed after 20 min, and the solution was heated in a microwave. The solution was filtered and diluted with deionized water to 100 mL in a volumetric flask after the vessels were cooled to room temperature (20 °C).

Figure 2. *Salix alba* at sampling site 14 near the KTPPs.

2.4. Determination of Total Phenols

Powdered and dried plant material was used to extract phenolic compounds (0.1 g) with 80% (v/v) CH_3OH in an ultrasonic bath for 30 min at 4 °C [32,33]. After that, the methanolic extracts were centrifuged at 12,000 rpm for 15 min, and the supernatants were used for the determination of phenolic compound and chlorophyll contents, as well as antioxidant activities. The total phenolic contents were determined by mixing plant methanolic extracts with Folin–Ciocalteu reagent and 0.7 M Na_2CO_3 [34]. The samples were incubated at 50 °C for 15 min, then cooled at room temperature. The absorbance was measured spectrophotometrically at 765 nm. The total phenolic content data are expressed as milligrams of gallic acid equivalent (GAE) per gram of dry weight (mg GAE g^{-1} DW).

2.5. Determination of Total Flavonoids

The assay described by Zhishen et al. (1999) was used to determine the total flavonoid content in plant extracts. Properly diluted extract was mixed with 5% $NaNO_2$ and allowed to react for 5 min. After that, 10% $AlCl_3$ was added and the mixture stood for another 5 min. Finally, 1 M NaOH and distilled water were added to the reaction mixture and the absorbance was measured spectrophotometrically at 510 nm [35]. Total flavonoid data are expressed as milligrams of catechin equivalent (CE) per gram of dry weight (mg CE g^{-1} DW).

2.6. Determination of Total Chlorophylls and Carotenoids

The chlorophyll and carotenoid contents in plant extracts were determined using the method of Lichtenthaler and Wellburn (1983) [36]. The absorbance of an acetone extract was measured at 666 nm (chlorophyll a), 653 nm (chlorophyll b), and 470 nm (carotenoids). The chlorophyll and carotenoid concentrations are expressed as mg chlorophyll/carotenoid per gram of dry weight (mg g^{-1} DW) [36].

2.7. Cupric-Reducing Antioxidant Capacity (CUPRAC)

The CUPRAC assay of plant extracts was carried out using the method of Apak et al. (2004). The reaction mixture consisted of plant extract, 10 mM $CuCl_2$, 7.5 mM neocuproine, and 1 M CH_3COONH_4 buffer (pH 7.0). The samples were incubated for 30 min at room temperature and the absorbance was measured at 450 nm. The molar extinction coefficient of Trolox (ε_{535} = 1.67 × 10^4 $Lmol^{-1}{\cdot}cm^{-1}$) was used for CUPRAC determination. The CUPRAC values of plant extracts are expressed as micromoles of Trolox equivalent (TE) per gram of dry weight (μmol TE g^{-1} DW).

2.8. Ferric-Reducing Antioxidant Power (FRAP)

The FRAP assay of plant extracts was conducted according to the method of Benzie and Strain (1996) [37]. The FRAP reagent consisted of 300 mM CH_3COONa buffer (pH 3.6), 10 mM TPTZ (2,4,6-tri(2-pyridyl)-s-triazine) in 40 mM HCl, and 20 mM $FeCl_3$ (10:1:1, *v*/*v*/*v*). An aliquot of plant extract was mixed with FRAP reagent and incubated for 30 min at 37 °C under dark conditions. The absorbance of samples was measured at 593 nm. Data are expressed as micromoles of Fe^{2+} equivalent per gram of dry weight (μM Fe^{2+} g^{-1} DW).

2.9. DPPH Radical Scavenging Activity

Using the method of Brand-Williams et al. (1995), we determined the ability of plant extracts to scavenge the stable free radical 2,2-diphenyl-1-picrylhydrazyl (DPPH). The reaction mixture consisted of plant extract and 0.25 mM DPPH in CH_3OH. The extract was substituted with CH_3OH in the control sample. The DPPH scavenging reaction was performed in the dark for 10 min at room temperature, and the decrease in absorbance was measured at 517 nm. Data are expressed as micromoles of Trolox equivalent (TE) per gram of dry weight (μM TE g^{-1} DW) [38].

2.10. Reagents and Equipment

The amounts of metals in the sample were determined by inductively coupled plasma optical emission spectrometry (ICP-OES), using a Perkin Elmer Optima 2100 DV (Optical Emission Spectrometer), which was set up and optimized according to the manufacturer's instructions. Blank samples were prepared in the same manner. A high-quality KERN ABJ analytical balance was used for weight measurement, and samples were milled in a Retsch miller (GM 200; plant samples). The samples were dried in an oven (Mega Term M-160) and heated in a Berghof Microwave (MWS 3+).

The following solvents (of analytical grade) were used: methanol (Sigma-Aldrich, Steinheim, Germany) and acetone (Carlo Erba, Milan, Italy). The following chemicals (of analytical grade) were used: Folin–Ciocalteu reagent (Carlo Erba, Milan, Italy); and sodium carbonate, sodium acetate, sodium nitrite, sodium hydroxide, ammonium acetate, aluminum chloride, cupric chloride, neocuproine, catechin hydrate, (±)-6-hydroxy-2,5,7,8-tetramethylchromane-2-carboxylic acid, gallic acid, 2,4,6-tri(2-pyridyl)-s-triazine, ferric chloride, and 2,2-diphenyl-1-picrylhydrazyl (Sigma-Aldrich, Steinheim, Germany). Spectrophotometric analyses of antioxidant parameters were carried out using a SpectraMax 190 Microplate Reader (Molecular Devices Corp., Sunnyvale, CA, USA) with SoftMax Pro (v. 5.4.1) software.

2.11. Statistical Analysis

All parameters were analyzed using the data analysis tools in VassarStats (http://vassarstats.net/ accessed on 4 February 2022) and Microsoft Excel 2016. A Pearson correlation test was used to analyze the correlation between variables in soil and *Salix alba* bark samples. The values indicate statistical significance at $p < 0.05$. The obtained data from the analysis of total phenols, flavonoids, and chlorophylls and carotenoids are expressed as mean values ± SD (n = 3). The correlations between the analyzed parameters were calculated according to Pearson's correlation coefficient, and details are presented in the Results and Discussion sections. Principal component analysis (PCA) was performed to determine the variance of selected parameters. Cluster analysis (CA) was conducted to determine the similarity and variance of influencing factors across various data sets [39].

3. Results

3.1. Selected Elements (Metals/Metalloids) in Soil

The content of heavy metals in soil is influenced by a variety of factors. Due to its negative ecological impacts, the heavy metal contamination of soil is a major environmental concern. Heavy metals occur naturally in low concentrations in soils; however, due to their acute and chronic toxicity, they are considered soil pollutants at high concentrations [40].

On both sides of the Sitnica river are several rivers and ravines, some of which pass near the KTPPs, having an obvious impact on it being one of the most polluted rivers in Kosovo. Water samples were taken at points from 30 km along the Sitnica river (approximately 15 km in both directions from the source of pollution: KTPPs), and were analyzed for the accumulation of aluminum, arsenic, cadmium, chromium, copper, iron, nickel, lead, and zinc. The values (mg L^{-1}) of selected elements in water samples varied, depending on the location from which the samples were taken and the distance from the main source of pollution (i.e., KTPPs). The results are presented in Table 1, showing the minimum, maximum, and average values for the analyzed metals/metalloids.

Table 1. Mean, minimum, and maximum contents (mg L^{-1}) of analyzed metals/metalloids in Sitnica river.

	Al	As	Cr	Cu	Fe	Ni	Zn	Pb
Mean	0.346	0.01	0.011	0.014	0.314	0.049	0.093	0.417
Minimum	0.163	0.0019	0.006	0.011	0.069	0.029	0.071	0.299
Maximum	0.99	0.022	0.037	0.024	1.571	0.078	0.122	0.642
Count	30	30	30	30	30	30	30	4

The mean value of aluminum in the analyzed water samples was 0.346 mg L^{-1}, with the highest concentration in water (W) sample W6 (0.99 mg L^{-1}) and the lowest in sample W11 (0.163 mg L^{-1}). The average amount of arsenic found in sampling sites was 0.01 mg L^{-1} (ranging from 0.0019 mg L^{-1} (W23) to 0.022 mg L^{-1} (W18)), while cadmium concentrations in the analyzed samples were all below the detection limit of the method used (i.e., 0.001 mg L^{-1}). Chromium values in all analyzed Sitnica river water samples were in the interval from 0.006 mg L^{-1} (W13) to 0.037 mg L^{-1} (W1). The mean copper value in the selected water samples was 0.014 mg L^{-1}, with a minimum concentration of 0.011 mg L^{-1} in sample W9 and the highest concentration in sample W24 (0.024 mg L^{-1}). The mean concentration of iron in the analyzed water samples from Sitnica river was 0.314 mg L^{-1}, with values ranging from 0.069 mg L^{-1} in sample W27 to 1.571 mg L^{-1}, detected in sample W6. Nickel is a very hazardous and dangerous element, and its content in the study area varied from 0.029 mg L^{-1} to 0.078 mg L^{-1}. During this research, traces of lead were determined only in four water samples, with concentrations ranging from 0.299 mg L^{-1} (W15) to 0.642 mg L^{-1} (W12). The mean detected value in water samples from Sitnica river for zinc was 0.093 mg L^{-1}, with values ranging from 0.071 mg L^{-1} in sample W28 to the maximum concentration in sample W2 (0.122 mg L^{-1}) [41]. These results indicate that the mean levels (expressed in mg L^{-1}) of the selected metals/metalloids in water samples covering 30 km of Sitnica river (which passes near the KTPPs) were in the following order: Pb > Al > Fe > Zn > Ni > Cu > Cr > As.

Soils are not considered contaminated, from human, animal, and plant standpoints, until a threshold concentration is surpassed and the metal(s) begin to influence biological processes [42]. The maximum permissible concentrations of heavy metals in soil are defined differently, mainly based on regulations at the national level, with some regional differences. All findings from soil samples collected along the Sitnica river are expressed on the basis of mg kg^{-1} dry mass; the data are summarized in Table 2. Selected mineral and trace elements were determined in close proximity to *Salix alba* roots at the same sampling sites. The soil pH was generally alkaline, ranging from 7.4 in sample S1 (S: soil sample) to 8.5 in sample S30, with a mean pH value of 8.08.

Table 2. Mean, minimum, and maximum concentrations (mg kg^{-1}) of analyzed metals/metalloids in soil.

	Al	As	Cd	Cr	Cu	Fe	Ni	Pb	Zn
Mean	15,698.4	37.67	14.85	24.27	58.85	16,975.67	95.04	185.22	226.7
Minimum	12,021.88	20.29	9.83	18.07	29.88	13,216.97	66.74	84.2	99.23
Maximum	19,279.32	78.36	22.23	35.85	100.02	20,399	129.59	667.97	948.71
Count	30	30	30	30	30	30	30	30	30

Aluminum was the largest constituent of inorganic soil fractions, typically ranging from 1% to 30% (i.e., 10,000 to 300,000 mg kg^{-1}) [43,44]. The obtained data suggest that the average aluminum concentration in analyzed samples in the KTPP region was in line with these criteria, where the highest concentration was observed in sample S9 and the lowest concentration in sample S29. The arsenic concentration ranged from 20.29 mg kg^{-1} (sample S18) to the highest determined level of 78.36 mg kg^{-1} in sample S10, which was collected within 10 km of the KTPPs. This is higher than the results obtained for the Anadrinia area (Kosovo), where the arsenic values ranged from 16.49 mg kg^{-1} to 62.44 mg kg^{-1} [45]. Cadmium is one of the most environmentally hazardous metals, harmfully impacting all biological processes in living organisms [42]. The cadmium concentration varied from the lowest content of 9.83 mg kg^{-1}, which was practically identical to the lowest content for the Anadrinia area (9.36 mg kg^{-1}) [45], to the highest content of 22.23 mg kg^{-1}, which is higher than the intervention value when compared with data of Dutch Target and Intervention Values (2000). The mean cadmium content was estimated to be 14.85 mg kg^{-1}, and the highest and lowest concentrations appeared at the same sampling points, presenting the highest and lowest concentrations of arsenic (sample S10 and S18), respectively. In all soil samples, the chromium content was less than the target value (100 mg kg^{-1}) [46], ranging from 18.07 mg kg^{-1} in sample S1 to 35.85 mg kg^{-1} in sample S22. The average copper concentration was measured to be 58.85 mg kg^{-1}, with the lowest (29.88 mg kg^{-1}) and highest (100.02 mg kg^{-1}) concentrations in sampling locations close to each other (samples S27 and S29). The average concentration of copper in the samples taken along Sitnica river was about 2.5 times higher compared with the average values of 20 to 30 mg kg^{-1} reported by Hooda (2010) [47]. Iron was the heavy metal with the highest average concentration (16,975.67 mg kg^{-1}) in the analyzed soil samples along Sitnica river and, at the same time, was the element with the highest concentration in all types of analyzed samples. The lowest concentration (13,216.97 mg kg^{-1}) was detected in sample S18, which was within 10 km of the source of pollution, while the highest concentration (20,399 mg kg^{-1}) was detected in sample S28. Nickel is most commonly found in soil and plants absorb nickel quickly and easily [48]. An average value of 25 mg kg^{-1} for nickel has been reported in the region surrounding Tunçbilek thermal power plant in Turkey, which is lower than the average value (95.04 mg kg^{-1}) found in samples around the KTPPs [49]. The measured lead content ranged from 84.20 mg kg^{-1} (sample S18) to 667.97 mg kg $^{-1}$ (sample S29), thus ranging from the target value's margin (85 mg kg^{-1}) to above the intervention value (530 mg kg^{-1}) [50]. This content was higher than that reported in records for the Anadrinia region, which ranged from 16.27 mg kg^{-1} to 42.58 mg kg^{-1} [45]. In terms of distance from the source of contamination, there were also differences in lead concentration between sampling points, with the highest concentration at sample point S29, located north of the KTPPs. Zinc, which is toxic to plants, occurs in soil mainly due to industrial contamination. The mean content in soil samples taken along the Sitnica river was more than four times higher, compared with data from the U.S., where the zinc concentration in agricultural soil ranges from 1.5 to 264 mg kg^{-1} [51]. The highest concentration of zinc (948.71 mg kg^{-1})

in the analyzed soil samples was detected in the same sample with the highest level of cadmium (sample S10), whereas the lowest concentration was observed in sample S27. Based on the average concentrations of the selected elements in soil samples taken along the Sitnica river, the following order was observed: Fe > Al > Zn > Pb > Ni > Cu > As > Cr > Cd.

3.2. Selected Elements (Metals/Metalloids) in Willow Bark Samples of *Salix alba*

Different types of plants accumulate different minerals and trace elements, based on the environmental state, type of element, and plant potential for heavy metals/metalloids [52]. Table 3 shows the concentrations of selected elements in willow bark samples obtained in the KTPP-polluted area along the Sitnica river, with a summary of mean, maximum, and minimum concentrations of the analyzed elements.

Table 3. Mean, minimum, and maximum contents (mg kg^{-1}) of analyzed metals/metalloids in *Salix alba* bark samples.

	Al	Ca	Cd	Cr	Cu	Fe	Mg	Mn	Ni	Zn
Mean	371.19	12,191.66	0.62	1.28	9.1	385.47	1306	123.13	7.69	87.77
Minimum	67.79	5260	0.09	0.89	5.09	66.79	840	19.68	6.49	56.39
Maximum	894.77	22,280	4.49	1.89	28.66	910.75	1680	392.75	10.09	140.94
Count	30	30	30	30	30	30	30	30	30	30

All 10 of the selected mineral and trace elements—aluminum, calcium, cadmium, chromium, copper, iron, magnesium, manganese, nickel, and zinc—were found in the *Salix alba* bark samples, while arsenic, lead, and molybdenum contents were below the detection limits for the analysis methods used.

It has been shown that aluminum is more soluble and that its bioavailability increases in acidic soil and water (pH < 5.5), while it tends to be biologically inactive in neutral alkaline media (pH 5.5–8.0); however, its solubility increases in a strongly alkaline medium (pH > 8.0), but its bioavailability in plants is not known [53]. Aluminum was found in all plant samples, ranging from 67.79 mg kg^{-1} in sample P29 (P: plant) to 894.77 mg kg^{-1} in sample P17; standard recorded values range from 10,000 to 300,000 mg kg^{-1} [42]. Calcium was detected in relatively high quantities, ranging from 5260 mg kg^{-1} to 22,280 mg kg^{-1}, in all plant samples tested [53]. The presence of calcium in high amounts restricts plant communities on calcareous soils [54]. The highest concentration of calcium was detected in sample P10 of willow bark, taken on the southern side of the KTPPs, while the lowest concentration of calcium was detected in sample P27, taken north of the KTPPs along the Sitnica river [55]. It has been shown that the genus *Salix* can accumulate large amounts of toxic elements, especially cadmium [56]. The cadmium content varied from 0.09 mg kg^{-1} to 4.49 mg kg^{-1}, with maximum detected concentrations similar to those reported by Vandecasteele et al. (2002) [57] (1.3–3.6 mg kg^{-1}), and significantly higher than the WHO-recommended level (<0.02 mg kg^{-1}) [46]. Cd affects the biochemical, physical, and genetic levels in plants, and is highly toxic even at low concentrations, but it has an affinity for concentrating in plants and marine species [58]. Interest in the study of chromium has recently increased, mainly due to the importance of chromium as an essential trace element. The soluble content of chromium in the soil controls the amount of chromium in plants. In plants, the concentration of chromium varied from 0.006 to 18 mg kg^{-1} d.w. and in the analyzed samples, the level of chromium ranged from 0.85 mg kg^{-1} (in sample P27) to 1.89 mg kg^{-1} (in sample P17), and was mostly (in 20 willow samples) within the WHO recommendation for plants (1.30 mg kg^{-1}) [46]; namely, the mean content of chromium (1.28 mg kg^{-1}) was within the suggested concentration. The copper concentration ranged from 5.09 mg kg^{-1} (sample P12) to 28.66 mg kg^{-1} (sample P7), the latter being higher than the WHO-recommended acceptable amount (10 mg kg^{-1}) in medicinal plants. Copper is an important micronutrient, for which the level of critical deficiency ranges from 1 to 5 mg kg^{-1}, the suitable level is 6–12 mg kg^{-1}, and the toxic level is above 20–30 mg kg^{-1} dry weight [59]. In this study, the mean content of copper (9.1 mg kg^{-1}) was within the range of suitable values for plants. The iron level in the willow bark samples ranged from 66.79 mg kg^{-1} (in sample P29) to 910.75 mg kg^{-1} (in sample P10), with an average concentration of 385.47 mg kg^{-1}. These findings were higher than the WHO-recommended iron concentration (20 mg kg^{-1}) in medicinal plants. The concentration of magnesium in plant samples ranged from 840 mg kg^{-1} (in the starting point sample, P1) to 1680 mg kg^{-1} (in sample P7). The values of Mg in willow bark, with an average concentration of 1360 mg kg^{-1}, were higher than the acceptable level (200 mg g^{-1}) for plants, according to the WHO report (1996) [46]. Exposure to magnesium for long periods may inhibit the accumulation of large amounts of ammonium in plants, so the Mg/Mn ratio can also be taken as an indicator of manganese toxicity in plants [60]. The manganese content in willow bark samples varied between 19.68 mg kg^{-1} (in sample P13, the entering point near the KTPPs) and 392.75 mg kg^{-1} (from the ending sample point, P30), which was above the acceptable level of manganese in medicinal plants recommended by the WHO (200 mg kg^{-1}). The mean content of manganese (123.13 mg kg^{-1}) was within the allowable range. The nickel content in willow bark samples varied from 6.49 mg kg^{-1} (in sample P27) to 10.09 mg kg^{-1} (in sample P11), which was very close to the highest allowed concentration (10 mg/kg) recommended by the WHO [61]. The acceptable concentration of nickel in plant tissues ranged between 0.5 and 5 mg kg^{-1}, and the mean content observed in willow bark samples (7.69 mg kg^{-1}) exceeded this limit [62]. The zinc content in all analyzed willow bark samples was higher than the recommended content in medicinal plants (50 mg kg^{-1}), ranging from 56.39 mg kg^{-1} (sample P22) to 140.94 mg kg^{-1} (P11). Sample P11, where the maximum content of zinc was detected, was collected very close to the KTPPs. The special motility system of zinc possessed by plants allows for its absorption from the soil, while a high content of zinc limits the growth of roots [63,64]. The data indicate that the concentrations of the selected elements in willow bark were in the following order (mean results in mg kg^{-1}): Ca > Mg > Fe > Al > Mn > Zn > Cu > Ni > Cr > Cd.

3.3. Phenolic Content and Antioxidant Properties of *Salix alba*

Plants are potential sources of antioxidants, such as tannins, terpenoids, saponins, flavonoids, phenols, ascorbic acid, and many more compounds having the potential to scavenge free radicals [65]. In Table 4, the values obtained at all the sampling points, in terms of the minimum, maximum, and mean values for the total phenols, total flavonoids, and chlorophylls and carotenoids, as well as mean values for CUPRAC, FRAP, and DPPH, are presented.

Table 4. Summary of results regarding total phenols, total flavonoids, chlorophylls and carotenoids, and antioxidant activities (CUPRAC, FRAP, and DPPH) of *Salix alba* bark.

Variable	Observations	Minimum	Maximum	Mean	Std. Deviation
TP	30	12.633	80.537	39.282	17.953
TF	30	9.252	67.479	28.222	15.333
CHLa	30	13.733	86.287	38.099	18.496
CHLb	30	26.068	71.504	49.240	10.564
CAR	30	42.264	152.585	94.976	27.190
CUPRAC	30	92.927	454.651	228.537	100.480
FRAP	30	89.616	504.368	317.189	116.852
DPPH	30	31.280	164.263	82.742	39.155

Total phenolic content was determined using the Folin–Ciocalteu method and are expressed as milligrams of gallic acid equivalent (GAE) per gram of dry weight. The range of total phenolics was from 12.63 to 78.53 mg GAE g^{-1} DW (Figure 3).

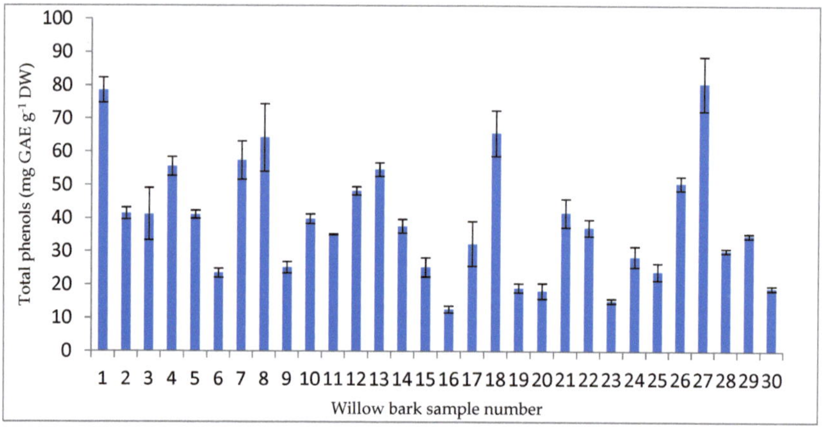

Figure 3. Total phenols in willow bark samples (mg GAE g^{-1} DW).

The highest total phenolic content (80.537 mg GAE g^{-1} DW) was determined at sample point P27, while the lowest content was determined at sample point P16 (12.633 mg GAE g^{-1} DW). The mean value of total phenolics was 39.282 mg GAE g^{-1} DW, higher than the values reported by Tahirović and Bašić (2017) in *Salix alba* leaves (8.75 ± 0.006 mg GAE g^{-1}) and *Salix alba* stem (7.02 ± 0.007 mg GAE g^{-1}) [66].

It has been estimated that willow bark contains up to 20% of condensed tannins and flavonoids [67]. In willow bark samples, the flavonoids are expressed as milligrams of catechin equivalent (CE) per gram of dry weight (mg CE g^{-1} DW), and the results ranged from 9.25 mg CE g^{-1} DW (sample point P16) to 67.47 mg CE g^{-1} DW (sample point P1). The lowest result for phenolics and flavonoids in willow bark was detected at the same sample point (P16), which was the sampling site nearest to the KTPPs (Figure 4).

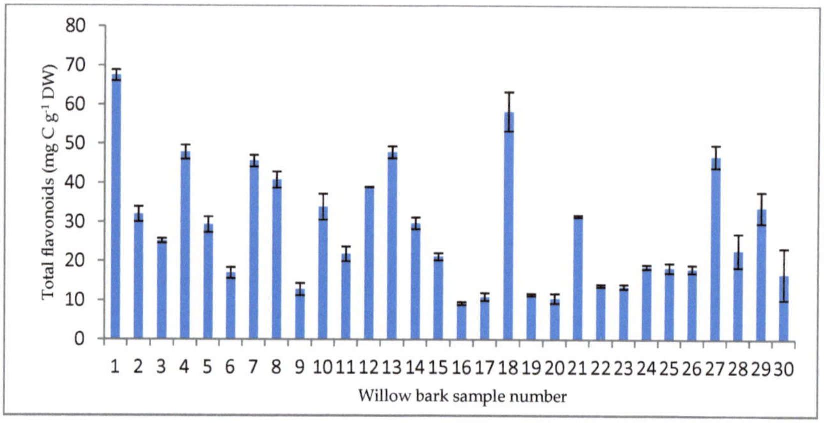

Figure 4. Total flavonoids in willow bark samples (mg CE g^{-1} DW).

The concentrations of CHLa, CHLb, and CARs are presented as mg chlorophyll/carotenoid per gram of dry weight (mg g^{-1} DW). CHLa concentrations were in the range 13.73–86.28 mg g^{-1} DW (sample points P15 and P21, respectively), CHLb concentrations were in the range 26.06–71.50 mg g^{-1} DW (sample points P9 and P21, respectively), and carotenoids were in the range 42.26–152.58 mg g^{-1} DW (sample points P9 and P1, respectively); see Figure 5.

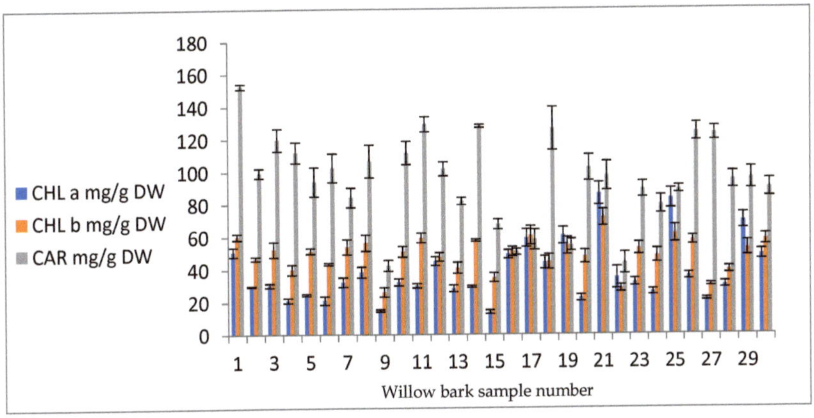

Figure 5. CHLa, CHLb, and carotenoids (CARs) in willow bark samples (mg g^{-1} DW).

3.4. *Antioxidant Activities of Salix alba*

At present, no method can be considered reliable for determining antioxidant activity/capacity; therefore, a series of analyses is typically needed to assess the antioxidant potential. The antioxidant activity/capacity of the willow bark extracts were investigated using three methods: CUPRAC, FRAP, and DPPH. The CUPRAC method is an uncomplicated and flexible antioxidant capacity assay that can be used to measure phenolic acids, flavonoids, carotenoids, and other polyphenols [28]. It is a more favorable method due to the fact that, compared to other electron transfer methods, the working pH is physiological (pH = 7), compared to the acidic pH used in the FRAP method and the alkaline pH used in the Folin method [68]. The results of CUPRAC are expressed as μmol TE g^{-1} DW, and ranged from 92.927 μmol TE g^{-1} DW (sample point P16) to 454.651 μmol TE g^{-1} DW (sample point P27), as presented in Figure 6.

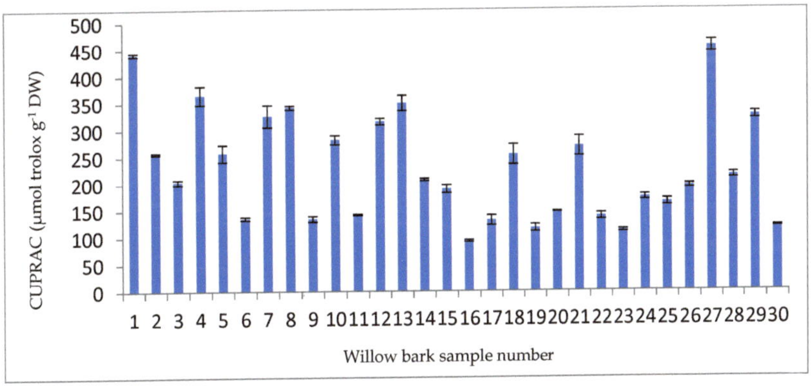

Figure 6. CUPRAC results for willow bark samples (μmol TE g^{-1} DW).

The FRAP assay is an uncomplicated automated test that measures the ferric-reducing potential of plasma, and has been presented as a new method for assessing antioxidant power. This assay measures an antioxidant's ability to reduce ferric iron. At a low pH, a complex of ferric iron and 2,3,5-triphenyl-1,3,4-triaza-2-azoniacyclopenta-1,4-diene chloride (TPTZ) is reduced to the ferrous form. The FRAP results are presented as micromoles of Fe^{2+} equivalent per gram of dry weight (μM Fe^{2+} g^{-1} DW) and, as presented in Figure 7, ranged from 89.61 μmol Fe^{2+} g^{-1} DW (sample P16) to 504.36 μmol Fe^{2+} g^{-1} DW (sample P18).

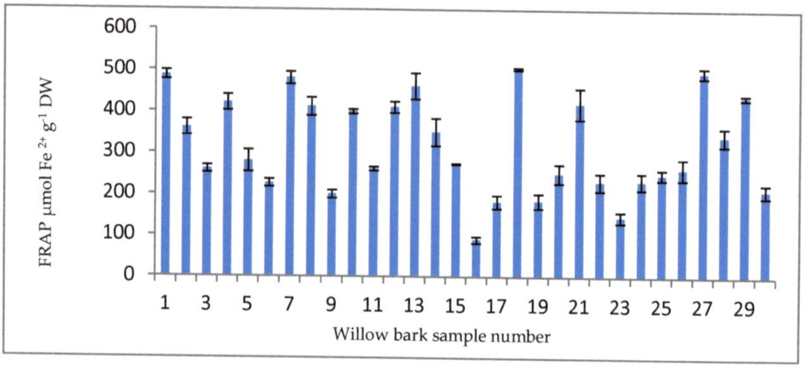

Figure 7. FRAP results for willow bark samples (µmol Fe^{2+} g^{-1} DW).

The DPPH method spectrophotometrically measures the capacity of antioxidants to reduce 2,2-diphenyl-1-picrylhydrazyl (DPPH), an additional uncommon radical determined in biological systems. The final results are expressed in terms of µmol of Trolox equivalent (TE) per gram of dry sample. The DPPH scavenging activity, as presented in Figure 8, ranged from 31.28 µmol TE g^{-1} DW (sample P16) to 164.26 µmol TE g^{-1} DW (sample P1).

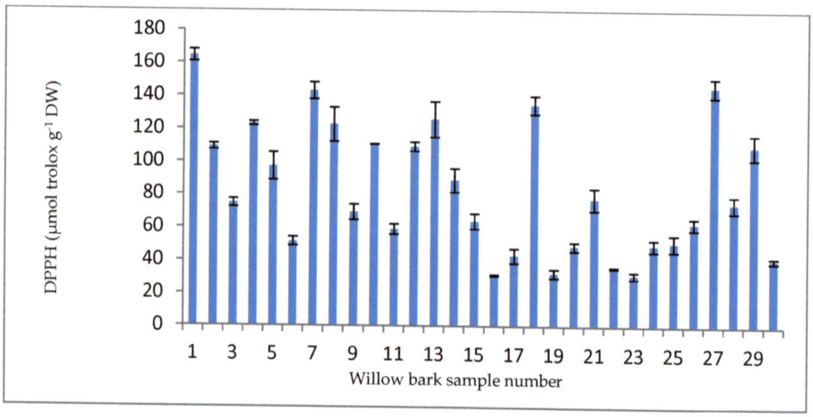

Figure 8. DPPH results for willow bark samples (µmol TE g^{-1} DW).

Compared with the mean values presented by Tahirović and Bašić (2017), the mean value in our willow bark samples (82.742 µmol Trolox g^{-1} DW) was higher than in leaves (31.25 ± 0.19 µmol TE g^{-1}) and *Salix alba* stem (30.13 ± 0.21 µmol TE g^{-1} DW). Furthermore, the mean value of DPPH was lower than that reported by Sulaiman et al. (2013) for *Salix alba* bark [69].

The antioxidant activity decreased in the following order: FRAP > CUPRAC > DPPH. The lowest results were detected mostly in samples taken from the area very close to the KTPPs (willow bark sample P16 for TP, TF, CUPRAC, FRAP, and DPPH, and CHLa in sample P15), and at sample point P9 (for CHLb and CAR). Meanwhile, the highest results were detected at sample P1 (TF, CAR, and DPPH), sample P18 (FRAP), sample P21 (CHLa and CHLb), and sample P27 (TP and CUPRAC).

3.5. Correlation Analysis between Selected Elements in Soil and Salix alba Bark

Pearson correlation analysis indicated moderate correlations between copper in water with nickel in soil (r = 0.382), as well as nickel in water to arsenic (r = 0.400), cadmium (r = 0.439), and chromium (r = 0.392) in soil. Regarding the correlations between the analyzed elements in soil and *Salix alba* bark, aluminum showed a moderate correlation with calcium in willow bark samples (r = 0.436); arsenic in soil with cadmium in willow bark (r = 0.653); cadmium in soil with its content in willow bark (r = 0.561); chromium in soil showed moderate correlations with copper and manganese in willow bark (r = 0.343 and r = 0.385, respectively); iron in soil with calcium (r = 0.513), magnesium (r = 0.449), and manganese (r = 0.438) in willow bark; nickel in soil showed moderate positive correlations with zinc (r = 0.558) and nickel (r = 0.340) in willow bark; and zinc in soil with cadmium in willow bark (r = 0.642).

The Pearson correlation matrix, provided in Table 5, shows significant correlations between most of the metals/metalloids determined in willow bark. Note the strong positive correlations between aluminum and chromium (r = 0.893), nickel (r = 0.883), and iron (r = 0.880); chromium and nickel (r = 0.801) and iron (r = 0.793); and the moderate negative correlation between calcium and manganese (r = −0.398).

Table 5. Correlations between selected elements in willow bark.

	Al	Ca	Cd	Cr	Cu	Fe	Mg	Mn	Ni	Zn
Al	1									
Ca	0.316	1								
Cd	-	0.458	1							
Cr	0.893	0.321	-	1						
Cu	-	-	-	-	1					
Fe	0.880	0.444	0.392	0.793	0.145	1				
Mg	-	0.564	-	-	0.307	0.313	1			
Mn	-	−0.398	-	-	-	-	-	1		
Ni	0.883	-	-	0.801	-	-	-	-	1	
Zn	-	-	-	-	-	-	-	-	0.327	1

The values indicate significance at $p < 0.05$.

3.6. *Correlation Analysis between Selected Elements, Production of Photosynthetic Pigments and Phenolic Compounds, and Antioxidant Activities in Salix alba Bark Samples*

The Pearson correlation matrix revealed significant positive correlations between Al, Cr, Fe, and Ni contents, as well between Cd, Fe, and Ca contents. Furthermore, Ca content was significantly positively correlated with Mg content, but negatively correlated with Mn content. With respect to the phenolic compounds, TP and TF contents were positively correlated with the CAR contents, as well as with the antioxidant activity assays (CUPRAC, FRAP, and DPPH). However, phenolic compounds—particularly TF—were significantly negatively correlated with Fe, Ni, and Mn contents. Regarding the photosynthetic pigments, CHLa content was negatively correlated with Fe and Ca contents. Antioxidant activity assays were significantly negatively correlated with the Ni and Mn contents. In addition, the FRAP assay was negatively correlated with Al, Cr, and Fe contents.

The data in Table 6 indicate significant positive correlations (Pearson coefficients) between TF and TP (r = 0.889); CHLb and CHLa (r = 0.665); CAR and TP, TF, and CHLb (r = 0.594, r = 617, and r = 0.332, respectively); CUPRAC and TP, TF, and CAR (r = 0.882, r = 0.902, and r = 0.540, respectively); FRAP and TP, TF, CAR, and CUPRAC (r = 0.854, r = 0.922, r = 0.577, and r = 0.910, respectively); and DPPH and TP, TF, CAR, CUPRAC, and FRAP (r = 0.886, r = 0.952, r = 0.568, r = 0.939, and r = 0.933, respectively).

Table 6. Correlation analysis between selected elements, production of photosynthetic pigments and phenolic compounds, and antioxidant activities in *Salix alba* bark samples.

r	Al	Cd	Cr	Cu	Fe	Ni	Zn	Ca	Mg	Mn	TP	TF	CHL a	CHL b	CAR	CUPRAC	FRAP
Cd	0.195																
Cr	**0.893**	0.154															
Cu	0.061	0.136	0.140														
Fe	**0.880**	**0.392**	**0.793**	0.145													
Ni	**0.883**	0.043	**0.801**	0.152	**0.790**												
Zn	0.253	−0.132	0.109	−0.210	0.149	0.327											
Ca	0.316	**0.458**	0.321	−0.018	**0.444**	0.166	0.190										
Mg	0.177	0.229	0.205	0.307	0.313	0.141	0.060	**0.564**									
Mn	0.097	−0.058	0.157	0.054	0.106	0.199	−0.276	**−0.398**	−0.048								
TP	−0.184	−0.033	−0.171	0.009	−0.293	−0.359	−0.176	0.106	−0.226	**−0.441**							
TF	−0.266	0.022	−0.327	−0.067	**−0.365**	**−0.424**	−0.081	0.177	−0.293	**−0.546**	**0.889**						
CHL a	−0.256	−0.094	−0.267	−0.096	**−0.361**	−0.287	0.012	**−0.401**	−0.339	0.274	−0.076	−0.006					
CHL b	0.074	0.086	0.176	0.045	−0.078	0.034	0.078	−0.047	−0.215	0.138	−0.053	0.008	**0.665**				
CAR	−0.070	0.081	−0.049	−0.229	−0.162	−0.125	−0.042	0.226	−0.246	−0.265	**0.594**	**0.617**	−0.058	0.332			
CUPRAC	−0.281	0.026	−0.329	−0.012	**−0.379**	**−0.432**	−0.165	0.137	−0.257	**−0.482**	**0.882**	**0.902**	−0.026	−0.059	**0.540**		
FRAP	**−0.361**	0.089	**−0.423**	0.058	**−0.411**	**−0.429**	−0.077	0.112	−0.205	**−0.512**	**0.845**	**0.922**	0.030	−0.023	**0.577**	**0.910**	
DPPH	−0.274	0.106	−0.320	0.042	−0.321	**−0.432**	−0.125	0.234	−0.156	**−0.590**	**0.886**	**0.952**	−0.111	−0.077	**0.568**	**0.939**	**0.933**

r: Pearson's coefficient; TP: total phenolic compounds; TF: total flavonoids; CHL a: chlorophyll a; CHL b: chlorophyll b; CAR: carotenoids; CUPRAC: cupric-reducing antioxidant capacity; FRAP: ferric-reducing antioxidant power; DPPH: DPPH radical scavenging activity. The values in bold indicate significance at $p < 0.05$.

4. Discussion

High concentrations of toxic elements in surface soil and groundwater, compared to normal and permissible values, are an increasing problem, considering their threat to the environment and human health. This is primarily a result of the development of many branches of chemical-based industries and the large discharges from them. Except for two of the samples, the aluminum content in water samples was higher than the level reported by Oram B (2020) [70]. The average arsenic, chromium, and copper concentration in all samples along Sitnica river were below the permissible metal limit reported by Brent L [71]. The iron levels in several sampling sites exceeded the average values reported for the presence of iron in rivers (0.5–1 mg L^{-1}) [72]. The obtained values for nickel were higher than those reported by Shala et al. (2015), detected at the source of Sitnica river (0.015–0.033 mg L^{-1}) [72]. Although lead was determined in only four samples, the values indicate that, in the samples in which it was detected, the lead concentration was higher than the value allowed by the WHO (0.05 mg L^{-1}), and the zinc content in all water samples was under the limit described by the WHO (5 mg L^{-1}) [46]. In terms of maximum detected concentrations, most of the selected elements showed the same tendency, in that the maximum values were observed in samples taken before the river approaches the KTPPs (Al, W6; Cr, W1; Fe, W6; Ni, W10; and Zn, W_2), while the maximum arsenic value was detected in a sample taken near the KTPPs (W15). Conversely, the maximum copper concentration was detected in a sample (W24) at a distance of about 10 km from the KTPPs, in the flow of the Sitnica river to the north. Of the analyzed elements, only arsenic showed a tendency toward a greater concentration in samples taken in the vicinity of (i.e., within 5 km in both directions) the source of pollution; that is, the KTPPs.

The aluminum concentrations in all soil samples collected along the Sitnica river were within the typically reported values of Lindsay (1979) and Dragun (1988) [43,44]. The arsenic concentration in most samples was above the values reported in the Anadrinia region in Kosovo [45]. Cadmium, as one of the greatest ecotoxic metals, showed content in the analyzed samples that was higher than intervention values of the Dutch list (12 mg kg^{-1}) [50]. The chromium content in all samples was higher than results from samples around the Tunçbilek thermal power plant, Turkey [49], but presented lower concentrations when compared with data from the Anadrinia region [45]. The copper content in soil samples was above the average values of 20–30 mg kg $^{-1}$ [47], and mostly above the target value presented in the Dutch list (36 mg kg^{-1}) [50]. The iron content was within the permissible range of 7000–500,000 mg/kg [71]. Nickel content was above average values, compared with data from the area surrounding Tunçbilek thermal power plant, Turkey (25 mg kg^{-1}) [49]. In terms of lead content, the intervention value (Dutch list; 530 mg kg) was only exceeded at sampling point S29. The zinc mean value of 226.7 mg kg^{-1} was higher than the natural zinc concentration in soil (10–100 mg kg^{-1}) [73]. Compared with data in the FAO Soil Bulletins (Tietjen, 1975) for the tolerable total content of some elements in soils, cadmium and lead exceeded the tolerable values [74]. Arsenic and cadmium showed the same tendency, with maximum and minimum values at the same sampling sites: the maximum content was determined for both of these elements in sample S10 and the minimum value was in sample S18. Furthermore, the highest value of zinc was determined in sample S10 and, in nearby sample S9, the highest value of aluminum was observed. Copper, iron, and lead showed a similar tendency, in that the highest value was detected in samples from sampling sites in the north direction of the Sitnica river (S29, Cu and Pb; S28, Fe). In sample S18, in addition to arsenic and cadmium, minimum values for iron and lead were also detected, while aluminum, copper, and zinc showed the same tendency of lower values being presented in the northern part of the sampling points (S29, Al; S27, Cu and Zn). Of the analyzed elements, copper and nickel showed a tendency toward greater concentrations in samples taken in the area in close vicinity (i.e., within 5 km in both directions) to the source of pollution: KTPPs. In samples collected within 5–10 km from the source of pollution, arsenic, cadmium, chromium, and zinc presented higher concentrations; however, at a distance of more than 10 km from the pollution source on both sides, aluminum, iron, and lead tended to accumulate more.

The arsenic, lead, and molybdenum contents in *Salix alba* bark were below the detection limits for the methods used. The aluminum content was within the limits reported in various plants, which range up to 1000 mg kg^{-1}. A significantly high content of calcium was detected in all willow bark samples, of which the highest content (2.2%) was within the range of acceptable values reported by the WHO (1996), and the cadmium content in all willow bark samples was higher than the value recommended for plants by the WHO (<0.02 mg kg^{-1}) [46]. The chromium level in ten willow bark samples was higher than the acceptable level (1.30 mg kg^{-1}) suggested by the WHO for plants, while copper in seven willow bark samples exceeded the content suggested by the WHO (10 mg kg^{-1}). In all of the willow bark samples, the iron content exceeded the value recommended by the WHO for plants (20 mg kg^{-1}). The magnesium content in all analyzed willow bark samples also exceeded the range suggested as acceptable by the WHO for plants (200 mg kg^{-1}). The nickel content in willow bark was above the acceptable concentration for nickel in plant tissue (0.5–5 mg kg^{-1}) recommended by Salt et al. (2000) [58]. Finally, the zinc content was higher than the recommended level for plants (50 mg kg^{-1}) [46].

Considering critical heavy metal values, it was observed that levels of cadmium, iron, nickel, lead, and zinc exceeded these values in soil, while only iron exceeded the critical value reported for plants in the willow bark samples. Compared with reported average data for *Salix alba*, the copper and manganese mean content in our study area were higher than WHO-reported data (Cu: 9.1 mg kg^{-1} vs. 8.89 mg kg^{-1}; Mn: 123.13 mg kg^{-1} vs. 43.94 mg kg^{-1}), but the value was lower in the case of zinc (87.77 mg kg^{-1} vs. 94.10 mg/kg) [75].

In terms of the maximum reported content of selected elements in willow bark samples, with the exception of manganese (P30), in the area including samples P7–P17 the maximum content was observed for all other elements. In willow bark sample P10, the maximum contents of Ca, Cd, and Fe were observed; Ni and Zn were highest in P11; Cu and Mg were highest in sample P7; and Al and Cr were highest in sample P17. The lowest contents were mostly detected in samples P27 (Ca, Cr, Ni), P29 (Al, Fe), P12 (Cd, Cu), P13 (Mn), and P18 (Ca). Between toxic and essential elements, there was a moderate correlation between aluminum and calcium content (r = 0.316); strong correlations between aluminum and iron (r = 0.880), nickel (r = 0.883), and chromium (r = 0.893); and a moderate correlation between cadmium and iron (r = 0.392) in *Salix alba* bark samples. Because the metal transmission from soil to *Salix alba* bark is observed in the following order of significant importance, zinc > copper > cadmium > nickel [41], the results point toward the importance of studying and applying *Salix alba* in the process of phytoremediation. Sustainable decision-making processes regarding ecological restoration using phytoremediation processes should find the few most favorable factors (human, earth, ethical issues, etc.) [76].

The differences in mineral elements and trace composition, the production of photosynthetic pigments and phenolic compounds, and antioxidant activities between the thirty *Salix alba* bark samples were analyzed using principal component analysis (PCA) and hierarchical agglomerative cluster analysis (HACA). These statistical tools were utilized to determine the variance in mineral element and heavy metal contents (Ca, Mg, Zn, Mn, Al, Cd, Cr, Cu, Fe, and Ni), the production of photosynthetic pigments (CHLa, CHLb, and CAR), the accumulation of phenolic compounds (TP) and flavonoids (TF), and antioxidant activities (CUPRAC, FRAP and DPPH) within *Salix alba* bark samples, as well as to cluster the samples depending on the analyzed variables (see Figures 9 and 10).

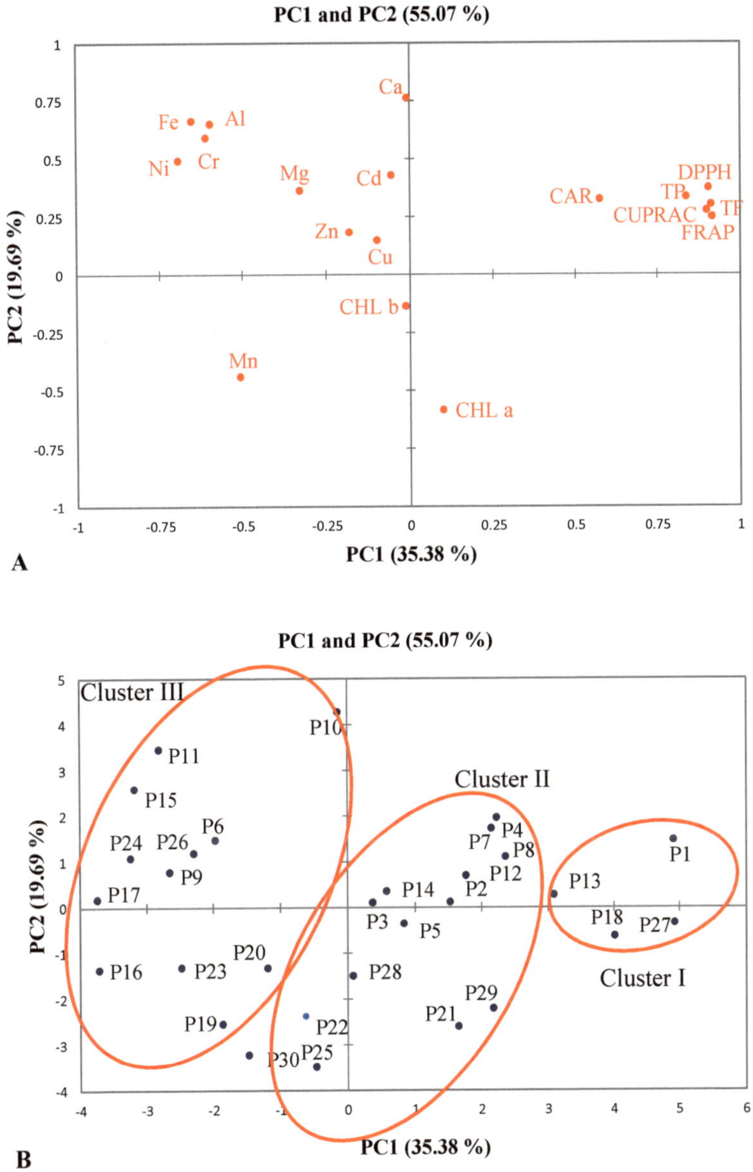

Figure 9. Loading plot (**A**) and score plot (**B**) of principal component analysis for analyzed parameters of *Salix alba* bark samples.

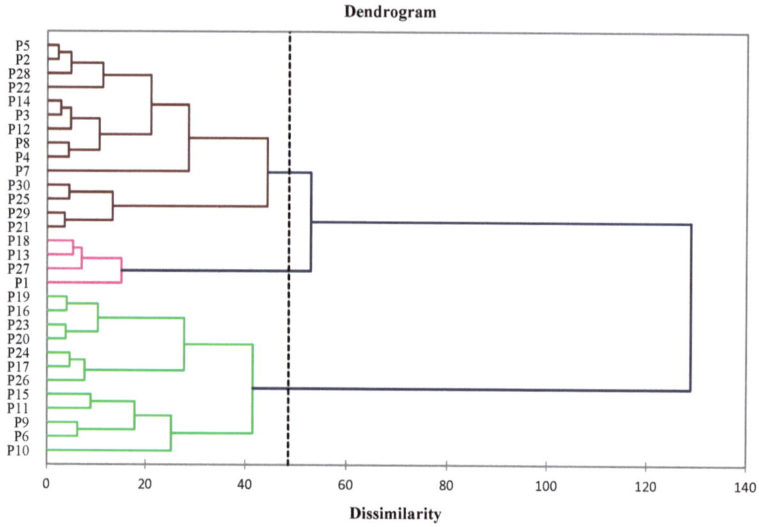

Figure 10. Cluster analysis of analyzed parameters for *Salix alba* bark samples.

Two main principal components (PCs) were used to characterize the mineral element and heavy metal composition, the production of photosynthetic pigments and phenolic compounds, and the antioxidant activities of the thirty *S. alba* samples. The PCA data revealed that PC1 and PC2 explained 55.07% of the total variation. The PCA loading plot (Figure 9A) indicates that PC1 explained 35.38% of the variance, which was connected positively to TP, TF, CUPRAC, FRAP, and DPPH, but negatively to Ni, Fe, Cr, and Al content. Meanwhile, PC2 explained 19.69% of the variance, and was positively related to Ca content.

The results for the mineral and heavy metal composition, the production of photosynthetic pigments and phenolic compounds, and the antioxidant activities showed that samples of *S. alba* bark were separated with respect to PC1 and grouped into three clusters (Figures 9B and 10).

Cluster I was represented by four samples (P_1, P_{13}, P_{18}, and P_{27}) with the largest positive scores in PC1. These samples were described as having a high production of phenolic and flavonoid compounds, as well strong antioxidant activities. In contrast, cluster III included 12 samples (P6, P9–P11, P15–P17, P19, P20, P23, P24, and P26) with the largest negative scores in PC1, which were characterized by the high accumulation of Ni, Fe, Cr, and Al. However, P10 and P11 samples were separated from this cluster, due to their high positive scores in PC2, and were therefore characterized by the increased accumulation of Ca. Cluster II was represented by 14 *S. alba* bark samples (P2–P5, P7, P8, P12, P14, P21, P22, P25, and P28–P30) with intermediate PC1 scores and which were characterized by average values for the production of phenolic and flavonoid compounds, antioxidant activities, and mineral element and heavy metal composition.

5. Conclusions

In this study, we showed that the group of chemical elements selected for analyses presented high contents in soil for the region around the KTPPs. As such, continuous and close monitoring—especially of arsenic, cadmium, copper, nickel, and lead—as well as their potential impacts on the environment and human health should be taken into account. Compared with WHO-recommended values, the mean cadmium content in *Salix alba* bark was approximately thirty times higher, while copper also exceeded the recommended values at several sampling points. Iron, magnesium, and zinc contents throughout the sampling area exceeded WHO values. The correlation between the cadmium and nickel contents in soil and their respective contents in *Salix alba* bark was demonstrated. Based on transfer factor data, it can be concluded that willow can be successfully used for phytoremediation, as it is able to extract and remove important quantities of heavy metals from soil, especially in terms of zinc, copper, cadmium, and nickel.

The results in this study indicate that the examined bark of *Salix alba* contained significant amounts of phenolic compounds and photosynthetic pigments with a strong positive correlation, proving to be sources of antioxidants with great potential. From the correlative point of view, negative moderate correlations were revealed between total phenolic compounds and iron, nickel, and manganese; photosynthetic compounds with iron and calcium; and antioxidant activity assays with nickel and manganese. In addition, the FRAP assay was negatively correlated with aluminum, chromium, and iron content in *Salix alba* bark. Further research is recommended, in order to analyze the impact of environmental pollution on the antioxidant capacity of plants, especially regarding *Salix alba* bark in regions where heavy metal/metalloid levels exceed recommended values. As already-published data have established that phenolic and flavonoid compounds are important components in drug design, isolated active components from *Salix alba* bark should be employed in future drug discovery research. Another aspect to be considered by future studies is the analysis of the uncertainty factors involved in making correct decisions to analyze the impact of heavy metals in *Salix alba* and its antioxidant potential.

Author Contributions: Conceptualization, D.B. and B.B.; methodology, D.B and L.Z.; software, D.B.; validation, D.B., B.B. and L.Z.; formal analysis, L.Z.; investigation, D.B.; resources, D.B.; data curation, D.B. and L.Z.; writing—original draft preparation, D.B.; writing—review and editing, D.B.; visualization, B.B.; supervision, B.B. and L.Z.; project administration, D.B.; funding acquisition, D.B. and L.Z. All authors have read and agreed to the published version of the manuscript.

Funding: This research received no external funding.

Institutional Review Board Statement: Not applicable.

Informed Consent Statement: Not applicable.

Data Availability Statement: Not applicable.

Conflicts of Interest: The authors declare no conflict of interest.

References

1. Tawfeek, N.; Mahmoud, M.F.; Hamdan, D.I.; Sobeh, M.; Farrag, N.; Wink, M.; El-Shazly, A.M. Phytochemistry, Pharmacology and Medicinal Uses of Plants of the Genus Salix: An Updated Review. *Front. Pharmacol.* **2021**, *12*, 593856. [CrossRef] [PubMed]
2. Barnes, J.; Anderson, L.A.; Phillipson, J.D. *Herbal Medicines*; Pharmaceutical Press: London, UK, 2007; pp. 598–600.
3. Schmid, B.; Tschirdewahn, B.; Kötter, I.; Günaydin, I.; Lüdtke, R.; Selbmann, H.; Schaffner, W.; Heide, L. Analgesic effects of willow bark extract in osteoarthritis: Results of a clinical double-blind trial. *Focus Altern. Complement. Ther.* **2010**, *3*, 186. [CrossRef]
4. Saller, R.; Melzer, J.; Felder, M. Pain relief with a proprietary extract of Willow bark in rheumatology. *Open Trial. Schweiz. Zschr. Ganzheitsmed. Medizin.* **2008**, *20*, 156–162. [CrossRef]
5. Praciak, A.; Pasiecznik, N.M.; Sheil, D.; van Heist, M.; Sassen, M.; Correia, C.S.; Dixon Ch Fyson, G.E.; Rushforth, K.; Teeling, C. *The CABI Encyclopedia of Forest Trees*; CABI: Oxfordshire, UK, 2013.
6. Dickmann, D.I.; Kuzovkina, J. Poplars and Willows in the World. In *Poplars and Willows: Trees for Society and the Environment*; Isebrands, J.G., Richardson, J., Eds.; Chapter: Poplars and Willows in the World Publisher; FAO: Rome, Italy, 2014; pp. 8–91.
7. Isebrands, J.G.; Richardson, J. *Poplars and Willows: Trees for Society and the Environment CABI*; FAO: Rome, Italy, 2014.
8. Larison, J.R.; Likens, E.; Fitzpatrick, J.W.; Crock, J.G. Cadmium toxicity among wildlife in the Colorado Rocky Mountains. *Nature* **2000**, *406*, 181–183. [CrossRef]
9. Takáč, P.; Szabová, T.; Kozáková, L.; Benková, M. Heavy metals and their bioavailability from soils in the long-term polluted Central Spis Region of SR. *Plant Soil Environ.* **2009**, *55*, 167–172. [CrossRef]
10. Casarett and Doull's. *Toxicology, the Basic Science of Poisons*, 7th ed.; McGraw-Hill: New York, NY, USA, 2008; pp. 1103–1104.
11. Tlustoš, P.; Szková, J.; Vysloužilová, M.; Pavlíková, D.; Weger, J.; Javorská, H. Variation in the uptake of Arsenic, Cadmium, Lead, and Zinc by different species of willows *Salix* spp. grown in contaminated soils. *Cent. Eur. J. Biol.* **2007**, *2*, 254–275. [CrossRef]
12. Viehweger, K. How plants cope with heavy metals. *Bot. Stud.* **2014**, *55*, 35. [CrossRef]
13. Ekmekçi, Y.; Tanyolaç, D.; Ayhan, B. Effects of cadmium on antioxidant enzyme and photosynthetic activities in leaves of two maize cultivars. *J. Plant Physiol.* **2008**, *165*, 600–611. [CrossRef]

14. Zeneli, L.; Daci-Ajvazi, M.; Daci, N.M.; Hoxha, D.; Shala, A. Environmental Pollution and Relationship Between Total Antioxidant Capacity and Heavy Metals (Pb, Cd, Zn, Mn, and Fe) in *Solanum tuberosum* L. and *Allium cepa* L. *Hum. Ecol. Risk Assess. Int. J.* **2013**, *19*, 1618–1627. [CrossRef]
15. Zeneli, L.; Daci, N.; Paçarazi, H.; Daci, A.M. Impact of Environmental Pollution on Human Health of Population Which Lives in Nearby Kosovo Thermopower Plants. *Indoor Build Environ.* **2011**, *20*, 479–482. [CrossRef]
16. Pietarila, H.; Varjoranta, R. *Dispersion of Exhaust Gases from Kosovo B Power Plant in Obilic, Kosovo*; Finnish Meteorological Institute, Air Quality Research: Helsinki, Finland, 2005.
17. Halliwell, B.; Gutteridge, J.M. The definition and measurement of antioxidants in biological systems. *Free Radic. Biol. Med.* **1995**, *18*, 125–126. [CrossRef]
18. Halliwell, B. Biochemistry of oxidative stress. *Biochem. Soc. Trans.* **2007**, *35*, 1147–1150. [CrossRef] [PubMed]
19. Justesen, U.; Knuthsen, P. Composition of flavonoids in fresh herbs and calculation of flavonoid intake by use of herbs in traditional Danish dishes. *Food Chem.* **2001**, *73*, 245–250. [CrossRef]
20. Yadav, A.; Kumari, R.; Yadav, A.; Mishra, J.P.; Srivatva, S.; Prabha, S.H. Antioxidants and its functions in human body—A Review. *Res. Environ. Life Sci.* **2016**, *9*, 1328–1331.
21. Anbudhasan, P.; Surendraraj, A.; Karkuzhali, S.; Sathishkumaran, S. Natural antioxidants and its benefits. *Int. J. Food Nutr. Sci.* **2014**, *3*, 225–232.
22. Gupta, D. Methods for Determination of Antioxidant capacity: A Review. *Int. J. Pharm. Sci. Res.* **2015**, *6*, 546–566.
23. Sin, H.P.Y.; Liu, D.T.L.; Lam, D.S.C. Life style Modification, Nutritional and Vitamins Supplements for Age-Related Macular Degeneration. *Acta Ophthalmol.* **2013**, *91*, 6–11. [CrossRef]
24. Willis, L.M.; Shukitt-Hale, B.; Joseph, J.A. Recent advances in berry supplementation and age-related cognitive decline. *Curr. Opin. Clin. Nutr. Metab. Care* **2009**, *12*, 91–94. [CrossRef]
25. Kaneria, M.; Kanani, B.; Chanda, S. Assessment of effect of hydroalcoholic and decoction methods on extraction of antioxidants from selected Indian medicinal plants. *Asian Pac. J. Trop. Biomed.* **2012**, *2*, 195–202. [CrossRef]
26. Huang, D.; Ou, B.; Prior, R.L. The Chemistry behind Antioxidant Capacity Assays. *J. Agric. Food Chem.* **2005**, *53*, 1841–1856. [CrossRef]
27. Antolovich, M.; Prenzler, P.D.; Patsalides, E.; Mc-Donald, D.; Robards, K. Methods for testing antioxidant activity. *Analyst* **2002**, *127*, 183–198. [CrossRef] [PubMed]
28. Özyürek, M.; Güçlü, K.; Tütem, E.; Başkan, K.S.; Erçağ, E.; Celik, S.E.; Baki, S.; Yıldız, L.; Karaman, S.; Apak, R. A comprehensive review of CUPRAC methodology. *Anal. Methods* **2011**, *3*, 2439–2453. [CrossRef]
29. Apak, R.; Gorinstein, S.; Böhm, V.; Schaich, K.M.; Özyürek, M.; Güçlü, K. Methods of measurement and evaluation of natural antioxidant capacity/activity (IUPAC Technical Report). *Pure Appl. Chem.* **2013**, *85*, 957–998. [CrossRef]
30. Moharram, H.A.; Youssef, M.M. Methods for Determining the Antioxidant Activity: A Review. *Alex. J. Food Sci. Technol.* **2014**, *11*, 31–42.
31. Ansari, A.Q.; Ahmed, S.A.; Waheed, M.A.; Juned, A.S. Extraction and determination of antioxidant activity of Withania somnifera Dunal. *Eur. J. Exp. Biol.* **2013**, *3*, 502–507.
32. Gadzovska, S.; Maury, S.; Delaunay, A.; Spasenoski, M.; Hagège, D.; Courtois, D.; Joseph, C. The influence of salicylic acid elicitation of shoots, callus, and cell suspension cultures on production of naphtodianthrones and phenylpropanoids in *Hypericum perforatum* L. *Plant Cell Tissue Organ Cult.* **2013**, *113*, 25–39. [CrossRef]
33. Gadzovska, S.; Maury, S.; Delaunay, A.; Spasenoski, M.; Joseph, C.; Hagege, D. Jasmonic acid elicitation of *Hypericum perforatum* L. cell suspensions and effects on the production of phenylpropanoids and naphtodianthrones. *Plant Cell Tissue Organ Cult.* **2007**, *89*, 1–13. [CrossRef]
34. Singleton, V.L.; Rossi, J.A. Colorimetry of total phenolics with phosphomolybdic-phosphotungstic acid reagents. *Am. J. Enol. Vitic.* **1965**, *16*, 144–158.
35. Zhishen, J.; Mengcheng, T.; Jianming, W. The determination of flavonoid contents in mulberry and their scavenging effects on superoxide radicals. *Food Chem.* **1999**, *64*, 555–559. [CrossRef]
36. Lichtenthaler, H.K.; Wellburn, A.R. Determinations of total carotenoids and chlorophylls a and b of leaf extracts in different solvents. *Biochem. Soc. Trans.* **1983**, *11*, 591–592. [CrossRef]
37. Benzie, I.F.; Strain, J.J. The ferric reducing ability of plasma (FRAP) as a measure of "antioxidant power": The FRAP assay. *Anal. Biochem.* **1996**, *239*, 70–76. [CrossRef] [PubMed]
38. Brand-Williams, W.; Cuvelier, M.E.; Berset, C.L.W.T. Use of a free radical method to evaluate antioxidant activity. *LWT-Food Sci. Technol.* **1995**, *28*, 25–30. [CrossRef]
39. Wang, Q.; Cui, Y.; Lin, X.M.; Dong, Y.Y.; Christie, P. Soil contamination and plant uptake of heavy metals at polluted sites in China. *J. Environ. Sci. Health A Tox. Hazard. Subst. Environ. Eng.* **2003**, *38*, 823–838. [CrossRef] [PubMed]
40. Han, S.-H.; Kim, D.-H.; Shin, S.-J. Bioaccumulation and Physiological, Response of Five Willows to Toxic Levels of Cadmium and Zinc. *Soil Sediment Contam.* **2013**, *22*, 241–255. [CrossRef]
41. Bajraktari, D.; Bauer, B.; Kavrakovski, Z.; Zeneli, L. Correlation between Environmental Pollution and Metals Accumulation in *Salix alba* L. (Fam. *Salicaceae*). *Agric. Conspec. Sci.* **2019**, *84*, 95–101.

42. Kabata-Pendias, A. *Trace Elements in Soil and Plants*, 4th ed.; Taylor & Francis Group: Boca Raton, FL, USA, 2011.
43. Lindsay, W.L. *Chemical Equilibria in Soils*; John Wiley and Sons: New York, NY, USA, 1979.
44. Dragun, J. Element adsorption and mobility in soil. In *The Soil Chemistry of Hazardous Materials*; Hazardous Materials Control Research Institute: Silver Spring, MD, USA, 1988.
45. Shehu, I.; Bajraktari, N.; Demaku, S.; Bekolli, A.; Malsiu, A. The Study of Absorption of Heavy Metals from the Soil at Some Vegetables in Anadrinia Region in Kosovo. *Int. J. Pharm. Sci. Rev. Res.* **2016**, *40*, 324–329.
46. WHO. *Permissible Limits of Heavy Metals in Soil and Plants*; WHO: Geneva, Switzerland, 1996.
47. Hooda, P.S. *Trace Elements in Soils*; Wiley: Hoboken, NJ, USA, 2010; Volume 28, p. 443.
48. Güne, A.; Alpaslan, M.; Ina, L.A. *Plant Growth and Fertilizer*; Pub. No: 1539; Univ. Agriculture Ankara: Ankara, Turkey, 2004.
49. Cicek, A.; Koparal, A.S. Accumulation of sulfur and heavy metals in soil and tree leaves sampled from the surroundings of Tuncbilek Thermal Power Plant. *Chemosphere* **2004**, *57*, 1031–1036. [CrossRef]
50. Dutch Target and Intervention Values, 2000 (the New Dutch List). Available online: https://www.esdat.net/environmental%20standards/dutch/annexs_i2000dutch%20environmental%20standards.pdf (accessed on 10 January 2022).
51. Holmgren, G.G.S.; Meyer, M.W.; Chaney, R.L.; Daniels, R.B. Cadmium, lead, zinc, copper, and in agricultural soils of the United States of America. *J. Environ. Qual.* **1993**, *22*, 335–348. [CrossRef]
52. Lokeshwari, H.; Chandrappa, G.T. Impact of heavy metal contamination of Bellandur Lake on soil and cultivated vegetation. *Curr. Sci.* **2006**, *91*, 622–627.
53. Sparling, D.W.; Lowe, T.P. *Environmental Hazards of Aluminum to Plants, Invertebrates, Fish and Wildlife. Reviews of Environmental Contamination and Toxicology*; Springer: New York, NY, USA, 1996; Volume 145.
54. Marschner, H. *Mineral Nutrition of Higher Plants*; Academic Press: London, UK, 1995.
55. Bajraktari, D.; Petrovska, B.B.; Zeneli, L.; Dimitrovska, A.; Kavrakovski, Z. Soil chemical evaluation and power plant ash impact on chemical properties of *Salix alba* L. (Fam. *Salicaceae*): The impact of bioaccumulation. *Toxicol. Res. Appl.* **2020**, *4*, 1–8. [CrossRef]
56. Klang-Westin, E.; Eriksson, J. Potential of Salix as phytoextractor for Cd on moderately contaminated soils. *Plant Soil* **2003**, *249*, 127–137. [CrossRef]
57. Vandecasteele, B.; De Vos, B.; Tack, F.M.G. Cadmium and Zinc uptake by volunteer willow species and elder rooting in polluted dredged sediment disposal sites. *Sci. Total Environ.* **2002**, *299*, 191–205. [CrossRef]
58. Benavides, M.P.; Gallego, S.M.; Tomaro, M.L. Cadmium toxicity in plants. *Braz. J. Plant Physiol.* **2005**, *17*, 21–34. [CrossRef]
59. Morrison, R.S.; Brooks, R.D.; Reeves, R.D.; Malaise, F.; Iiorowitz, P.; Aronson, M.; Merrian, G.R. The diverse chemical forms of heavy metals in tissue extracts of metallophytes from Shaba province, Zaire. *Photochem* **1981**, *20*, 155–158. [CrossRef]
60. Le Bot, J.; Goss, M.; Carvalho, M.J.G.P.R.; Van Beusichem, M.L.; Kirkby, E.A. The significance of the magnesium to manganese ratio in plant tissues for growth and alleviation of manganese toxicity in tomato (*Lycopersicon esculentum*) and wheat (*Triticum aestivum*) plants. *Plant Soil* **1990**, *124*, 205–210. [CrossRef]
61. Dt, O.; Aa, A.; Oe, O. Heavy Metal Concentrations in Plants and Soil along Heavy Traffic Roads in North Central Nigeria. *J. Environ. Anal. Toxicol.* **2015**, *5*, 6. [CrossRef]
62. Salt, D.E.; Kato, N.; Kramer, U.; Smith, R.D.; Raskin, I. The role of root exudates in nickel hyperaccumulation and tolerance in accumulator and nonaccumulator species of Thlaspi. In *Phytoremediation of Contaminated Soil and Water*; Terry, N., Banuelos, G., Eds.; Elsevier: Amsterdam, The Netherlands, 2000; pp. 189–200.
63. Herrero, E.M.; López-Gonzálvez, A.; Ruiz, M.A.; Lucas-García, J.A.; Barbas, C. Uptake and distribution of zinc, cadmium, lead and copper in Brassica napus var. oleífera and Helianthus annus grown in contaminated soils. *Int. J. Phytoremediat.* **2003**, *3*, 153–167. [CrossRef]
64. Choi, J.M.; Pak, C.H.; Lee, C.W. Micronutrient toxicity in French marigold. *J. Plant Nutr.* **1996**, *2019*, 901–916. [CrossRef]
65. Krishnaiah, D.; Sarbatly, R.; Bono, A. Phytochemical antioxidants for health and medicine—A move towards nature. *Biotechnol. Mol. Biol.* **1996**, *1*, 97–104.
66. Tahirović, A.; Bašić, N. Determination of phenolic content and antioxidant properties of methanolic extracts from Viscum album ssp. album Beck. *Bull. Chem. Technol. Bosnia Herzeg.* **2017**, *49*, 25–30.
67. Shao, Y. Phytochemischer Atlas der Schweizer Weiden. Ph.D. Thesis, University of Zürich (CH), Zürich, Switzerland, 1991.
68. Apak, R.; Güçlü, K.; Demirata, B.; Özyürek, M.; Çelik, S.E.; Bektaşoğlu, B.; Berker, K.I.; Özyurt, D. Comparative Evaluation of Various Total Antioxidant Capacity Assays Applied to Phenolic Compounds with the CUPRAC Assay. *Molecules* **2007**, *12*, 1496–1547. [CrossRef] [PubMed]
69. Sulaiman, G.M.; Hussien, N.N.; Marzoog, T.R.; Awad, H.A. Phenolic content, Antioxidant, Antimicrobial and Cytotoxic activities of ethanol extract of *Salix alba*. *Am. J. Biochem. Biotechnol.* **2013**, *9*, 41–46. [CrossRef]
70. Oram, B. Metals in the Environment, Water Rsearch Center. 2020. Available online: https://water-research.net/index.php/metals-in-the-environment (accessed on 15 January 2022).
71. Dasa, B.H.; Swarnakar, A.K. Mobility of Heavy Metals in Soils and Wastewater from Landfill, Sarona, Raipur, Chhattisgarh, India. *J. Univ. Shanghai Sci. Technol.* **2020**, *21*, 372–381.
72. Shala, A.; Sallaku, F.; Shala, A.; Ukaj, S. The effects of industrial and agricultural activity on the water quality of the Sitnica River (Kosovo). *Geoadria* **2015**, *20*, 13–21. [CrossRef]
73. Mertens, J.; Smolders, E. *Zinc. Heavy Metals in Soils*; Environmental Pollution Book Series (EPOL); Springer: Berlin/Heidelberg, Germany, 2012; Volume 22, pp. 465–493.

74. FAO Soils Bulletins. Available online: https://www.fao.org/3/x5872e/x5872e0i.htm (accessed on 20 December 2021).
75. Popoviciu, D.R.; Ticuţa, N. Copper, manganese and zinc bioaccumulation in three common woody species from black sea coastal area. *UPB Sci. Bull. Ser. B* **2018**, *80*, 49–56.
76. Băbău, A.M.C.; Micle, V.; Damian, G.E.; Sur, I.M. Sustainable Ecological Restoration of Sterile Dumps Using Robinia pseudoacacia. *Sustainability* **2021**, *13*, 14021. [CrossRef]

horticulturae

Article

Larvicidal Activity and Phytochemical Profiling of Sweet Basil (*Ocimum basilicum* L.) Leaf Extract against Asian Tiger Mosquito (*Aedes albopictus*)

Chee Aun Chan, Lai Yee Ho and Nam Weng Sit *

Department of Allied Health Sciences, Faculty of Science, Universiti Tunku Abdul Rahman, Kampar 31900, Malaysia; zian_chan@hotmail.com (C.A.C.); holy@utar.edu.my (L.Y.H.)
* Correspondence: sitnw@utar.edu.my or sitnw06@gmail.com

Abstract: Applying larvicides to interrupt a mosquito's life cycle is an important strategy for vector control. This study was conducted to evaluate the larvicidal properties of the hexane extract of sweet basil (*Ocimum basilicum* L.; family Lamiaceae) leaves against the wild strain of Asian tiger mosquito, *Aedes albopictus* (Skuse). Third instar larvae (20 larvae/replicate, n = 3) were exposed to different concentrations of the extract (6.25–200 μg/mL), and the mortality rate was recorded. Probit analysis showed that the median lethal concentration and 95% lethal concentration of the extract were 16.0 (10.9–22.1) and 53.0 (34.6–136.8) μg/mL, respectively, after 24 h exposure. Only the fractions F3, F4, and F5 from the column chromatography displayed high mortality rates of 91.7–100% at 25.0 μg/mL after 24 h exposure. Subsequent column chromatography from the pooled fraction yielded two active subfractions, H-F345-S2 and H-F345-S3, with mortality rates of 100% and 98.3 ± 2.9%, respectively, at 12.5 μg/mL. Gas chromatography–mass spectrometry analysis unveiled that methyl chavicol, 2-(2-butoxyethoxy)ethanol, cedrelanol, methyl eugenol, 2,4,di-tert-butylphenol, and phytol were the major components in both subfractions with some of them being reported as larvicidal compounds. The results suggest that sweet basil has substantial larvicidal activity against *Ae. albopictus* mosquito and is a potential source of naturally derived larvicide.

Keywords: *Aedes albopictus*; gas-chromatography–mass spectrometry; median lethal concentration; 95% lethal concentration; *Ocimum basilicum*; probit analysis

Citation: Chan, C.A.; Ho, L.Y.; Sit, N.W. Larvicidal Activity and Phytochemical Profiling of Sweet Basil (*Ocimum basilicum* L.) Leaf Extract against Asian Tiger Mosquito (*Aedes albopictus*). *Horticulturae* **2022**, *8*, 443. https://doi.org/10.3390/horticulturae8050443

Academic Editors: Milica Aćimović and Ivan Salamon

Received: 22 April 2022
Accepted: 13 May 2022
Published: 16 May 2022

1. Introduction

The Asian tiger mosquito, *Aedes albopictus* (Skuse), which derives from the subgenus Stegomyia of the family Culicidae, is an invasive species based on its ever-increasing global distribution. For the past few decades, the *Ae. albopictus* has spread from its native range of Southeast Asia to several new regions as they have strong ecological plasticity, which allows them to adapt progressively to anthropogenic influences [1]. Due to its high disease vector potential and its substantial biting activities, *Ae. albopictus* is considered one of the important mosquito species responsible for several pandemic-prone viral diseases. Formerly known as a secondary vector, the *Ae. albopictus* mosquitoes have evolved to be the primary vectors in several parts of the world [2]. They are the known vectors to transmit dengue and chikungunya viruses, and in recent years, they have been reported to be correlated to the Zika virus outbreaks as well [3,4]. The burden of these diseases is typically highest in tropical and subtropical areas, affecting the most impoverished populations [5].

Being a naturally tree-hole mosquito, *Ae. albopictus* are able to establish themselves and survive in non-urbanized and rural localities that lack artificial containers. Although they inhabit densely vegetated rural areas, their ecological flexibility also allows them to colonize a wide range of artificial water containments and also in urban localities. In Malaysia, there were a sizeable amount of evidence which pointed to the *Ae. albopictus* becoming a dominant species over *Ae. aegypti* (Linnaeus) in certain urban areas due to the higher environmental temperature and precipitation, and adaptability to various breeding habitats [1]. The biting rates of female adult *Ae. albopictus* mosquitoes can be as high as 30 to 45 bites per hour, giving them the name "aggressive daytime biters". They are usually found biting during the early mornings and late afternoons, preferring to bite outdoors rather than indoors. Although they may feed on a variety of hosts, including humans as well as mammals, reptilians, amphibians, and birds, they prefer mammalian blood [6]. Their anthropophilic behavior, frequent biting, and aptitude to create habitats near human dwellings, have led the *Aedes* mosquitoes to be the most effective vectors for arboviruses [5].

In the year 2020, Kinta district of Perak was declared a dengue epidemic area. Out of the 3226 reported dengue cases, Kinta topped the list with 968 cases. Kampar district came in fourth, with 298 reported cases [7]. The constant usage of insecticides has led to the development of resistance among the *Aedes* population in Kampar [8,9].

Rising public health concerns over the mosquito population and the arboviral diseases transmitted by them have intensified the need for adequate mosquito control strategies. The use of synthetic insecticides is usually recommended during outbreaks of any vector-borne diseases in order to reduce the vector population as quickly as possible [10]. Nonetheless, the use of plant-based products that have insecticidal activity has intensified in an effort to combat the development of insecticide resistance in disease-related insect vectors [2]. Until today, various kinds of plant-based products have been identified to possess potential insecticidal activities, with pyrethrum, neem, rotenone, and essential oils being considered as the four major types to control insects [11,12]. In contrast to synthetic insecticides, botanicals and botanical components with potential insecticidal activity have the features of low toxicity to mammals and easy biodegradability, causing less or no harm to beneficial insects, yet still show promising effects that are able to control the targeted insects [13,14].

Sweet basil, *Ocimum basilicum* (Linnaeus), is the most common species in the Lamiaceae family. There are about 50 to 150 members being reported under the *Ocimum* genus based on variation in morphological characteristics such as growing habit, leaf and flower color, size and shape, and aromatic composition [15]. It is the largest among the Lamiaceae family, with many of them considered to be rich sources of essential oils as they have been known to express various biological activities [16]. Other significant representatives of the *Ocimum* genus include *O. americanum* L. (syn. *O. hispidulum* Schum.) (hoary basil), *O. tenuiflorum* L. (syn. *O. sanctum* L.) (holy basil), and *O. gratissimum* L. (syn. *O. suave* Willd.) (clove basil) [17,18].

Ocimum basilicum is incredibly aromatic, and because of its popularity, it is often referred to as the "king of the herbs". Along with its pleasant taste, it is a versatile culinary ingredient, used as a common spice especially in Italian and Southeast Asian cuisines. It is native to Asia, Africa, Central and South America but it has been widely cultivated in many countries, especially in the Mediterranean region. Similar to other aromatic plants grown for fresh herbs, basil is mainly cultivated as a short-lived annual crop, preferably in an organic production system, including hydroponics [19]. It is a green herb that can reach about 90 cm in height, displaying lanceolate leaves, which are glossy and fragrant [20]. The *O. basilicum* has several varieties that differ in the general morphological structure and texture, and in the chemical contents as well [21].

The essential oils from some *Ocimum* spp. have been shown to repel insects and exhibited larvicidal activity against houseflies, blue bottle flies, and mosquitoes [22]. In Cuba, *O. basilicum* is commonly used as insect repellent in greenhouses [23]. In two separate reports, Erler et al. [24] noted the repellency activities of the essential oils extracted from dried foliage of *O. basilicum* against female adult *Culex pipiens*, whereas Adam et al. [21] reported on the repellency properties of *O. basilicum* essential oils against *Anopheles* mosquitoes. Besides having substantial repellency properties, some reports have also noted the larvicidal effects of the *Ocimum* spp. against dengue vectors. Ghosh et al. [25] discovered that the nanoemulsion of essential oils of *O. basilicum* has shown a dose-dependent larvicidal effect against wild strain *Aedes aegypti* larvae. In another study, treatment with granulated formulations of *O. tenuiflorum* led to up to 90% mortality of the *Ae. aegypti* larvae at 6000 ppm [26].

According to Krüger et al. [27], the main components of essential oils distilled from various basil cultivars contain alcohols (linalool), oxides (1,8-cineole), phenols (eugenol, methyl eugenol, methyl isoeugenol, thymol), esters (methyl cinnamate), aldehydes (citral), and camphor. Among all of them, the 1,8-cineole, methyl cinnamate, methyl chavicol, and linalool are the major constituents that are primarily responsible for the distinct aroma of basil plants [28]. The known insecticidal compounds found in the volatile oils of some *Ocimum* spp. are methyl cinnamate, methyl chavicol, and eugenol. Other chemicals which have been reported to have insect repellent activity are thymol, carvacrol, camphor, caryophyllene oxide, cineole, limonene, and myrcene [29]. In spite of that, the chemical components and essential oil levels of *Ocimum* may vary between the different species and cultivars, and under different growing conditions. Various parameters such as genotype, cropping seasons, and geographical properties would also influence the biochemical components of the plants [17].

Increasing number of studies have reported on the toxicity of plant extracts against mosquitoes, indicating that they exhibit potent larvicidal, ovicidal, and repellent activities against these vectors of diseases [30–32]. In this present study, the larvicidal efficacy of the *O. basilicum* leaf extract and its corresponding fractions were investigated against a wild strain of *Ae. albopictus*. The results from the present study highlighted that the hexane extract of the leaves of *O. basilicum* contained many secondary metabolites which were active against the *Ae. albopictus* larvae.

2. Materials and Methods

2.1. Chemicals and Reagents

The following chemicals were purchased from the respective manufacturers: dimethyl sulfoxide (DMSO) from Fisher Scientific Ltd., Loughborough, UK; absolute ethanol, n-hexane (chromatography grade), and silica gel 60 F_{254} from Merck, Darmstadt, Germany; acetone, hexane, and methanol (analytical grade) from Prochem Chemicals Inc., High Point, NC, USA; and temephos from Vector Control Research Unit, Universiti Sains Malaysia, Penang, Malaysia.

2.2. Extraction of Plant Material

The fresh leaves of *O. basilicum* L. were harvested in January 2015 from the Agriculture Park of Universiti Tunku Abdul Rahman, Kampar, Malaysia. The plant species name was authenticated by the ethnobotanist Dr. Hean Chooi Ong, a former professor affiliated with the Faculty of Science, Universiti Malaya, Malaysia. The leaves were cleaned under running tap water to remove soil and dirt, and air-dried under shade for 24 h. The leaves (1.3 kg) were then ground into small pieces using a stainless-steel blender prior to extraction. The blended leaves were soaked in hexane for two days at room temperature and agitated at 120 rpm using an orbital shaker (Orbit LS, Labnet International Inc., Woodbridge, NJ, USA). The maceration was repeated for five cycles. The filtrate was concentrated at 40 °C to dryness using a rotary evaporator (Rotavapor R-200, Buchi Labortechnik AG, Flawil, Switzerland) and the dried extract was stored at −20 °C pending bioassay.

2.3. Mosquito Sampling and Larvae Culturing

Sixty ovitraps (H: 9.0 cm × D: 7.8 cm) were randomly set up outdoors surrounding the West Lake Garden (Kampar, Malaysia) from February 2015 to June 2015 for mosquito oviposition. The garden is a suburban area consisting of moderately populated houses and an ex-mining lake. The ovitraps were hung on the trees or bamboo plants approximately 20–30 m apart and 1 m above the ground. A hardboard paddle (H: 10.0 cm × L: 2.5 cm × W: 0.3 cm) was inserted diagonally into each ovitrap filled with dechlorinated tap water. The paddles and the water in the ovitrap were replaced every two to three days. The egg-containing paddles were brought back to the laboratory and air-dried for two days before immersing them in distilled water for hatching. The emerged larvae were fed with ground cat food (Cuties Catz, Perfect Companion (M) Sdn. Bhd., Kuala Lumpur, Malaysia) until they developed into pupae and adults. Temperature and relative humidity in the laboratory were maintained within the range of 22.0 °C to 27.7 °C and 67% to 78%, respectively [33]. The emerged adults were used for species identification [34] using a stereomicroscope (Zoom 2000, Leica Microsystems Inc., Buffalo, NY, USA). Only the third instar larvae of the species *Ae. albopictus* were used for larvicidal bioassay.

2.4. *Larvicidal Bioassay*

The larvicidal bioassay was performed according to the World Health Organization guidelines [35] with slight modifications. Each extract was dissolved in a dimethyl sulfoxide-ethanol mixture (3:2, v/v), sonicated for 5 min using an ultrasonic water bath (Elmasonic E 100 H, Elma Schmidbauer GmbH, Singen, Germany), and filtered using 0.45 μm nylon syringe filters to produce a stock solution of 60 mg/mL. Appropriate volumes of the stock solution were pipetted into round plastic containers (H: 6.0 cm × D: 9.2 cm) filled with dechlorinated tap water to produce six different concentrations (6.25, 12.5, 25, 50, 100, and 200 μg/mL) for the assay. The final volume for each container was 150 mL. Later, 20 third instar larvae of *Ae. albopictus* were collected in separate containers and after a one hour holding period they were introduced into each container and the larval mortality was observed and recorded at 24 and 48 h post-treatment. It was ensured that the DMSO-ethanol content in the assay medium was ≤1% to reduce DMSO/ethanol toxicity to the larvae. Food was not given to the larvae during the assay. The larvae were considered dead if they were unable to reach the water surface and did not move when prodded with a needle in the siphon or cervical region. A 1% DMSO-ethanol mixture was used as negative control while temephos at 1 μg/mL was used as positive control. Temephos is a synthetic, organophosphate larvicide widely used in the vector control programs against mosquitoes in Malaysia. The assay was conducted in triplicate. The percentage mortality was calculated. The morphological changes that occurred on the treated larvae were observed under the stereomicroscope and recorded.

2.5. *Column Chromatography*

Fractionation of the hexane extract was carried out using gravity column chromatography. A glass column (H: 50.0 cm × D: 3.5 cm) was packed with silica gel 60 F_{254}. Two grams of extract was dissolved completely in hexane and introduced into the column. The elution began with 600 mL each of 100% hexane, followed by different compositions of hexane-acetone mixture (9:1, 8:2, 7:3, 6:4, 5:5, 4:6, 3:7, 2:8, and 1:9 v/v), 100% acetone, and lastly with a different composition of acetone-methanol mixture (8:2, 6:4, and 5:5 v/v). The collected eluents were dried at 40 °C using the rotary evaporator. A total of 14 dried fractions (F1–F14) were obtained and subjected to the larvicidal bioassay at 12.5 and 25.0 μg/mL, as described in Section 2.4, to identify active fractions. Subsequently, the three most active fractions, F3, F4, and F5 were pooled together (1.1 g) as H-F345 was subjected to another round of gravity column chromatography using different compositions of hexane-acetone mixture (9:1, 8:2, 7.5:2.5, 7:3, 6:5:3.5, 6:4, and 5:5, v/v). Seven subfractions (H-F345-S1 to H-F345-S7) were collected from the column and together with the pooled fraction H-F345 were evaluated again for larvicidal activity at 12.5 μg/mL.

2.6. *Thin-Layer Chromatography*

Thin-layer chromatography (TLC) was performed to determine the compound profile in the active fractions F3, F4, and F5. Ten microliters of each fraction (1 mg/mL) was spotted onto a silica gel 60 F_{254} TLC plate (10 cm × 4 cm; Merck KGaA, Darmstadt, Germany) using a microcapillary tube. The TLC plate was developed using hexane-acetone mixture (8:2, v/v) as a mobile phase. The plate was then viewed at 366 nm using the TLC Visualizer (CAMAG, Muttenz, Switzerland) and the photo was captured using the software winCATS version 1.4.4.6337 (CAMAG, Muttenz, Switzerland).

2.7. Phytochemical Profiling Using Gas Chromatography–Mass Spectrometry (GC-MS)

Profiling of phytochemicals in the most active subfractions, H-F345-S2 and H-F345-S3, was performed using a gas chromatography–mass spectrometry (QP2010 Plus, Shimadzu Corporation, Kyoto, Japan). The components were separated on a BPX-5 capillary column, 30 m × 0.25 mm × 0.25 μm (SGE International Pty Ltd., Ringwood, Victoria, Australia). Helium gas (purity > 99.999%) was used as a carrier gas with a flow rate of 0.65 mL/min and a linear velocity of 29.9 cm/s. The injector port was maintained at 250 °C. The sample was prepared at 1 mg/mL in chromatography grade n-hexane and 1 μL was injected into the column using split mode with a split ratio of 10:1. The column temperature was set at 100 °C for 3 min, then ramped at a rate of 5 °C/min to 300 °C and maintained for 3 min. The total program time for each analysis was 46 min. For the mass spectrometer, the ion source and interface temperatures were maintained at 200 °C and 300 °C, respectively. Electron ionization at 70 eV was used. Each ion peak was scanned from 40 to 600 m/z. The data acquisition was recorded using the GCMS Solution software version 4.45 SP1 (Shimadzu Corporation, Kyoto, Japan). The possible identity of each component was made by comparison of the mass spectrum obtained with the NIST 14 Mass Spectral Library and Search Software (National Institute of Standards & Technology, Gaithersburg, MD, USA).

2.8. Data Analysis

The larval mortality rate was expressed as the mean ± standard deviation of three replicates. The data were subjected to probit analysis to determine median lethal concentration (LC_{50}), 95% lethal concentration (LC_{95}), upper confidence limit, lower confidence limit, and regression coefficient. The percentages of larval mortality were tested for significance using one-way analysis of variance (ANOVA), followed by Duncan's multiple range test as the post hoc test. The analysis was performed using the IBM SPSS Statistics for Windows Version 23.0 (IBM Corp, Armonk, NY, USA). The significance level was set at $\alpha = 0.05$.

3. Results

In this study, the hexane extract of the leaves of *O. basilicum* was evaluated against third instar larvae of *Ae. albopictus* using six concentrations ranging from 6.25 to 200 μg/mL for 24 and 48 h of exposure. As shown in Figure 1, the larval mortality rates increased from 3.3% at 6.25 μg/mL to 100% at 100 and 200 μg/mL after 24 h of exposure. Further exposure for another 24 h resulted in higher larval mortality rates for 12.5 μg/mL (51.7% to 66.7%), 25.0 μg/mL (68.3% to 78.3%), and 50 μg/mL (91.7% to 100%). Statistical comparisons of the larval mortality performed using one-way ANOVA revealed that there were significant differences between groups for both 24 h ($F(6, 14) = [360.205]$, $p = 0.00$) and 48 h of exposures ($F(6, 14) = [277.130]$, $p = 0.00$).

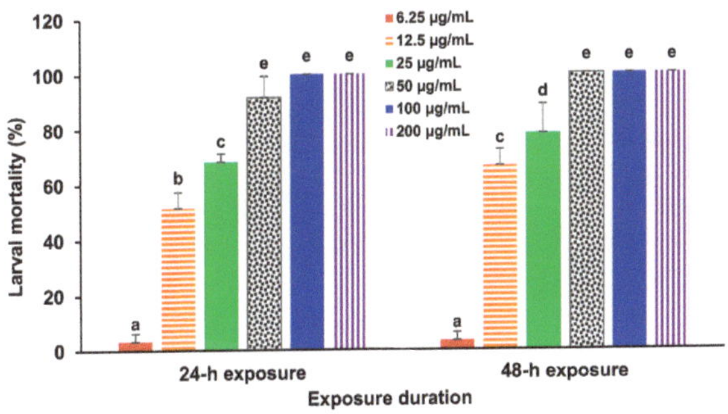

Figure 1. Percentages of mortality of third instar *Ae. albopictus* larvae treated with various concentrations of the hexane extract of *O. basilicum* leaves for 24 and 48 h exposure. The percentages are expressed as mean ± standard deviation of three replicates. Mean values which are statistically not significant ($p > 0.05$) using one-way ANOVA are indicated with the same superscript above the bars.

Probit analysis revealed that the median lethal concentration (LC_{50}) and 95% lethal concentration (LC_{95}) of the extract were 16.0 and 53.0 μg/mL, respectively, after 24 h exposure, and 12.8 and 32.7 μg/mL, respectively, after 48 h exposure (Table 1 and Figure 2). This clearly demonstrated that the larvicidal effect of a plant extract was dependent on the extract concentration and larval exposure period. In terms of morphological changes, the *Ae. albopictus* larva treated with the hexane extract of the leaves of *O. basilicum* showed an extended cephalo-thoracic junction and blackening of the midgut or abdomen compared to the healthy untreated larva (Figure 3).

Table 1. Probit analysis of the larvicidal activity of the hexane extract of *O. basilicum* leaves against third instar *Ae. albopictus* larvae after 24 and 48 h exposure periods.

Extract	24 h Exposure	48 h Exposure
Median lethal concentration (LC_{50}, μg/mL) (LCL-UCL)	15.98 (10.95–22.12)	12.82 (7.64–19.22)
95% lethal concentration (LC_{95}, μg/mL) (LCL-UCL)	53.00 (34.61–136.82)	32.66 (21.11–151.96)
Regression coefficient ± Standard error	3.159 ± 0.312	4.051 ± 0.442
Chi Square, X^2	10.292	17.151
Degree of freedom, df	4	4
p-value	0.036	0.002

LCL denotes lower confidence limit; UCL denotes upper confidence limit.

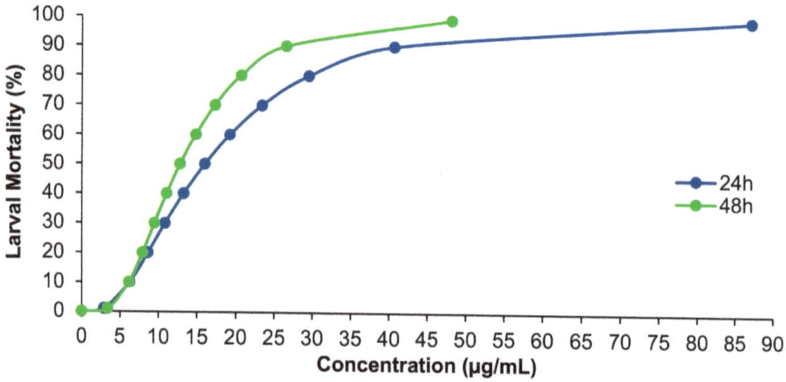

Figure 2. Probit analysis of the larvicidal activity of the hexane extract of *O. basilicum* leaves against third instar *Ae. albopictus* larvae after 24 and 48 h exposure periods.

Fourteen fractions (F1–F14) were obtained from the column chromatography and their larvicidal activity was evaluated at 12.5 and 25.0 μg/mL for 24 and 48 h. As shown in Figure 4, the larvicidal compounds were found to be mainly present in the fractions F3, F4, and F5 as their mortality rates shown were 80.0–100% at 12.5 μg/mL and 91.7–100% at 25.0 μg/mL. The fractions were eluted using the hexane-acetone mixture with the compositions of 8:2, 7:3, and 6:4 *v*/*v*, respectively. Seven fractions (F1, F7, F8, F10, F12, F13, and F14) were devoid of larvicidal activity at both concentrations. The three active fractions were pooled together based on their similar TLC profiles (Figure S1) into a single fraction (H-F345) and subjected to another round of column chromatography. Seven subfractions were obtained and tested for larvicidal activity at 12.5 μg/mL. Out of these, only the subfractions H-F345-S2, H-F345-S3, and H-F345-S4 exhibited larvicidal activity with mortality rates of 100%, 98.3 ± 2.9%, and 75.0 ± 5.0%, respectively, after 24 h of exposure. The latter two subfractions achieved a mortality rate of 100% at 48 h after exposure. These three subfractions were eluted using the hexane-acetone mixture with the compositions of 8:2, 7.5:2.5, and 7:3 *v*/*v*, respectively.

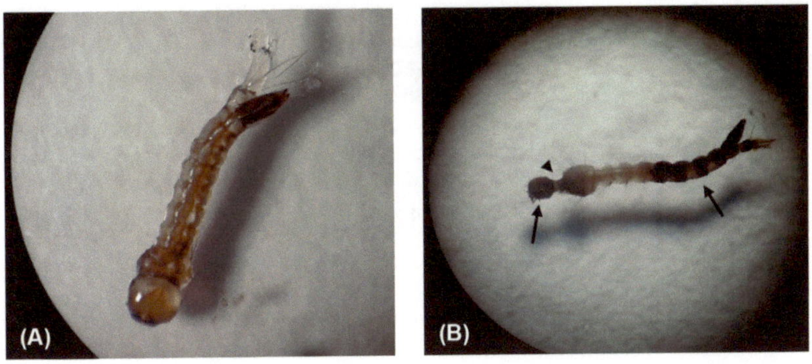

Figure 3. Morphology of third instar larvae of *Ae. albopictus* (40X magnification). (**A**) Normal, untreated larva; (**B**) Larva treated with 200 μg/mL of the hexane extract of *O. basilicum* leaves after 24 h exposure. Deformities such as blackening abdomen (arrow) and extended cephalo-thoracic junction (arrowhead) were noticed on the larva.

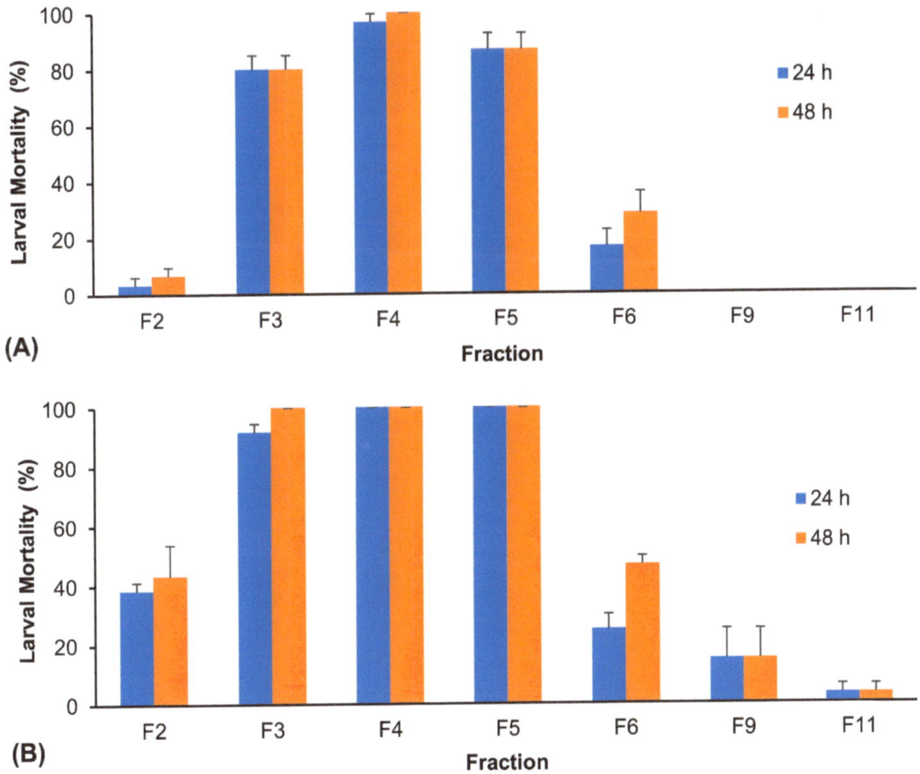

Figure 4. Percentage of mortality of third instar larvae of *Ae. albopictus* treated with fractions of the hexane extract of *O. basilicum* leaves for 24 and 48 h at 12.5 μg/mL (**A**) and 25.0 μg/mL (**B**). The fractions were eluted with different compositions of hexane and/or acetone. The percentage of mortality for each fraction is shown as mean ± standard deviation of triplicate.

In order to gain some insights into the larvicidal compounds, the two most active subfractions (H-F345-S2 and H-F345-S3) were selected for phytochemical profiling using gas chromatography–mass spectrometry (GC-MS). The subfraction H-F345-S2 contained a higher number of compounds than H-F345-S3. In total, 14 out of 31 peaks detected in the H-F345-S2 were successfully identified and they represented 91.52% of the total peak area or 88.67% of the total peak height of the chromatogram (Table 2). While for the subfraction H-F345-S3, 19 peaks were registered in the chromatogram, and nine of them were successfully identified, accounting for 95.16% of the total peak area or 92.69% of the total peak height (Table 3). The major components present in both subfractions are quite similar, i.e., methyl chavicol, 2-(2-butoxyethoxy)ethanol, cedrelanol, methyl eugenol, 2,4-di-tert-butylphenol, and phytol. The main compound, methyl chavicol constituted 57.67% and 64.34% of the peak area for H-F345-S2 and H-F345-S3, respectively.

Table 2. Phytochemical profiling of the subfraction H-F345-S2 derived from the hexane extract of the leaves of *O. basilicum* using gas chromatography–mass spectrometry.

Peak	Retention Time (min)	Peak Area (%)	Peak Height (%)	Compound	MW	Chemical Formula	Similarity (%)
1	13.811	10.92	8.86	2-(2-Butoxyethoxy)ethanol	162	$C_8H_{18}O_3$	95
2	14.544	57.67	47.12	Methyl chavicol	148	$C_{10}H_{12}O$	93
3	15.918	0.53	0.58	-			
4	16.961	0.18	0.24	-			
5	19.158	0.20	0.28	-			
6	19.316	0.45	0.63	Allyl methallyl ether	112	$C_7H_{12}O$	91
7	19.864	0.40	0.64	beta-Elemene	204	$C_{15}H_{24}$	86
8	19.971	4.35	6.10	Methyl eugenol	178	$C_{11}H_{14}O_2$	92
9	20.922	0.28	0.40	-			
10	21.019	0.49	0.66	-			
11	22.001	0.42	0.62	Humulene	204	$C_{15}H_{24}$	86
12	22.668	2.39	3.16	2,4-Di-tert-butylphenol	206	$C_{14}H_{22}O$	91
13	23.334	1.23	1.81	Germacrene D	204	$C_{15}H_{24}$	88
14	24.443	0.53	0.91	1-Dodecene	168	$C_{12}H_{24}$	90
15	24.851	0.28	0.41	-			
16	25.035	0.25	0.37	-			
17	25.176	0.94	1.37	-			
18	25.408	0.31	0.45	-			
19	25.908	0.31	0.49	-			
20	26.004	1.53	1.89	-			
21	26.589	9.92	13.54	Cedrelanol	222	$C_{15}H_{26}O$	88
22	26.951	0.46	0.63	-			
23	27.140	0.63	0.85	-			
24	27.225	0.55	0.74	-			
25	27.447	0.55	0.44	-			
26	29.068	0.22	0.41	cis-3-Tridecene	182	$C_{13}H_{26}$	88
27	31.949	0.78	1.34	Methyl palmitate	270	$C_{17}H_{34}O_2$	90
28	35.569	0.61	1.01	9,12,15-Octadecatrienal	262	$C_{18}H_{30}O$	88
29	35.669	1.63	2.52	Phytol	296	$C_{20}H_{40}O$	89
30	37.411	0.21	0.34	-			
31	40.552	0.76	1.19	-			

'-' denotes unidentified as the similarity with the NIST 14 Mass Spectral Library is <86%.

Table 3. Phytochemical profiling of the subfraction H-F345-S3 derived from the hexane extract of the leaves of *O. basilicum* using gas chromatography–mass spectrometry.

Peak	Retention Time (min)	Peak Area (%)	Peak Height (%)	Compound	MW	Chemical Formula	Similarity (%)
1	13.816	11.03	9.08	2-(2-Butoxyethoxy)ethanol	162	$C_8H_{18}O_3$	95
2	14.537	64.34	54.66	Methyl chavicol	148	$C_{10}H_{12}O$	94
3	19.867	0.33	0.49	-			
4	19.982	4.13	6.00	Methyl eugenol	178	$C_{11}H_{14}O_2$	91
5	21.038	0.42	0.57	-			
6	22.677	2.00	2.81	2,4-Di-tert-butylphenol	206	$C_{14}H_{22}O$	88
7	23.347	0.96	1.44	Germacrene D	204	$C_{15}H_{24}$	86
8	24.456	0.37	0.67	1-Dodecene	168	$C_{12}H_{24}$	88
9	25.184	0.70	1.12	-			
10	26.010	1.12	1.51	-			
11	26.589	9.30	13.13	Cedrelanol	222	$C_{15}H_{26}O$	88
12	26.958	0.28	0.43	-			
13	27.145	0.43	0.65	-			
14	27.242	0.31	0.48	-			
15	31.952	0.76	1.29	Methyl palmitate	270	$C_{17}H_{34}O_2$	87
16	35.577	0.36	0.66	-			
17	35.674	2.27	3.61	Phytol	296	$C_{20}H_{40}O$	89
18	40.557	0.61	0.95	-			
19	42.595	0.29	0.45	-			

'-' denotes unidentified as the similarity with the NIST 14 Mass Spectral Library is <86%.

4. Discussion

The hexane extract of *O. basilicum* leaves is considered to have strong larvicidal activity as its LC_{50} values were less than 50 µg/mL [36]. Our results are consistent with the findings from other reported studies. The essential oils or extracts from the leaves and/or stems of *O. basilicum* have been shown to possess larvicidal activity against several disease-carrying mosquitoes, including *Culex quinquefasciatus* [37,38], *Cx. pipiens* [39,40], *Culex tritaeniorhynchus*, *Anopheles subpictus* [41], *Anopheles stephensi* [42], *Anopheles culicifacies* [43], and *Ae. aegypti* [44,45]. The leaf essential oils of *O. basilicum* are also active against the larvae of *Ae. albopictus* with LC_{50} values of 11.97–107.7 µg/mL [41,46]. On the other hand, the methanol extract of the aerial part of *O. basilicum* showed a weak larvicidal effect against *Ae. albopictus* with an LC_{50} value of 755.13 µg/mL after 24 h of exposure [43]. Our study using the hexane extract (LC_{50} = 16.0 µg/mL) suggested that the larvicidal compounds from the leaves of *O. basilicum* are likely to be volatile or non-polar compounds.

In this study, the larvae of *Ae. albopictus* treated with the hexane extract showed deformities such as an extended cephalo-thoracic junction and blackening of the mid/hind gut. Similar deformities have also been reported when *Ae. albopictus* larvae were treated with the acetone leaf extract of *Ipomoea cairica* and the ethanol leaf and fruit extracts of *Piper nigrum* [47,48]. This suggests that the mid/hind gut of mosquito larvae is a common target organ for the larvicidal effect of plant extracts. These morphological deformities are likely caused by the interference of hormonal control or interruption of chitin synthesis from the bioactive phytochemicals in the extracts [49].

Analysis using GC-MS showed that the leaves of *O. basilicum* contained many phytochemicals or different classes of secondary metabolites. Methyl chavicol (also known as estragole) and methyl eugenol belong to the phenylpropene group of secondary metabolites. Both phenylpropenes have been reported from the leaves or aerial parts of *O. basilicum* [40,50–52]. Methyl chavicol, isolated from the essential oils of *Clausena anisata*, has been shown to exhibit larvicidal activities against *An. stephensi*, *Ae. aegypti*, and *Cx. quinquefasciatus* with LC_{50} values of 11.01–14.01 µg/mL [53]. Methyl chavicol also exhibited larvicidal activity against *Ae. albopictus* but with a much higher LC_{50} value of 58.7 µg/mL [46]. Similarly, methyl chavicol and methyl eugenol obtained from the root essential oils of *Asarum heterotropoides* possess larvicidal activity against *Ae. aegypti*, *Cx. pipiens*, and *Ochlerotatus togoi* with LC_{50}

values of 46.39–58.52 μg/mL and 53.30–67.02 μg/mL, respectively [54,55]. The LC_{50} values recorded from these studies strongly suggested that methyl chavicol was not the sole active larvicidal compound for *O. basilicum* albeit it is the major component in both active subfractions.

Cedrelanol (tau-cadinol) is a natural oxygenated sesquiterpene that has been documented in some of the *Ocimum* species, such as *O. basilicum*, *O. carnosum*, *O. gratissimum*, and *O. tenuiflorum* [56–58]. It can also be found in the essential oils of aromatic plants such as *Salvia aratocensis*, *Lippia americana*, *Melissa officinalis*, and *Thymus* spp. [59–61]. Several biological activities of cedrelanol have been reported in the literature, these include cytotoxic against human cancer cells, antibacterial, antifungal, antiparasitic, spasmolytic, and immunomodulatory activities [62–65]. Cedrelanol isolated from the bark of *Swartzia polyphylla* killed all the larvae of *Cx. quinquefasciatus* at 300 μg/mL [66]. However, the larvicidal activity of cedrelanol against *Aedes* and other species of mosquitoes remains to be investigated.

2-4-Di-tert-butylphenol is a natural lipophilic phenol and a toxin produced by many living organisms. This toxin has been documented in at least 169 species of bacteria, fungi, diatom, liverwort, pteridophytes, gymnosperms, dicots, monocots, and animals [67]. The GC-MS results of the present study revealed that this toxin is also detected for the first time from a species of *Ocimum* genus, *O. basilicum*. Wang et al. [68] studied the larvicidal effect of 2-4-di-tert-butylphenol and reported LC_{50} values of 1.98–3.90 μg/mL against the larvae of *Cx. pipiens*, *Ae. aegypti*, *Ae. albopictus*, and *Anopheles sinensis*. Our study suggested that the toxin works synergistically with other phytochemicals of the plant in exerting the killing effect on *Ae. albopictus* larvae.

Phytol (3,7,11,15-tetramethylhexadec-2-en-1ol) is a diterpene or isoprenoid alcohol with 20 carbon atoms and a double bond. It is a constituent of plant and phytoplankton chlorophyll. Besides the synthesis of chlorophyll, phytol is involved in the production of tocopherol, phylloquinol, and fatty acid phytyl ester in plants [69]. Hence, it is not surprising to find the presence of phytol in the leaves of many *Ocimum* spp., such as *O. basilicum* [50,70], *O. gratissimum* [71], *O. obovatum* [72], and *O. tenuiflorum* [73]. A comprehensive review of published literature highlighted the diverse biological activities of phytol, including antibacterial, antifungal, antiparasitic, cytotoxic, antimutagenic, antioxidant, anti-inflammatory, insulin-sensitizing, and lipid-lowering activities [74]. Despite these, further study is needed to ascertain whether phytol possesses any larvicidal activities against mosquito larvae.

2-(2-Butoxyethoxy)ethanol or diethylene glycol monobutyl ether is a colorless liquid with a low odor and high boiling point. It is an inert ingredient or a stabilizer in plant protection products, pesticides, or fertilizers, or a common solvent in many washing and cleaning household products, cosmetics, and personal care products [75]. To the authors' knowledge, this compound has not been isolated from plants. Hence, it is unlikely a bioactive product from the secondary metabolism of *O. basilicum*.

In addition to the major components of the two subfractions, some minor components are also believed to play their roles in contributing to the larvicidal activity of *O. basilicum* leaves. Germacrene D, α-humulene (α-caryophyllene), and β-elemene are sesquiterpenes that accounted for 0.40–1.23% of the subfraction H-F345-S2. Germacrene D was also detected (0.96%) in the subfraction H-F345-S3. These sesquiterpenes have also been detected and quantified from the essential oils of *O. basilicum* cultivated in Romania [56], Algeria [39], Greece [46], Serbia [51], Jordan [76], and Saudi Arabia [52]. Germacrene D is able to kill the larvae of *Ae. aegypti*, *An. stephensi*, and *Cx. quinquefasciatus* with LC_{50} values of 16.95–63.6 μg/mL [53,77]. Similarly, α-humulene possesses larvicidal activities against *Ae. aegypti*, *Ae. albopictus*, *An. subpictus*, and *Cx. tritaeniorhynchus* with LC_{50} values of 6.19–90.50 μg/mL [78,79]. β-Elemene is the major isomer of elemene, which is a naturally occurring compound in the herb *Curcuma wenyujin*. It is used as an herbal extract in traditional Chinese medicine for treating various types of cancers owing to its broad-spectrum anticancer properties [80]. The β-elemene isolated from *Syzygium zeylanicum* was found to be active against the larvae of *Ae. albopictus* with an LC_{50} value of 11.15 μg/mL [78].

5. Conclusions

The hexane extract and fractionations obtained from the leaves of *O. basilicum* have demonstrated significant larvicidal effects against wild strain *Ae. albopictus* that are both concentration and time dependent. GC-MS analysis revealed the presence of methyl chavicol, methyl eugenol, cedrelanol, and 2,4-di-tert-butylphenol as the major components in the active subfractions with reported larvicidal activities against disease-carrying mosquitoes, including *Ae. albopictus*. In addition, the larvicidal effects were also contributed by germacrene D, α-humulene, or β-elemene, which were minor components with known larvicidal property. These results are indicative of the use of *O. basilicum* leaves as a potential source of natural mosquitocidal agents in integrated vector management programs. Application of these natural larvicides alongside the reduction of usage of synthetic larvicides may serve to decrease the population of *Aedes* mosquitoes, and eventually the transmission of arboviruses to humans without the decremental effects to the environment. Further investigations using laboratory strains of both dengue vectors are necessary to substantiate the herb's efficacy.

Supplementary Materials: The following supporting information can be downloaded at: https://www.mdpi.com/article/10.3390/horticulturae8050443/s1, Figure S1: Thin-layer chromatography (TLC) profiles of the hexane fractions of the leaves of *Ocimum basilicum*.

Author Contributions: Conceptualization, L.Y.H. and N.W.S.; Data curation, C.A.C. and N.W.S.; Formal analysis, C.A.C. and N.W.S.; Funding acquisition, N.W.S.; Investigation, C.A.C.; Methodology, L.Y.H. and N.W.S.; Project administration, N.W.S.; Resources, L.Y.H. and N.W.S.; Supervision, L.Y.H. and N.W.S.; Visualization, N.W.S.; Writing—original draft, L.Y.H. and N.W.S.; All authors have read and agreed to the published version of the manuscript.

Funding: The APC was funded by Universiti Tunku Abdul Rahman (UTAR) Financial Support for Journal Publication and UTAR Research Publication Scheme (Vote: 6251/S02).

Institutional Review Board Statement: Not applicable.

Informed Consent Statement: Not applicable.

Data Availability Statement: All the data generated in this study have been provided in the main text.

Acknowledgments: The authors thank Saravanan Sivasangaran and Shin Wei Tie for their technical assistance.

Conflicts of Interest: The authors declare no conflict of interest.

References

1. Misbah, S.; Low, V.L.; Mohd Rahim, N.F.; Jaba, R.; Basari, N.; Ya'cob, Z.; Abu Bakar, S. Mitochondrial diversity of the Asian tiger mosquito *Aedes albopictus* (Diptera: Culicidae) in Peninsular Malaysia. *J. Med. Entomol.* **2022**, *59*, 865–873. [CrossRef] [PubMed]
2. Shoukat, R.F.; Shakeel, M.; Rizvi, S.A.H.; Zafar, J.; Zhang, Y.; Freed, S.; Xu, X.; Jin, F. Larvicidal, ovicidal, synergistic, and repellent activities of *Sophora alopecuroides* and its dominant constituents against *Aedes albopictus*. *Insects* **2020**, *11*, 246. [CrossRef] [PubMed]
3. Wong, P.S.; Li, M.Z.; Chong, C.S.; Ng, L.C.; Tan, C.H. *Aedes* (Stegomyia) *albopictus* (Skuse): A potential vector of Zika virus in Singapore. *PLoS Negl. Trop. Dis.* **2013**, *7*, e2348. [CrossRef] [PubMed]
4. Grard, G.; Caron, M.; Mombo, I.M.; Nkoghe, D.; Ondo, S.M.; Jiolle, D.; Fontenille, D.; Paupy, C.; Leroy, E.M. Zika virus in Gabon (Central Africa)–2007: A new threat from *Aedes albopictus*? *PLoS Negl. Trop. Dis.* **2014**, *8*, e2681. [CrossRef]
5. Kraemer, M.U.; Sinka, M.E.; Duda, K.A.; Mylne, A.Q.; Shearer, F.M.; Barker, C.M.; Moore, C.G.; Carvalho, R.G.; Coelho, G.E.; Van Bortel, W.; et al. The global distribution of the arbovirus vectors *Aedes aegypti* and *Ae. albopictus*. *eLife* **2015**, *4*, e08347. [CrossRef]
6. Kweka, E.J.; Baraka, V.; Mathias, L.; Mwang'onde, B.; Baraka, G.; Lyaruu, L.; Mahande, A.M. Ecology of *Aedes* mosquitoes, the major vectors of arboviruses in human population. In *Dengue Fever. A Resilient Threat in the Face of Innovation*; Falcón-Lezama, J.A., Betancourt-Cravioto, M., Tapia-Conyer, R., Eds.; IntechOpen: London, UK, 2019.
7. Zahratulhayat, M.A.; Kinta District Declared Dengue Epidemic Area. The Star. Available online: https://www.nst.com.my/news/nation/2020/01/553440/kinta-district-declared-dengue-epidemic-area (accessed on 7 May 2022).
8. Ho, L.Y.; Loh, T.S.; Yam, L.Y. Surveillance and resistance status of *Aedes* population in two suburban residential areas in Kampar town, Perak, Malaysia. *Trop. Biomed.* **2014**, *31*, 441–448.

9. Rajendran, D.; Najwa, F.; Ummi, A.; Besar, A.; Yusoff, M.; Zuharah, W.F. Status of insecticide resistance on *Aedes aegypti* (L.) and *Aedes albopictus* (Skuse) in Kampar, Perak, Malaysia. *Serangga* **2021**, *26*, 245–254.
10. National Pesticides Information Center. Pesticides Used in Mosquitoes Control. Available online: http://npic.orst.edu/pest/mosquito/mosqcides.html (accessed on 19 July 2019).
11. Park, Y.L.; Tak, J.H. Essential oils for arthropod pest management in agricultural production systems. In *Essential Oils in Food Preservation, Flavor and Safety*; Preedy, V.R., Ed.; Elsevier: San Diego, CA, USA, 2015; pp. 61–70.
12. Khan, B.A.; Freed, S.; Zafar, J.; Farooq, M.; Shoukat, R.F.; Ahmad, K.W.; Li, S.; Zhang, Y.; Hua, Y.; Shoukat, R.F. Efficacy of different entomopathogenic fungi on biological parameters of pulse beetle *Callosobruchus chinensis* L. (Coleoptera: Bruchidae). *J. Entomol. Zool. Stud.* **2018**, *6*, 1972–1976.
13. Mdoe, F.P.; Cheng, S.S.; Lyaruu, L.; Nkwengulila, G.; Chang, S.T.; Kweka, E.J. Larvicidal efficacy of *Cryptomeria japonica* leaf essential oils against *Anopheles gambiae*. *Parasites Vectors* **2014**, *7*, 426. [CrossRef]
14. Shoukat, R.F.; Freed, S.; Ahmad, K.W. Evaluation of binary mixtures of entomogenous fungus and botanicals on biological parameters of *Culex pipiens* (Diptera: Culicidae) under laboratory and field conditions. *Int. J. Mosq. Res.* **2016**, *3*, 17–24.
15. Makri, O.; Kintzios, S. *Ocimum* sp. (Basil): Botany, cultivation, pharmaceutical properties, and biotechnology. *J. Herbs Spices Med. Plants* **2008**, *13*, 123–150. [CrossRef]
16. Avetisyan, A.; Markosian, A.; Petrosyan, M.; Sahakyan, N.; Babayan, A.; Aloyan, S.; Trchounian, A. Chemical composition and some biological activities of the essential oils from basil *Ocimum* different cultivars. *BMC Complement Altern. Med.* **2017**, *17*, 60. [CrossRef]
17. Jirovetz, L.; Buchbauer, G.; Shafi, M.P.; Kaniampady, M.M. Chemotaxonomical analysis of the essential oil aroma compounds of four different *Ocimum* species from southern India. *Eur. Food Res. Technol.* **2003**, *217*, 120–124. [CrossRef]
18. Shahrajabian, M.H.; Sun, W.; Cheng, Q. Chemical components and pharmacological benefits of basil (*Ocimum basilicum*): A review. *Int. J. Food Prop.* **2020**, *23*, 1961–1970. [CrossRef]
19. Bączek, K.; Kosakowska, O.; Gniewosz, M.; Gientka, I.; Węglarz, Z. Sweet basil (*Ocimum basilicum* L.) productivity and raw material quality from organic cultivation. *Agronomy* **2019**, *9*, 279. [CrossRef]
20. Filip, S. Basil (*Ocimum basilicum* L.) a source of valuable phytonutrients. *Int. J. Clin. Nutr. Diet.* **2017**, *3*, 118. [CrossRef]
21. Adam, A.A.; Ahmed, A.S.; Mohamed, T.A.; Azrag, R.A.; Mustfa, S.E.; Hamdi, O.A.A. Evaluation of repellent activities of the essential oil of *Ocimum basilicum* against *Anopheles* mosquito and formulation of mosquitoes repellent cream. *Biomed. Res. Clin. Pract.* **2019**, *4*, 2–5. [CrossRef]
22. Abo-Elseoud, M.A.; Sorhan, M.M.; Omar, A.E.; Helal, M.M. Biocides formulation of essential oils having antimicrobial activity. *Arch. Phytopathol. Plant Prot.* **2005**, *38*, 175–184. [CrossRef]
23. de la Parte, E.M.; Pérez-Vicente, L.; Bernal, B.; García, D. First report of *Peronospora* sp. on sweet basil (*Ocimum basilicum*) in Cuba. *Plant Pathol.* **2010**, *59*, 800. [CrossRef]
24. Erler, F.; Ulug, I.; Yalcinkaya, B. Repellent activity of five essential oils against *Culex pipiens*. *Fitoterapia* **2006**, *77*, 491–494. [CrossRef]
25. Ghosh, V.; Mukherjee, A.; Chandrasekaran, N. Formulation and characterization of plant essential oil based nanoemulsion: Evaluation of its larvicidal activity against *Aedes aegypti*. *Asian J. Chem.* **2013**, *25*, S321–S323.
26. Ikhsanudin, A.; Lolita, L.; Ramadani, Z.S. Larvicidal activity of granulated pharmaceutical products using Indonesian holy basil leaf extract. *Int. J. Public Health* **2021**, *10*, 934–941. [CrossRef]
27. Krüger, H.; Wetzel, S.B.; Zeiger, B. The chemical variability of *Ocimum* species. *J. Herbs Spices Med. Plants* **2002**, *9*, 335–344. [CrossRef]
28. Klimankova, E.; Holadová, K.; Hajšlová, J.; Cajka, T.; Poustka, J.; Koudela, M. Aroma profile of five basil (*Ocimum basilicum* L.) cultivars grown under conventional and organic conditions. *Food Chem.* **2008**, *107*, 464–472. [CrossRef]
29. Chokechaijaroenporn, O.; Bunyapraphatsara, N.; Kongchuensin, S. Mosquito repellent activities of *Ocimum* volatile oils. *Phytomedicine* **1994**, *1*, 135–139. [CrossRef]
30. Kumar, S.; Mishra, M.; Wahab, N.; Warikoo, R. Larvicidal, repellent, and irritant potential of the seed-derived essential oil of *Apium graveolens* against dengue vector, *Aedes aegypti* L. (Diptera: Culicidae). *Front. Public Health* **2014**, *2*, 147. [CrossRef]
31. Thomas, A.; Mazigo, H.D.; Manjurano, A.; Morona, D.; Kweka, E.J. Evaluation of active ingredients and larvicidal activity of clove and cinnamon essential oils against *Anopheles gambiae* (sensu lato). *Parasites Vectors* **2017**, *10*, 411. [CrossRef]
32. Andrade-Ochoa, S.; Sánchez-Aldana, D.; Chacón-Vargas, K.F.; Rivera-Chavira, B.E.; Sánchez-Torres, L.E.; Camacho, A.D.; Nogueda-Torres, B.; Nevárez-Moorillón, G.V. Oviposition deterrent and larvicidal and pupaecidal activity of seven essential oils and their major components against *Culex quinquefasciatus* Say (Diptera: Culicidae): Synergism–antagonism effects. *Insects* **2018**, *9*, 25. [CrossRef]
33. Tuetun, B.; Choochote, W.; Kanjanapothi, D.; Rattanachanpichai, E.; Chaithong, U.; Chaiwong, P.; Jitpakdi, A.; Tippawangkosol, P.; Riyong, D.; Pitasawat, B. Repellent properties of celery, *Apium graveolens* L., compared with commercial repellents, against mosquitoes under laboratory and field conditions. *Trop. Med. Int. Health* **2005**, *10*, 1190–1198. [CrossRef]
34. Rueda, L. Pictorial Keys for the Identification of Mosquitoes (Diptera: Culicidae) Associated with Dengue Virus Transmission. Available online: http://www.mosquitocatalog.org/files/pdfs/wr385.pdf (accessed on 20 March 2015).
35. World Health Organization. Guidelines for Laboratory and Field Testing of Mosquito Larvicides. Available online: https://apps.who.int/iris/handle/10665/69101 (accessed on 19 August 2014).
36. Pavela, R. Essential oils for the development of eco-friendly mosquito larvicides: A review. *Ind. Crops Prod.* **2015**, *76*, 174–187. [CrossRef]
37. Rajamma, A.J.; Dubey, S.; Sateesha, S.B.; Tiwari, S.N.; Ghosh, S.K. Comparative larvicidal activity of different species of *Ocimum* against *Culex quinquefasciatus*. *Nat. Prod. Res.* **2011**, *25*, 1916–1922. [CrossRef] [PubMed]

38. Manzoor, F.; Samreen, K.B.; Parveen, Z. Larvicidal activity of essential oils against *Aedes aegypti* and *Culex quinquefasciatus* larvae (Diptera: Culicidae). *J Anim. Plant Sci.* **2013**, *23*, 420–424.
39. Dris, D.; Tine-Djebbar, F.; Bouabida, H.; Soltani, N. Chemical composition and activity of an *Ocimum basilicum* essential oil on *Culex pipiens* larvae: Toxicological, biometrical and biochemical aspects. *S. Afr. J. Bot.* **2017**, *113*, 362–369. [CrossRef]
40. Hamad, Y.K.; Abobakr, Y.; Salem, M.Z.M.; Ali, H.M.; Al-Sarar, A.S.; Al-Zabib, A.A. Activity of plant extracts/essential oils against three plant pathogenic fungi and mosquito larvae: GC/MS analysis of bioactive compounds. *BioResources* **2019**, *14*, 4489–4511. [CrossRef]
41. Govindarajan, M.; Sivakumar, R.; Rajeswary, M.; Yogalakshmi, K. Chemical composition and larvicidal activity of essential oil from *Ocimum basilicum* (L.) against *Culex tritaeniorhynchus*, *Aedes albopictus* and *Anopheles subpictus* (Diptera: Culicidae). *Exp. Parasitol.* **2013**, *134*, 7–11. [CrossRef]
42. Murugan, K.; Aarthi, N.; Kovendan, K.; Pannerselvam, C.; Chandramohan, B.; Kumar, P.M.; Amerasan, D.; Paulpandi, M.; Chandirasekar, R.; Dinesh, D.; et al. Mosquitocidal and antiplasmodial activity of *Senna occidentalis* (Cassiae) and *Ocimum basilicum* (Lamiaceae) from Maruthamalai hills against *Anopheles stephensi* and *Plasmodium falciparum*. *Parasitol. Res.* **2015**, *114*, 3657–3664. [CrossRef]
43. Pandey, A.; Pandey, U.K.; Shrivastava, P.K. Studies on mosquito larvicidal efficacy of indigenous plant: Extracts. *J. Ecophysiol. Occup. Health* **2010**, *10*, 223–227.
44. Murugan, K.; Murugan, P.; Noortheen, A. Larvicidal and repellent potential of *Albizzia amara* Boivin and *Ocimum basilicum* Linn against dengue vector, *Aedes aegypti* (Insecta:Diptera:Culicidae). *Bioresour. Technol.* **2007**, *98*, 198–201. [CrossRef]
45. Nasir, S.; Batool, M.; Hussain, S.M.; Nasir, I.; Hafeez, F.; Debboun, M. Bioactivity of oils from medicinal plants against immature stages of dengue mosquito *Aedes aegypti* (Diptera: Culicidae). *Int. J. Agric. Biol.* **2015**, *17*, 843–847. [CrossRef]
46. Giatropoulos, A.; Kimbaris, A.; Michaelakis, A.; Papachristos, D.P.; Polissiou, M.G.; Emmanouel, N. Chemical composition and assessment of larvicidal and repellent capacity of 14 Lamiaceae essential oils against *Aedes albopictus*. *Parasitol. Res.* **2018**, *117*, 1953–1964. [CrossRef]
47. Ahbirami, R.; Zuharah, W.F.; Thiagaletchumi, M.; Subramaniam, S.; Sundarasekar, J. Larvicidal efficacy of different plant parts of railway creeper, *Ipomoea cairica* extract against dengue vector mosquitoes, *Aedes albopictus* (Diptera: Culicidae) and *Aedes aegypti* (Diptera: Culicidae). *J. Insect Sci.* **2014**, *14*, 180. [CrossRef]
48. Aziz, E.I.A.; Rahim, N.A.A.; Raduan, S.Z.; Safii, R. A preliminary study on larvicidal efficacy of *Piper nigrum* L. (Piperaceae) extracts against dengue vector, *Aedes albopictus* (Diptera: Culicidae). *Serangga* **2021**, *26*, 80–94.
49. Soonwera, M.; Phasomkusolsil, S. Effect of *Cymbopogon citratus* (lemongrass) and *Syzygium aromaticum* (clove) oils on the morphology and mortality of *Aedes aegypti* and *Anopheles dirus* larvae. *Parasitol. Res.* **2016**, *115*, 1691–1703. [CrossRef]
50. Hassanpouraghdam, M.B.; Hassani, A.; Vojodi, L.; Farsad-Akhtar, N. Drying method affects essential oil content and composition of basil (*Ocimum basilicum* L.). *J. Essent. Oil-Bear. Plants* **2010**, *13*, 759–766. [CrossRef]
51. Ilić, A.S.; Antić, M.P.; Jelačić, S.C.; Šolević Knudsen, T.M. Chemical composition of the essential oils of three *Ocimum basilicum* L. cultivars from Serbia. *Not. Bot. Horti Agrobot. Cluj Napoca* **2019**, *47*, 347–351. [CrossRef]
52. Anwar, F.; Alkharfy, K.M.; Mehmood, T.; Bakht, M.A.; Najeeb-ur-Rehman. Variation in chemical composition and effective antibacterial potential of *Ocimum basilicum* L. essential oil harvested from different regions of Saudi Arabia. *Pharm. Chem. J.* **2021**, *55*, 187–193. [CrossRef]
53. Govindarajan, M. Chemical composition and larvicidal activity of leaf essential oil from *Clausena anisata* (Willd.) Hook. f. ex Benth (Rutaceae) against three mosquito species. *Asian Pac. J. Trop. Med.* **2010**, *3*, 874–877. [CrossRef]
54. Perumalsamy, H.; Kim, N.J.; Ahn, Y.J. Larvicidal activity of compounds isolated from *Asarum heterotropoides* against *Culex pipiens pallens*, *Aedes aegypti*, and *Ochlerotatus togoi* (Diptera: Culicidae). *J. Med. Entomol.* **2009**, *46*, 1420–1423. [CrossRef]
55. Perumalsamy, H.; Chang, K.S.; Park, C.; Ahn, Y.J. Larvicidal activity of *Asarum heterotropoides* root constituents against insecticide-susceptible and-resistant *Culex pipiens pallens* and *Aedes aegypti* and *Ochlerotatus togoi*. *J. Agric. Food Chem.* **2010**, *58*, 10001–10006. [CrossRef]
56. Stefan, M.; Zamfirache, M.M.; Padurariu, C.; Truţă, E.; Gostin, I. The composition and antibacterial activity of essential oils in three *Ocimum* species growing in Romania. *Cent. Eur. J. Biol.* **2013**, *8*, 600–608. [CrossRef]
57. Ricarte, L.P.; Bezerra, G.P.; Romero, N.R.; Da Silva, H.C.; Lemos, T.L.G.; Arriaga, A.M.C.; Alves, P.B.; Santos, M.B.D.; Millitao, G.C.G.; Silva, T.D.S.; et al. Chemical composition and biological activities of the essential oils from *Vitex-agnus castus*, *Ocimum campechianum* and *Ocimum carnosum*. *An. Acad. Bras. Cienc.* **2020**, *92*, e20180569. [CrossRef]
58. Braglia, R.; Costa, P.; Di Marco, G.; D'Agostino, A.; Redi, E.L.; Scuderi, F.; Gismondi, A.; Canini, A. Phytochemicals and quality level of food plants grown in an aquaponics system. *J. Sci. Food Agric.* **2022**, *102*, 844–850. [CrossRef]
59. Dawson, B.S.W.; Franich, R.A.; Meder, R. Essential oil of *Melissa officinalis* L. subsp. *altissima* (Sibthr. et Smith) Arcang. *Flavour Fragr. J.* **1988**, *3*, 167–170. [CrossRef]
60. Bueno, J.; Escobar, P.; Martínez, J.R.; Leal, S.M.; Stashenko, E.E. Composition of three essential oils, and their mammalian cell toxicity and antimycobacterial activity against drug resistant-tuberculosis and nontuberculous mycobacteria strains. *Nat. Prod. Commun.* **2011**, *6*, 1743–1748. [CrossRef]
61. Trendafilova, A.; Todorova, M.; Ivanova, V.; Zhelev, P.; Aneva, I. Essential oil composition of five *Thymus* species from Bulgaria. *Chem. Biodivers.* **2021**, *18*, e2100498. [CrossRef]
62. Claeson, P.; Andersson, R.; Samuelsson, G. T-cadinol: A pharmacologically active constituent of scented myrrh: Introductory pharmacological characterization and high field ^{1}H- and ^{13}C-NMR data. *Planta Med.* **1991**, *57*, 352–356. [CrossRef]

63. Takei, M.; Umeyama, A.; Arihara, S. T-cadinol and calamenene induce dendritic cells from human monocytes and drive Th1 polarization. *Eur. J. Pharmacol.* **2006**, *537*, 190–199. [CrossRef]
64. Su, Y.C.; Hsu, K.P.; Wang, E.I.C.; Ho, C.L. Composition, in vitro cytotoxic, and antimicrobial activities of the flower essential oil of *Diospyros discolor* from Taiwan. *Nat. Prod. Commun.* **2015**, *10*, 1311–1314. [CrossRef]
65. dos Santos, A.L.; Amaral, M.; Hasegawa, F.R.; Lago, J.H.G.; Tempone, A.G.; Sartorelli, P. (-)-T-Cadinol—A sesquiterpene isolated from *Casearia sylvestris* (Salicaceae) displayed in vitro activity and causes hyperpolarization of the membrane potential of *Trypanosoma cruzi*. *Front. Pharmacol.* **2021**, *12*, 734127. [CrossRef]
66. Rojas, R.; Bustamante, B.; Ventosilla, P.; Fernadez, I.; Caviedes, L.; Gilman, R.H.; Lock, O.; Hammond, G.B. Larvicidal, antimycobacterial and antifungal compounds from the bark of the Peruvian plant *Swartzia polyphylla* DC. *Chem. Pharm. Bull.* **2006**, *54*, 278–279. [CrossRef]
67. Zhao, F.; Wang, P.; Lucardi, R.D.; Su, Z.; Li, S. Natural sources and bioactivities of 2,4-di-tert-butylphenol and its analogs. *Toxins* **2020**, *12*, 35. [CrossRef] [PubMed]
68. Wang, Z.Q.; Perumalsamy, H.; Wang, M.; Shu, S.; Ahn, Y.J. Larvicidal activity of *Magnolia denudata* seed hydrodistillate constituents and related compounds and liquid formulations towards two susceptible and two wild mosquito species. *Pest Manag. Sci.* **2016**, *72*, 897–906. [CrossRef] [PubMed]
69. Gutbrod, K.; Romer, J.; Dörmann, P. Phytol metabolism in plants. *Prog. Lipid Res.* **2019**, *74*, 1–17. [CrossRef] [PubMed]
70. Araújo, S.G.; Alves, L.F.; Pinto, M.E.A.; Oliveira, G.T.; Siquera, E.P.; Ribeiro, R.I.M.A.; Ferreira, J.M.S.; Lima, L.A.R.S. Volatile compounds of Lamiaceae exhibit a synergistic antibacterial activity with streptomycin. *Braz. J. Microbiol.* **2014**, *45*, 1341–1347. [CrossRef]
71. Chogo, J.B.; Crank, G. Chemical composition and biological activity of the Tanzanian plant *Ocimum suave*. *J. Nat. Prod.* **1981**, *44*, 308–311. [CrossRef]
72. Naidoo, Y.; Sadashiva, C.T.; Kasim, N.; Nicholas, A.; Naidoo, G. Chemical composition and antimicrobial activity of the essential oil of *Ocimum obovatum* E. Mey. Ex Benth. (Lamiaceae). *J. Essent. Oil-Bear. Plants* **2014**, *17*, 142–147. [CrossRef]
73. Selvaraju, R.; Sakuntala, P.; Jaleeli, K.A. Comparative analysis of chemical compounds and trace elements in the leaves of few *Ocimum* species plants by using different spectroscopic methods. *Rom. J. Biophys.* **2021**, *31*, 65–78.
74. Islam, M.T.; De Alencar, M.V.O.B.; Da Conceição Machado, K.; Da Conceição Machado, K.; De Carvalho Melo-Cavalcante, A.A.; De Sousa, D.P.; De Freitas, R.M. Phytol in a pharma-medico-stance. *Chem.-Biol. Interact.* **2015**, *240*, 60–73. [CrossRef]
75. European Chemicals Agency. 2-(2-Butoxyethoxy)Ethanol. Available online: https://echa.europa.eu/substance-information/-/substanceinfo/100.003.601 (accessed on 28 February 2022).
76. Aburjai, T.A.; Mansi, K.; Azzam, H.; Alqudah, D.A.; Alshaer, W.; Abuirjei, M. Chemical compositions and anticancer potential of essential oil from greenhouse-cultivated *Ocimum basilicum* leaves. *Indian J. Pharm. Sci.* **2020**, *82*, 179–184. [CrossRef]
77. Ravi Kiran, S.; Bhavani, K.; Sita Devi, P.; Rajeswara Rao, B.R.; Janardhan Reddy, K. Composition and larvicidal activity of leaves and stem essential oils of *Chloroxylon swietenia* DC against *Aedes aegypti* and *Anopheles stephensi*. *Bioresour. Technol.* **2006**, *97*, 2481–2484. [CrossRef]
78. Govindarajan, M.; Benelli, G. α-Humulene and β-elemene from *Syzygium zeylanicum* (Myrtaceae) essential oil: Highly effective and eco-friendly larvicides against *Anopheles subpictus*, *Aedes albopictus*, and *Culex tritaeniorhynchus* (Diptera: Culicidae). *Parasitol. Res.* **2016**, *115*, 2771–2778. [CrossRef]
79. França, L.P.; Amaral, A.C.F.; Ramos, A.S.; Ferreira, J.L.P.; Maria, A.C.B.; Oliveira, K.M.T.; Araujo, E.S., Jr.; Branches, A.D.S.; Silva, J.N.; Barros, G.A.; et al. *Piper capitarianum* essential oil: A promising insecticidal agent for the management of *Aedes aegypti* and *Aedes albopictus*. *Environ. Sci. Pollut. Res.* **2021**, *28*, 9760–9776. [CrossRef]
80. Tong, H.; Liu, Y.; Jiang, L.; Wang, J. Multi-targeting by β-elemene and its anticancer properties: A good choice for oncotherapy and radiochemotherapy sensitization. *Nutr. Cancer* **2020**, *72*, 554–567. [CrossRef]

horticulturae

MDPI

Article

Weather Conditions Influence on Lavandin Essential Oil and Hydrolate Quality

Milica Aćimović [1], Biljana Lončar [2,*], Jovana Stanković Jeremić [3], Mirjana Cvetković [3], Lato Pezo [4], Milada Pezo [5], Marina Todosijević [6] and Vele Tešević [6]

1 Institute of Field and Vegetable Crops Novi Sad—National Institute of the Republic of Serbia, Maksima Gorkog 30, 21000 Novi Sad, Serbia; acimovicbabicmilica@gmail.com
2 Faculty of Technology Novi Sad, University of Novi Sad, Bulevar Cara Lazara 1, 21000 Novi Sad, Serbia
3 Institute of Chemistry, Technology and Metallurgy, Njegoševa 12, 11000 Belgrade, Serbia; jovana_stankovic@hotmail.com (J.S.J.); miracvet13@gmail.com (M.C.)
4 Institute of General and Physical Chemistry, Studentski trg 10–12, 11000 Belgrade, Serbia; latopezo@yahoo.co.uk
5 Department of Thermal Engineering and Energy, "Vinča" Institute of Nuclear Sciences—National Institute of the Republic of Serbia, University of Belgrade, P.O. Box 522, 11001 Belgrade, Serbia; milada@vin.bg.ac.rs
6 Department of Organic Chemistry, Faculty of Chemistry, Studentski trg 16, 11000 Belgrade, Serbia; marinab@chem.bg.ac.rs (M.T.); vtesevic@chem.bg.ac.rs (V.T.)
* Correspondence: cbiljana@uns.ac.rs; Tel.: +381-(0)21-485-3708

Abstract: *Lavandula* sp. essential oil and hydrolate are commercially valuable in various industry branches with the potential for wide-ranging applications. This study aimed to evaluate the quality of these products obtained from *L. x intermedia* cv. 'Budrovka' for the first time cultivated on Fruška Gora Mt. (Serbia) during three successive seasons (2019, 2020, and 2021). Essential oil extraction was obtained by steam distillation, and the composition and influence of weather conditions were also assessed, using flowering tops. The obtained essential oils and hydrolates were analysed by gas chromatography with a flame ionization detector (GC-FID) and gas chromatography coupled to mass spectrometry (GC-MS). A linear regression model was developed to predict *L. x intermedia* cv. 'Budrovka' essential oil volatile compound content and hydrolate composition during three years, according to temperature and precipitation data, and the appropriate regression coefficients were calculated, while the correlation analysis was employed to analyse the correlations in hydrolate and essential oil compounds. To completely describe the structure of the research data that would present a better insight into the similarities and differences among the diverse *L. x intermedia* cv. 'Budrovka' samples, the PCA was used. The most dominant in *L. intermedia* cv. 'Budrovka' essential oil and hydrolate were oxygenated monoterpenes: linalool, 1,8-cineole, borneol, linalyl acetate, and terpinene-4-ol. It is established that the temperature was positively correlated with all essential oil and hydrolate compounds. The precipitations were positively correlated with the main compounds (linalool, 1,8-cineole, and borneol), while the other compounds' content negatively correlated to precipitation. The results indicated that Fruška Gora Mt. has suitable agro-ecological requirements for cultivating *Lavandula* sp. and providing satisfactory essential oil and hydrolate.

Keywords: lavandin; precipitations; temperature; volatile compounds

Citation: Aćimović, M.; Lončar, B.; Stanković Jeremić, J.; Cvetković, M.; Pezo, L.; Pezo, M.; Todosijević, M.; Tešević, V. Weather Conditions Influence on Lavandin Essential Oil and Hydrolate Quality. *Horticulturae* **2022**, *8*, 281. https://doi.org/10.3390/horticulturae8040281

Academic Editor: Alessandra Carrubba

Received: 2 March 2022
Accepted: 25 March 2022
Published: 27 March 2022

1. Introduction

The genus *Lavandula* (order Lamiales, family Lamiaceae) originates from the Mediterranean. It is cultivated worldwide as an ornamental and essential oil-bearing plant [1,2]. There are 39 species, but only three are commercially important: true or English lavender (*L. angustifolia* Mill.), spike or Spanish lavender (*L. latifolia* Medic.), and their hybrid known as lavandin (*L. x intermedia* Emeric ex Loisel). With more than 400 registered cultivars and hybrids, *Lavandula* sp. cultivation, essential oil production, and consumption rapidly has increased due to certain profits [3–6]. Major producing regions of *Lavandula* sp. are in Europe, and the dominant countries are Bulgaria (with 3700 ha) and France (with 3500 ha) [1,3], but it is crucial in Mediterranean countries, especially Greece, Spain, and Turkey, as well [7]. Croatia is traditionally associated with *Lavandula* sp. cultivation and essential oil production on the *Adriatic coast*, but continental parts have become a commercial cultivation region [8,9]. In Hungary, a country with a continental climate, the cultivation of *Lavandula* sp. has expanded considerably in recent years [10], and in Romania [5]. Serbia follows the same path, with more than 100 ha under this crop in 2021, with an increasing tendency for the future (Aćimović, personal communication).

The *Lavandula* sp. essential oil is used in cosmetic (soaps, bath, colognes, perfumes, skin lotions, and after-shaves), pharmaceutical (mild sedative and analgesic, rubefacient and wound healing, and antioxidant and antimicrobial agents), and food industries (flavourings in baked goods, beverages, puddings, ice creams, candies, and chewing gums), in household products (detergents and hygiene products) and aromatherapy (for decreasing stress and anxiety, as well as pain intensity and against migraine), but it is also used as a biopesticide [3,4,11–15]. However, the application of lavender depends on its chemical composition. The common criteria for determining *Lavandula* sp. essential oil quality are camphor, linalool, and linalyl acetate percentage [4]. Essential oil of *L. angustifolia* is highly valued due to the low content of camphor (up to 1.2% according to European Pharmacopoeia) and is much more expensive. Therefore, it is often mixed with cheaper oils of *L. latifolia* and *L. x intermedia* to achieve better quality that satisfies ISO 8902 standard [15,16]. However, *L. latifolia* and *L. x intermedia* achieve higher essential oil yields [17,18]. Essential oil content and composition are primarily determined by plant genotype. However, it may be influenced by environmental factors (climate, but also weather conditions during growth year, soil conditions, and nutrient supply), harvest time, post-harvest treatments, and extraction methods [10,19–21].

Hydrolates are accrued as by-products during the steam distillation of essential oils. Essential oils contain volatile, lipid-soluble, and partially water-soluble compounds, generally of lower density than water [22,23]. However, volatile water-soluble compounds remain in condensate water in a Florentine flask (oil–water separator) and give a specific fragrance. This water is called hydrolate, hydrosol, aromatic or floral water [24–27]. Since hydrolates contain only a small amount of dissolved essential oil components in water, the amount of volatile organic compounds affects their biological properties [28,29]. Principally, they are by-products of essential oil distillation that could be useful as raw material in many industries, such as the food, and beverage industry (for flavouring and preservation, as well as in soft drinks), cosmetics (replacement for water phase in cosmetics, lotions, creams, soaps, and tonics) and aromatherapy (skincare or as massage products, as facial and body sprays to feel relaxed and refreshed, as air fresheners) [24]. Although the quality standards for hydrolates are not defined, their global economic impact is increasing [30]. Therefore, it is necessary to develop norms and standards for plant-derived components as there is growing interest in using them. Furthermore, due to volatile organic compounds, *Lavandula* sp. hydrolate has a pleasant lavender aroma and biological activities. Therefore, it has the potential for wide-ranging applications [28,29,31].

Fruška Gora Mt. (in Latin Alma Mons) is situated in northern Serbia, at the confluence of the Danube and Sava Rivers. This is the oldest national park in Serbia (with 25,525 ha of protected area), with almost 90% of the linden, oak, and beech forest. The rims of the Fruška Gora Mt. are used for grape cultivation, and this tradition dates from the Roman period. However, medicinal and aromatic plants were not cultivated in this region until recently. As a result, many wild species of medicinal plants, are mainly collected by gatherers [32]. This region is suitable for organic cultivation, and previous research indicated promising results in organically grown *Lavandula* sp. through Europe [33–35]. *Lavandula* sp. is a perennial, heliophyte, and drought-resistant shrub [10]. It requires well-drained soil, but it can adapt to poor soils with low fertility [36]. Croatian cultivar of *L. x intermedia* 'Budrovka' is widely grown in the former Yugoslavia region. This cultivar is well adapted to the continental climate and low temperatures during winter, up to –20 °C [37,38].

Considering the great commercial importance of *Lavandula* sp. flowers and essential oil in different industry branches and the increasing trend of cultivation lavender and lavandin through Europe, this paper aimed to evaluate the quality of introduced *L. x intermedia* cv. 'Budrovka' on Fruška Gora Mt. (Serbia). The main goal of this investigation was the influence of microclimatic conditions in the foreground temperature and precipitations in a selected location on *L. x intermedia* essential oil and hydrolate quality. The second goal of this investigation is to evaluate the essential oil quality of cv. 'Budrovka' compared to other samples *L. x intermedia* reported in the literature and ISO standards. Further, results of the chemical composition of *L. x intermedia* 'Budrovka' hydrolates from this study and other from literature were used for setting ranges of the main compounds of *Lavandula* sp. hydrolate.

2. Materials and Methods

2.1. Plant Material

The plantation of *Lavandula x intermedia* Emeric ex Loisel cv. 'Budrovka' was established in autumn 2014, at a certified organic farm in the village of Bukovac (area Beljevo; 45.111744 N, 19.530732 E) on Fruška Gora region (Figure 1). One-year-old seedlings were planted at a 1.5 m interrow distance and 0.5 m between plants.

Figure 1. Field location and lavender field in full blooming with honeybees.

2.2. Soil Characteristics

Basic chemical properties of the soil where *L. x intermedia* was cultivated were determined before establishing the experiment (October 2014) in samples of the top layer (0 to 40 cm) and are shown in Table 1. The soil samples were analysed at the Agriculture Extension Service Novi Sad, following standardized methods adopted in Serbia [39].

Table 1. Basic chemical characteristics of the soil.

	pH		$CaCO_3$	Humus	Total Nitrogen	P_2O_5	K_2O
	1MKCl	H_2O		(%)		(mg·100 g^{-1} Soil)	
Hill	7.13	8.22	11.27	1.03	0.051	3.6	8.0
Middle	7.19	8.34	8.10	0.56	0.028	7.4	8.0
Valley	7.41	8.31	7.89	0.96	0.048	7.4	8.0

2.3. Weather Conditions

The climate of Fruška Gora, as an isolated mountain, significantly differs from its surroundings and modifies the local meteorological conditions. Although the mountain is small, the wooded slopes and the E–W direction of the ridge influence the passing air masses greatly [40]. The weather conditions during the three successive growing years are presented in Figure 2.

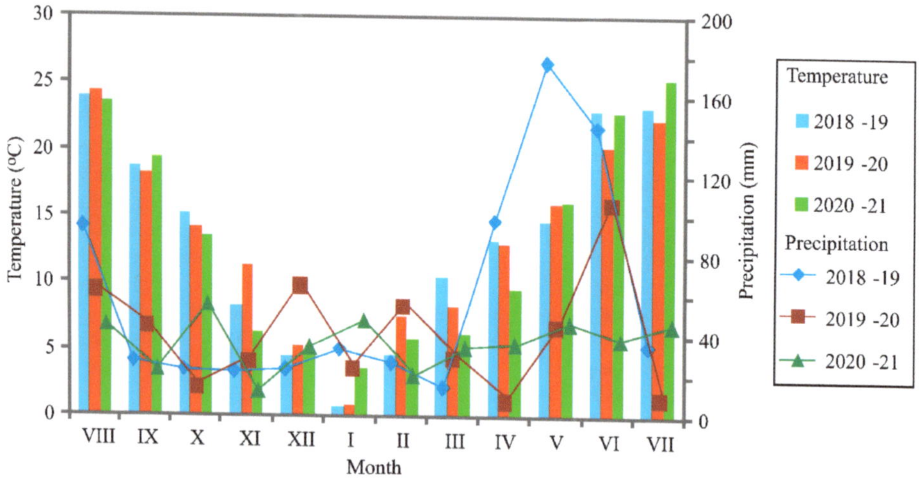

Figure 2. The weather conditions during the three successive growing seasons.

2.4. Essential Oil and Hydrolate Extraction

The small-scale distillation unit at the Institute of Field and Vegetable Crops Novi Sad, consists of a cylindrical distillation vessel with a conical bottom of stainless-steel pipe, which connects the vessel with a water-cooled condenser Florentine flask. The steam was produced externally, by a steam generator, and supplied the vessel with plant material via a pipe on the bottom. Our previous paper provides detailed information about capacity and conditions [41].

Briefly, 100 kg of fresh flowers of *L. x intermedia* cv. 'Budrovka' were placed in a distillation vessel, routed upwards through a plumbing system, and supplied with steam. After 20 min, the condensed vapours started to collect in the Florentine flask. After 2 h the distillation process was over. Accumulated essential oil floated on the water phase (hydrolate). The hydrolate was collected in sterile plastic bottles via filter paper, while essential oil was dried applying sodium sulphate and stored in amber glass bottles. The same extraction process was repeated three times.

2.5. Analysis of Volatile Compounds

The hydrolates before analysis were subjected to simultaneous distillation and extraction with dichloromethane via a Likens–Nickerson apparatus. Obtained essential oils and hydrolates samples were analysed by gas chromatography with flame ionization detector (GC-FID) and gas chromatography coupled to mass spectrometry (GC-MS). GC-FID/MS analyses were performed according to Acimovic et al. [42] with some modifications of the split ratio. Injection volume was 1 µL, split ratio 10:1 for essential oils and 50:1 for hydrolates.

2.6. Statistical Analysis

A linear regression model was developed to predict the *L. x intermedia* cv. 'Budrovka' essential oil active compound content and hydrolate composition during three years according to temperature and precipitation data, and the appropriate regression coefficients were calculated using the linear formula: $x = a + b_T \cdot T + b_P \cdot P$, where x was EO active compound content or hydrolate content, a was the intercept, b_T and b_P were temperature and precipitation coefficients, while T and P were temperature and precipitation amounts, respectively.

The collected data were processed statistically using the software package STATISTICA 10.0.

3. Results

3.1. Fresh Herb Yield and Essential Oil Content

This variety (Vouch No 2-0987, Herbarium BUNS, Serbia), obtained from local nursery garden, is a shrub ranging up to 150 cm, with linear-lanceolate to spatulate leaves, often tomentose. Its inflorescence stalk is branched, and flowers show a corolla with bilateral symmetry and vary in colour from lilac-purple to white, blooming from late June to July. Flowering tops were collected at the full bloom stage. Harvest dates and yields are given in Table 2.

Table 2. Date of harvest, fresh herb yield and essential oil content for *L. x intermedia* cv. 'Budrovka' cultivated in Fruška gora Mt. during 2019–2021.

	2019	2020	2021
Date of harvest	11th–17th July	27th June–2nd July	9th–15th July
Fresh herb yield (kg/ha)	4.840	5.040	5.670
Essential oil content (%)	1.26	1.19	1.03

3.2. Chemical Composition of Essential Oil

The essential oil from *L. x intermedia* cultivated on Fruška Gora Mt. contains 56 compounds (Table 3). The most abundant compounds were: linalool (33.3–42.0%), 1,8-cineole (12.9–19.0%), borneol (8.1–16.1%), linalyl acetate (5.3–8.8%), and terpinene-4-ol (3.5–6.6%). All these compounds belong to oxygenated monoterpenes, which are the dominant compound class in essential oils.

Table 3. *L. x intermedia* cv. 'Budrovka' essential oil composition during three growing years and related multiple correlation coefficients with the observed temperature and precipitation amounts.

No	Compound	RI	2019	2020	2021	Temp. Coeff.	Prec. Coeff.
			EO19	EO20	EO21		
6	*n*-hexanol [O]	861	0.2	-	-	0.70	−0.25
7	*α*-thujene [MT]	924	0.1	0.1	0.1	0.70	−0.25
8	*α*-pinene [MT]	931	0.9	0.6	1.4	0.71	−0.23
9	camphene [MT]	945	0.4	0.7	0.5	0.71	−0.24
12	sabinene [MT]	970	0.3	0.1	0.4	0.70	−0.24
14	*β*-pinene [MT]	974	1.6	0.9	2.1	0.71	−0.21
15	3-octanone [O]	984	0.1	-	-	0.70	−0.25
16	dehydro-1,8-cineole [OMT]	989	-	0.1	-	0.70	−0.25
17	myrcene [MT]	990	0.5	0.1	0.7	0.70	−0.24
18	*δ*-3-carene [MT]	1008	0.3	0.1	0.4	0.70	−0.24
19	hexyl acetate [O]	1010	0.1	-	-	0.70	−0.25
20	*α*-terpinene [MT]	1013	-	-	0.1	0.70	−0.25
22	*p*-cymene [MT]	1022	0.4	0.8	0.3	0.71	−0.24
23	limonene [MT]	1025	1.1	0.7	1.1	0.71	−0.23
24	1,8-cineole [OMT]	1028	12.9	16.4	19.0	0.82	0.11
25	*cis*-*β*-ocimene [MT]	1033	1.9	-	3.6	0.72	−0.21
27	*trans*-*β*-ocimene [MT]	1044	0.2	-	0.4	0.70	−0.24
28	*γ*-terpinene [MT]	1055	0.1	-	0.2	0.70	−0.25
29	*cis*-sabinene hydrate (IPP vs OH) [OMT]	1063	0.2	0.2	-	0.70	−0.25
30	*cis*-linalool oxide (furanoid) [OMT]	1069	0.1	2.4	-	0.71	−0.23
32	*trans*-linalool oxide (furanoid) [OMT]	1085	-	2.1	-	0.71	−0.23
33	terpinolene [MT]	1086	0.2	-	0.3	0.70	−0.24
34	linalool [OMT]	1101	41.8	33.3	42.0	0.98	0.59
36	*allo*-ocimene [MT]	1126	0.3	-	0.6	0.70	−0.24
38	*trans*-pinocarveol [OMT]	1134	-	0.3	-	0.70	−0.25
40	camphor [OMT]	1141	3.7	4.6	3.9	0.73	−0.16

RI—Retention index (relative to C8–C36 n-alkanes on HP-5MSI column).

Table 3. *Cont.*

No	Compound	RI	2019 EO19	2020 EO20	2021 EO21	Temp. Coeff.	Prec. Coeff.
41	hexyl isobutanoate [O]	1145	0.1	0.1	-	0.70	−0.25
43	borneol [OMT]	1164	11.4	16.1	8.1	0.79	0.01
45	*trans*-linalool oxide (pyranoid) [OMT]	1168	-	0.2	-	0.70	−0.25
46	terpinen-4-ol [OMT]	1175	6.6	3.5	5.3	0.74	−0.14
48	cryptone [O]	1180	-	0.3	-	0.70	−0.25
49	α-terpineol [OMT]	1188	0.8	0.5	-	0.70	−0.24
50	hexyl butanoate [O]	1189	0.4	0.4	0.5	0.70	−0.24
51	myrtenal [OMT]	1190	-	0.2	-	0.70	−0.25
55	hexyl 2-methyl butanoate [O]	1234	0.2	0.3	0.1	0.70	−0.24
56	cumin aldehyde [O]	1236	-	0.1	-	0.70	−0.25
57	carvone [OMT]	1238	-	0.1	-	0.70	−0.25
58	hexyl isovalerate [O]	1239	0.1	-	-	0.70	−0.25
61	linalyl acetate [OMT]	1254	6.5	8.8	5.3	0.75	−0.10
63	bornyl acetate [OMT]	1284	0.1	0.2	-	0.70	−0.25
64	lavandulyl acetate [OMT]	1289	0.6	1.2	0.3	0.71	−0.23
65	neryl acetate [OMT]	1363	0.1	-	-	0.70	−0.25
66	daucene [ST]	1377	0.1	-	-	0.70	−0.25
67	geranyl acetate [OMT]	1382	0.2	-	-	0.70	−0.25
68	hexyl hexanoate [O]	1384	0.2	0.2	-	0.70	−0.25
69	7-*epi*-sesquithujene [ST]	1388	0.1	-	-	0.70	−0.25
70	sesquithujene [ST]	1403	0.1	-	-	0.70	−0.25
71	α-santalene [ST]	1414	-	0.3	-	0.70	−0.25
72	*trans*-caryophyllene [ST]	1417	0.9	-	0.6	0.70	−0.24
73	*trans*-α-bergamotene [ST]	1433	0.1	-	-	0.70	−0.25
74	*trans*-β-farnesene [ST]	1455	2.0	0.3	1.1	0.71	−0.23
75	germacrene D [ST]	1480	0.3	-	0.1	0.70	−0.25
76	lavandulyl isovalerate [OST]	1508	0.3	0.4	-	0.70	−0.24
77	γ-cadinene [ST]	1513	tr	0.1	-	0.70	−0.25
79	caryophyllene oxide [OST]	1580	0.2	0.8	-	0.70	−0.24
80	*epi*-α-bisabolol [OST]	1682	0.1	-	-	0.70	−0.25
	Monoterpene hydrocarbons (MT)		8.3	4.1	12.2		
	Oxygenated monoterpenes (OMT)		85.0	90.2	83.9		
	Sesquiterpene hydrocarbons (ST)		3.6	0.7	1.8		
	Oxygenated sesquiterpens (OST)		0.6	1.2	-		
	Other (O)		1.4	1.4	0.6		
	Total identified		98.9	97.6	98.5		

RI—Retention index (relative to C8–C36 n-alkanes on HP-5MSI column).

Comparing rainfall among years (Figure 2), the growing season 2018/19 had significantly higher rainfall (694.2 mm) in comparison with the other two growing seasons (458.8 mm for 2019/20, and 416.9 mm for 2020/21). On the other hand, comparing average year temperatures for all three investigated seasons, the third growing season (2020/21) was slightly cooler (13.2 °C) in comparison to the other two (13.5 °C on average). Therefore, it could be said that temperature was in positive correlation with all essential oil compounds (temp. coeff. varied between 0.70 and 0.98), while precipitations were in negative correlation with a large number of essential oil compounds (prec. coeff. varied between −0.10 and −0.25), except linalool (0.59), 1,8-cineole (0.11), and borneol (0.01).

3.3. Chemical Composition of Hydrolate

There were 40E compounds identified in *L. x intermedia* cv. 'Budrovka' hydrolate (Table 4). The most abundant was linalool (21.9–32.1%), 1,8-cineole (12.7–26.2%), borneol (10.6–24.4%), terpinen-4-ol (6.4–12.2%), and *cis*- and *trans*-linalool oxides (1.4–11.5% and 1.3–10.9%, respectively).

Table 4. *L. x intermedia* cv. 'Budrovka' hydrolate composition during three growing years and related multiple correlation coefficients with the observed temperature and precipitation amounts.

No	Compound	RI	2019	2020	2021	Temp. Coeff.	Prec. Coeff.
			EO19	EO20	EO21		
1	3-methyl-2-butenal [O]	772	0.1	-	0.4	0.70	−0.24
2	hexanal [O]	798	-	-	0.2	0.70	−0.25
3	2,2-dimethyl-3(2H)-furanone [O]	829	-	-	0.1	0.70	−0.25
4	furfural [O]	832	-	-	0.1	0.70	−0.25
5	*cis*-3-hexenol [O]	847	-	-	0.3	0.70	−0.25
6	*n*-hexanol [O]	858	1.0	0.6	2.2	0.71	−0.22
10	4-methyl pent-2-enolide (impure) [O]	949	0.1	-	0.3	0.70	−0.25
11	2-ethenyltetrahydro-2,6,6-trimethyl-2H-pyran [O]	968	0.1	0.1	-	0.70	−0.25
13	1-octen-3-ol [O]	974	0.9	0.6	0.9	0.71	−0.23
15	3-octanone [O]	982	-	-	0.1	0.70	−0.25
16	dehydro-1,8-cineole [OMT]	990	tr	-	0.1	0.70	−0.25
21	1,4-cineole [OMT]	1014	0.1	-	-	0.70	−0.25
24	1,8-cineole [OMT]	1030	12.7	14.4	26.2	0.84	0.15
26	lavender lactone [O]	1036	-	0.2	0.1	0.70	−0.25
30	*cis*-linalool oxide (furanoid) [OMT]	1071	1.4	11.5	2.7	0.74	−0.13
31	camphenilone [NOMT]	1080	-	-	0.1	0.70	−0.25
32	*trans*-linalool oxide (furanoid) [OMT]	1088	1.3	10.9	2.4	0.74	−0.14
34	linalool [OMT]	1102	26.0	21.9	32.1	0.90	0.33
35	hotrienol [OMT]	1102	-	1.0	-	0.70	−0.24
37	nopinone [NOMT]	1136	0.2	0.2	0.2	0.70	−0.24
38	*trans*-pinocarveol [OMT]	1134	-	0.2	0.1	0.70	−0.25
39	*trans*-sabinol (trans for OH vs. IPP) [OMT]	1136	0.1	-	-	0.70	−0.25
40	camphor [OMT]	1143	7.1	4.5	6.3	0.74	−0.12
42	*neoiso*-3-thujanol [OMT]	1146	-	0.2	-	0.70	−0.25
43	borneol [OMT]	1166	24.4	16.8	10.6	0.81	0.10
44	*cis*-linalool oxide (pyanoid) [OMT]	1168	0.6	1.0	0.5	0.71	−0.23
45	*trans*-linalool oxide (pyanoid) [OMT]	1173	0.4	0.8	0.4	0.71	−0.24
46	terpinen-4-ol [OMT]	1177	12.2	6.4	9.7	0.77	−0.05
47	*p*-cymen-8-ol [O]	1179	-	0.2	0.2	0.70	−0.24
48	cryptone [O]	1180	-	1.0	0.3	0.71	−0.24
49	α-terpineol [OMT]	1190	6.0	4.5	1.4	0.73	−0.17
51	myrtenol [OMT]	1191	-	0.1	-	0.70	−0.25
52	verbenone [OMT]	1208	0.2	0.3	0.1	0.70	−0.24
53	*trans*-carveol [OMT]	1216	-	0.1	-	0.70	−0.25
54	nerol [OMT]	1222	-	0.3	-	0.70	−0.25
57	carvone [OMT]	1243	0.1	0.1	-	0.70	−0.25
60	geraniol [OMT]	1248	-	0.9	-	0.70	−0.24
61	linalyl acetate [OMT]	1248	-	-	0.2	0.70	−0.25
64	lavandulyl acetate [OMT]	1286	-	0.1	-	0.70	−0.25
78	*cis*-nerolidol [OST]	1526	-	-	0.2	0.70	−0.25
	Nor oxygenated monoterpenes (NOMT)		0.2	0.2	0.3		
	Oxygenated monoterpenes (OMT)		92.6	96.0	92.8		
	Sesquiterpene hydrocarbons (ST)		-	-	-		
	Oxygenated sesquiterpens (OST)		-	-	0.2		
	Other (O)		2.2	2.7	5.2		
	Total identified		95.0	98.9	98.5		

RI—Retention index (relative to C8–C36 n-alkanes on HP-5MSI column).

During all three years oxygenated monoterpenes were the dominant class, with 92.6–96.0%. Similar to essential oils, differences in climate conditions during tree investigated years have not impacted hydrolate composition variations. In addition, the temperature was in positive correlation with all hydrolate compounds (temp. coeff. varied between 0.70 and 0.84), while precipitations were in negative correlation with a large number of essential oil compounds (prec. coeff. varied between −0.10 and −0.25), except linalool (0.33), 1,8-cineole (0.15), and borneol (0.10).

3.4. Correlation between Chemical Compounds of Essential Oil and Hydrolate

The correlation analysis was employed to examine the relations in hydrolate and essential oil compounds of *L. x intermedia* cv. 'Budrovka' samples from three growing years (2019, 2020, and 2021), and the results were displayed in Figure 3. It can be noticed from the figure that the darker blue colour of the squares, which indicates the two active compounds 'content similarity, presents a more significant correlation linking observed active compounds. In contrast, the lighter tone indicates a particular dissimilarity active compound. Therefore, if the colour tone is lighter, consequently the correlation is lower. On the other hand, the red colour symbolizes a negative correlation between active compounds.

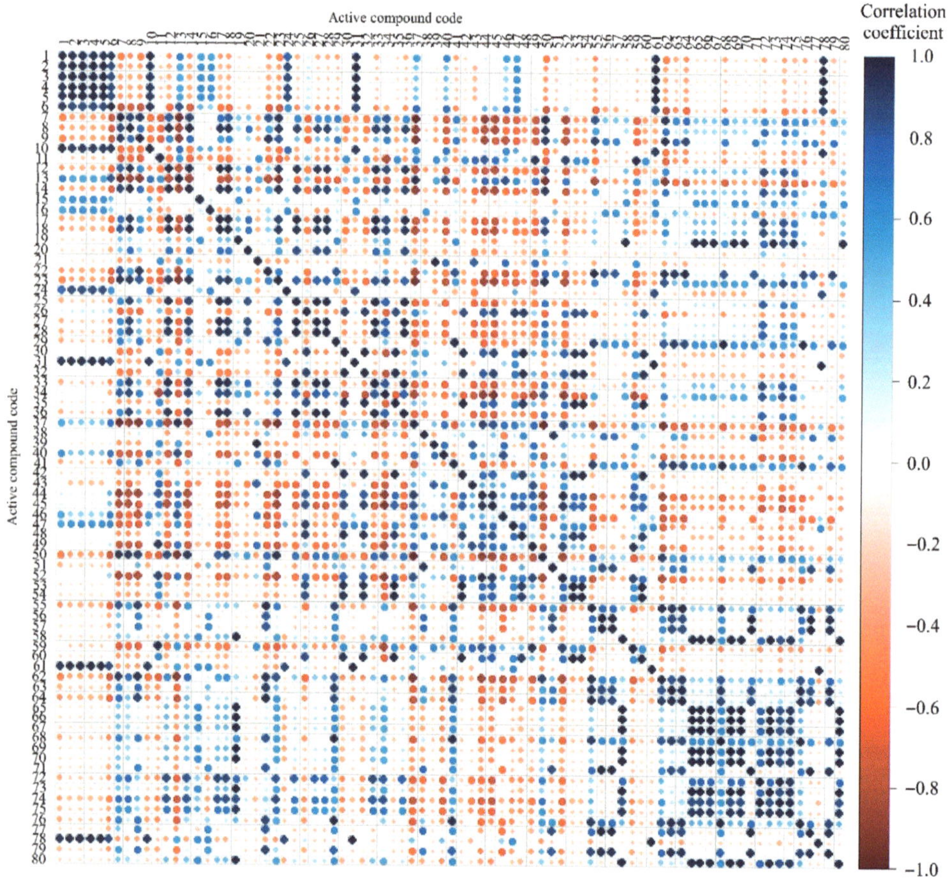

Figure 3. Correlation between active compounds content in hydrolate and essential oil of *L. x intermedia* cv. 'Budrovka' samples from 2019, 2020, and 2021 (active compounds codes are listed in Tables 3 and 4).

To thoroughly explain the structure of the research data that would provide a better perception of similarities and differences among diverse the *L. x intermedia* cv. 'Budrovka' samples from 2019, 2020, and 2021, PCA was used, and the results are presented in Figure 4. The first PC explained 43.46%, the second 19.10%, and the third 16.53% of the total variance within the experimental data. The separation between samples could be recognized from the PCA figures, where the samples from *L. x intermedia* cv. 'Budrovka' hydrolate composition during 2019, 2020, and 2021 are grouped on the right side of the graphic, while the samples from the *L. x intermedia* cv. 'Budrovka' essential oil composition during three growing years are grouped on the graphic's left side. Table 5 represents the correlation matrix among active compounds content in hydrolate and essential oil of *L. x intermedia* cv. 'Budrovka' samples from 2019, 2020, and 2021.

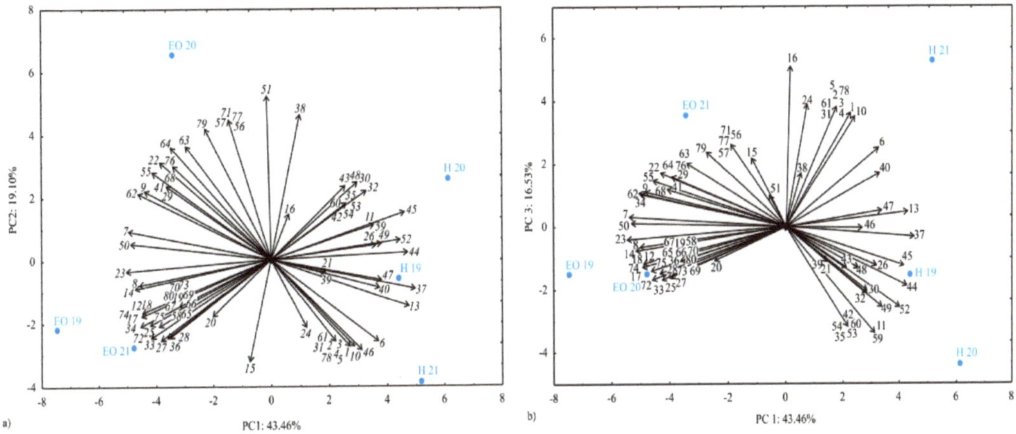

Figure 4. The PCA biplot diagram depicting the relationships among hydrolate and essential oil compounds of *L. x intermedia* cv. 'Budrovka' samples from three growing years: (**a**) projection of variables and cases in the PC1 and PC2 plane, (**b**) projection of variables and cases in PC1 and PC3 plane (active compounds codes and samples codes are listed in Tables 3 and 5).

Table 5. Correlation matrix between active compounds content in hydrolate and essential oil of *L. x intermedia* cv. 'Budrovka' samples from 2019, 2020, and 2021.

	EO20	EO21	H19	H20	H21
EO19	0.967	0.987	0.855	0.795	0.908
EO20		0.959	0.896	0.868	0.928
EO21			0.820	0.788	0.940
H19				0.893	0.874
H20					0.870

EO20—essential oil active compounds from 2020, EO21—essential oil active compounds from 2021, H19—hydrolate active compounds from 2019, H20—hydrolate active compounds from 2020, H21—hydrolate active compounds from 2021.

4. Discussion

4.1. Essential Oil

Investigations conducted with *L. latifolia* showed that the environment significantly affected the essential oil quality, especially altitude [43]. Further, in the case of *L. angustifolia*, agronomical practices such as irrigation [44], variety [19], and substrate [45] significantly impact essential oil composition. In addition, investigations show that during the blooming period, linalool content was influenced by temperature; although, rainfalls remarkably decreased its production [46]. As it can be seen from Figure 4, our long-term experiment (three successive years) shows that on accumulation linalool and linalyl acetate precipitation, and the temperature are in positive relations, while in the case of linalyl acetate, the temperature is in positive (0.75), and precipitation in negative (−0.10) correlations.

The isolated essential oil of *L. x intermedia* cv. 'Budrovka' cultivated in Fruška gora Mt. did not comply with international standards requirements for the lavandin (ISO 8902), similar is with the same cultivar grown in Croatia, which contained linalool (57.1%), linalyl acetate (9.8%), and 1,8-cineole (8.5%) as dominant compounds [9]. One of the main criteria for *Lavandula* sp. essential oil quality is the ratio between linalool and its ester form, linalyl acetate. A ratio lower than one indicates the high quality of essential oil [47]. In this study, the linalool: linalyl acetate ratio ranged from 3.78 (in 2020) and 6.43 (in 2019) to 7.92 (in 2021). On average, this value was 6.04, while in Croatia, this ratio was 5.81 for the same cultivar [9].

Table 6 shows 28 samples of *L. x intermedia* essential oil from literature and average values from this study (total 29 accessions) with ten compounds from ISO standard specification. As can be seen, no sample satisfies all values required for essential oil quality. However, it is known that *L. x intermedia* is a cheaper hybrid, 6–7 times than *L. angustifolia*, and usually is mixed with them to obtain a higher quality of essential oil [16]. In addition, the accumulation of main compounds in *Lavandula* sp. plants are genetically influenced, i.e., by a key gene involved in controlling the production of linalyl acetate, camphor, 1,8-cineole, and borneol [48].

Table 6. Comparison results of main *L. x intermedia* essential oil components and ISO 8902 standard.

No.	Cultivar	Origin	Limonene	1,8-cineole	Cis-B-Ocimene	Trans-B-Ocimene	Camphor	Linalool	Linalyl Acetate	Terpinene-4-Ol	Lavandulyl Acetate	Lavandulol	Reference
1	Hidcote Giant	Canada	1.4	17.1	7.8	12.2	13.3	23.8	0.3	4.1	0.0	1.6	[19]
2	Grosso	Canada	2.0	10.7	3.4	4.4	10.8	30.6	8.3	3.3	3.7	0.0	[19]
3	Super	Canada	2.7	13.1	3.5	6.2	5.5	28.3	10.0	0.0	3.4	0.0	[19]
4	OK-Farms Super	Canada	2.4	4.8	3.1	8.0	2.1	35.2	11.6	0.0	1.8	0.0	[19]
5	French Super	Canada	2.5	5.2	4.0	9.8	4.8	25.2	12.2	0.5	2.7	0.0	[19]
6	Super	Turkey	0.0	2.6	0.0	0.0	4.8	34.0	47.7	0.0	0.0	0.0	[49]
7	n.s.	Macedonia	0.0	6.7	2.3	0.8	6.6	39.0	2.1	5.1	1.9	1.7	[11]
8	Dutch	Turkey	1.1	7.6	0.0	0.0	11.3	44.8	8.2	1.5	0.0	0.1	[50]
9	Giant Hidcote	Turkey	1.3	12.0	0.0	0.0	6.6	41.6	3.9	1.5	0.0	0.2	[50]
10	Super A	Turkey	0.8	2.1	0.0	0.0	7.5	38.1	36.2	0.5	0.0	0.0	[50]
11	Super A	Turkey	0.4	3.2	1.5	6.6	43.7	24.6	5.4	0.0	0.0		[51]
12	Sumiens	Italy	0.7	12.0	1.9	1.0	7.1	40.4	9.9	0.2	0.3	0.0	[3]
13	Super A	Italy	0.6	6.9	1.3	0.7	6.6	36.2	18.4	3.3	4.5	0.1	[3]
14	Grosso	Italy	0.5	8.1	1.2	0.8	8.1	38.4	15.7	3.6	4.1	0.1	[3]
15	Super	Turkey	0.7	0.0	2.6	1.5	5.3	36.8	33.1	0.0	1.2	0.1	[4]
16	Grey Hedge	Turkey	2.3	0.0	1.8	5.0	6.4	28.5	4.6	6.9	0.8	0.5	[4]
17	Budrovka	Croatia	4.0	8.4	0.0	0.4	0.1	57.1	9.8	2.3	0.2	1.1	[9]
18	Grosso	Turkey	2.5	33.1	0.0	0.0	21.8	0.2	0.0	1.0	0.2	0.4	[18]
19	Dutch	Turkey	2.9	35.8	0.0	0.0	22.2	0.2	0.0	0.8	0.1	0.2	[18]
20	Abriel	Turkey	2.5	35.9	0.0	0.0	22.8	0.1	0.0	1.0	0.2	0.1	[18]
21	Grosso	Italy	1.0	5.2	0.9	0.7	6.0	41.6	23.0	4.8	3.2	0.0	[52]
22	n.s.	Serbia	0.4	14.6	0.4	0.0	16.3	23.1	10.0	0.7	2.6	0.9	[53]
23	Grappenhall	Hungary	1.4	9.0	9.3	3.4	2.8	46.8	3.3	3.0	0.9	1.2	[10]
24	Grosso	Hungary	0.7	3.0	4.1	1.9	14.8	55.2	4.4	0.9	1.2	1.4	[10]
25	n.s.	China	0.0	44.8	0.0	0.5	12.2	7.6	0.0	0.6	0.0	0.0	[21]
26	Abrialis	Italy	0.5	7.0	3.0	8.3	9.4	40.3	18.4	0.6	1.4	0.8	[54]
27	Rinaldi Cerioni	Italy	0.4	10.0	0.0	0.0	11.5	65.8	0.5	2.9	0.0	0.6	[54]
28	Sumiens	Italy	0.7	12.1	3.0	0.4	6.8	48.0	14.9	0.3	0.0	0.0	[54]
29	Budrovka	Serbia	1.0	16.1	1.8	0.2	4.1	39.0	6.9	5.1	0.7	0.0	TS
	AVERAGE		1.3	12.0	2.0	2.4	9.1	34.1	11.6	2.1	1.2	0.4	
	RANGE		≤4.0	≤44.8	≤9.3	≤12.2	0.1–22.8	0.1–65.8	≤47.7	≤6.9	≤4.5	≤1.7	
	ISO STANDARD		0.5–1.5	4.0–7.0	0.5–1.5	tr-1.0	6.0–8.0	24.0–35.0	28.0–38.0	1.5–5.0	1.5–3.0	0.2–0.8	

n.s.—not specified; TS—this study (average value for three investigated year).

The primary source of variability in chemical composition and oil yield among the various populations of *L. x intermedia* are differences in environmental conditions [14].

According to 29 accessions of *L. x intermedia* from the literature, unrooted cluster tree (Figure 5) shows the presence of four chemotypes: (1) linalool + linalyl acetate, (2) linalool + 1,8-cineole, (3) linalool + camphor, and (4) 1,8-cineole + camphor. The first one, with dominant linalool and linalyl acetate, could be divided into three subgroups according to linalool:linalyl acetate ratio: (1) ratio ranged between 0.7 and 1.1—all three accessions from Turkey [4,49,50]; (2) ratio 1.8—one accession from Turkey [51]; and (3) ratio ranged from 1.8 to 5.5—five accessions from Italy [3,52,54], and one from Turkey [50]. In the second chemotype, the dominant compounds were linalool, and 1,8-cineole, with two subgroups: (1) 39.0–46.8% of linalool and 6.7–16.1% of 1,8-cineole [3,10,11,50], and this study; and (2) 55.2–65.8% of linalool and 3.0–10.0% of 1,8-cineole [9,10,54]. Linalool+camphor chemotype is present in Canada (five accessions, [19]) as well as with one sample from Turkey and Serbia [4,53]. Chemotype with dominant 1,8-cineole, and camphor is noted in Turkey (three accessions, [18]) and China [21].

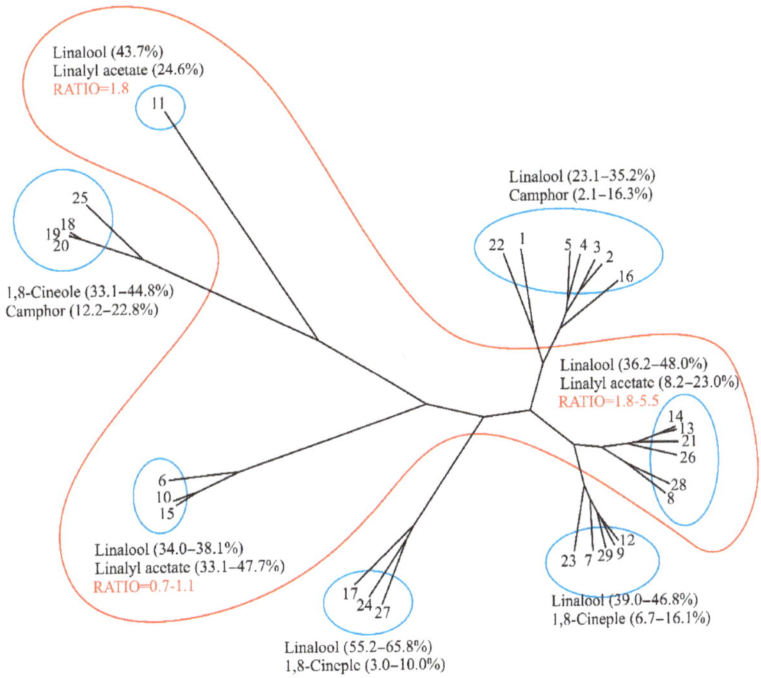

Figure 5. Unrooted cluster tree for chemical composition of *L. x intermedia* essential oil (samples are coded according to Table 6.

4.2. Hydrolate

Lavandula sp. hydrolates haves a characteristic delicate lavender scent [28]. However, the chemical composition of hydrolates and essential oil, depends on plant part (herb or flower), postharvest processing (fresh or dry plant material), and isolation technique [55,56]. The chemical composition of different *Lavandula* sp. hydrolates found in literature and from the study is shown in Table 7. As can be seen, all *Lavandula* sp. hydrolates contain linalool (ranged between 7.7% and 55.6%). The content of other significant compounds (average content more than 1.5% according to 19 samples from literature, and average value from this study) are: linalool oxides (≤67.3%), 1,8-cineole (≤52.9%), camphor (≤19.6%), borneol (≤17.3%), α-terpineol (≤13.0%), terpinen-4-ol (≤9.4%), and geraniol (≤5.0%). The most abundant group of compounds are oxygenated monoterpenes, mainly monoterpene alcohols [28]. The hydrosols were not found to contain linalyl acetate or sesquiterpenes, which are present in the essential oil [28,49].

Table 7. Chemical composition of different *Lavandula* sp. hydrolates according to literature and this study.

No.	Species/ Variety/ Cultivar	Origin	Extraction Technique/ Plant Material	Linalool	1,8-cineole	Linalool Oxides	α-terpineol	Camphor	Borneol	Terpinen-4-Ol	Geraniol	Reference
1	LI cv "Super"	Turkey	f.f., SD	55.6	9.8	6.0	0.0	13.4	13.5	0.0	1.6	[49]
2	LO	Morocco	n.s.	45.0	14.8	0.4	11.8	15.7	11.3	0.0	0.0	[57]
3	LA	Poland	HV400	39.2	0.0	19.2	7.1	1.3	4.8	4.6	2.9	[58]
4	LA	Poland	HV3200	29.0	0.0	19.1	12.7	3.0	9.3	6.9	0.0	[58]
5	LA	Poland	f.h.	53.0	2.9	0.4	8.3	0.9	5.3	3.9	4.0	[55]
6	LA	Poland	d.h.	48.0	2.7	0.6	8.8	1.5	5.8	6.6	5.0	[55]
7	LA	Poland	f.f.	43.6	4.4	1.0	7.5	0.0	6.6	5.6	3.4	[55]
8	LA	Poland	d.f.	39.2	0.0	19.2	7.0	1.3	4.8	4.6	2.9	[55]
9	LA	Poland	d.f., VH400	43.6	4.4	0.0	7.5	0.0	6.6	5.6	0.0	[28]
10	LA	Poland	d.f., VH800	44.9	2.0	0.0	8.5	1.3	5.2	5.4	3.4	[28]
11	LA	Poland	d.f., VH1200	43.9	4.0	1.2	5.8	2.7	4.0	3.9	4.5	[28]
12	LA	Poland	d.f., VH1600	25.7	2.7	1.4	4.1	1.0	4.6	3.7	1.1	[28]
13	LI	Italy	f.f.	43.8	25.4	0.1	1.8	12.8	4.3	4.5	0.0	[30]
14	LI	Italy	f.s.	34.4	28.9	0.0	2.2	15.4	4.0	2.7	0.2	[30]
15	LI	Serbia	d.f.	7.7	6.8	67.3	2.7	7.2	0.0	0.4	0.0	[53]
16	LA	Croatia	d.f., SD	7.9	20.6	21.0	10.4	0.4	0.0	1.1	1.0	[56]
17	LA	Croatia	d.f., HD	23.2	19.5	13.8	13.0	0.5	0.0	1.2	2.3	[56]
18	LI cv „Grosso"	Italy	n.s.	12.6	52.9	0.7	4.8	19.6	3.0	5.4	0.0	[29]
19	LA	Italy	f., SD	42.9	11.8	0.1	12.6	18.4	5.8	8.4	0.0	[31]
20	LI cv „Budrovka"	Serbia	f.f., SD	26.7	17.8	11.3	4.0	6.0	17.3	9.4	0.3	TS
	AVERAGE			35.5	11.6	9.1	7.0	6.1	5.8	4.2	1.6	
	RANGE			7.7–55.6	≤52.9	≤67.3	≤13.0	≤19.6	≤17.3	≤9.4	≤5.0	

LI—*Lavandula × intermedia*; LO—*Lavandula officinalis*; LA—*Lavandula angustifolia*; TS—This Study (average values for all three years); VH—Volume of Hydrosols in ml; SD—Steam Distillation; HD—Hydro Distillation; n.s.—not specified; f.f.—fresh flowers; f.h.—fresh herb; d.h.—dry herb; d.f.—dry flower; f.s.—fresh stem; f.—flowers.

Moroccan hydrolate of *L. officinalis* contained linalool as the dominant compound, followed by camphor, 1,8-cineole, α-terpineol, and borneol [57]. The main component of the volatile fraction of *L. angustifolia* from Poland was linalool, followed by α-terpineol, borneol, and geraniol [55], while in another study from Poland, it was linalool, followed by α-terpineol and terpinen-4-ol [28,58]. Hydrolate obtained from *L. angustifolia* buds from Croatia had the most significant proportion of 1,8-cineole, linalool oxide, and linalool [56]. Turkish *L. x intermedia* hydrolate had linalool, borneol, and camphor as the main compounds [49], while Italian *L. intermedia* samples contained linalool, 1,8-cineole, and camphor in different proportions [29–31]. Serbian *L. x intermedia* hydrolate had a high percentage of linalool oxides. A higher abundance of monoterpene alcohols oxides (*cis-* and *trans-*linalool) in hydrolate compared to the corresponding oil is probably due to their better solubility in water [53].

According to the unrooted cluster tree (Figure 6), it could be said that in *Lavandula* sp. hydrolate linalool dominated in almost all samples (in different proportions, from 25.7% to 55.6%), while in one sample, the dominant compound was linalool oxide [53], and in three it was 1,8-cineole [29,56]. Moreover, there is no difference between the chemical compositions of *L. angustifolia* and *L. x intermedia* hydrolates.

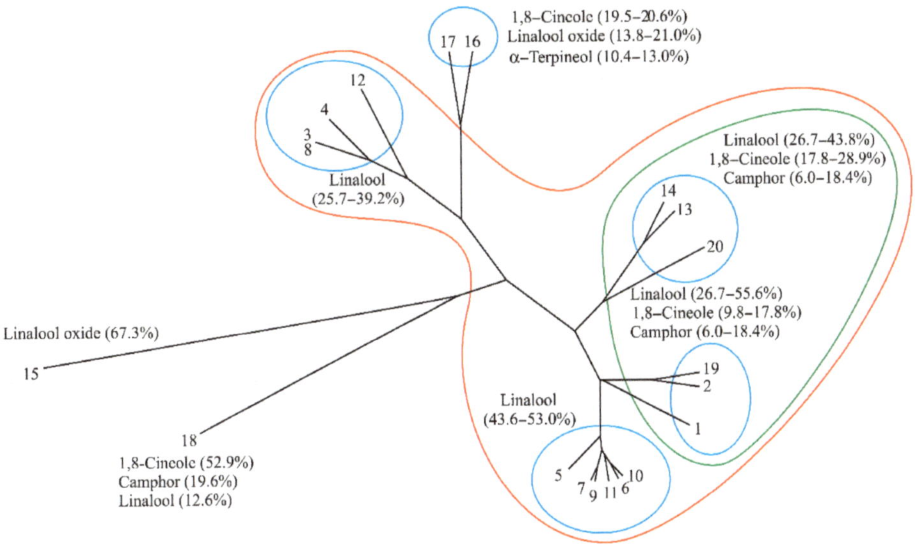

Figure 6. Unrooted cluster tree for chemical composition of *L. x intermedia* hydrolate according to literature and this study (samples are coded according to Table 7).

5. Conclusions

This research has proved that Fruška Gora (Serbia) has good agro-ecological conditions for cultivating *Lavandula* sp. and producing acceptable quality essential oil and hydrolate. Growing this species in this region could be recommended as Fruška Gora is a protected area suitable for organic farming. Since *Lavandula* sp. is a pollen and nectar-producing plant, it can positively affect the ecosystem's biodiversity. Honeybees are the most common visitors to *Lavandula* sp. fields and produce lavender honey. This type of unifloral honey has a high commercial value, which further supports the cultivation of this species. Tourism on Fruška Gora is quite developed due to many Orthodox monasteries, vineyards, and small wineries. Therefore, lavender fields could improve touristic content in this region following examples of other countries.

Author Contributions: Conceptualization, M.A. and B.L.; methodology, J.S.J.; software, L.P.; validation, M.C., M.P. and M.T.; formal analysis, J.S.J.; investigation, B.L.; resources, M.A.; data curation, L.P.; writing—original draft preparation, M.A.; writing—review and editing, V.T.; visualization, M.P.; supervision, B.L.; project administration, M.A.; funding acquisition, M.A., B.L. and M.P. All authors have read and agreed to the published version of the manuscript.

Funding: This research was funded by the Ministry of Education, Science and Technological Development of the Republic of Serbia, grant number 451-03-68/2022-14/200032, 451-03-68/2022-14/200134, 451-03-68/2022-14/200168, 451-03-68/2022-14/ 200051, 337-00-21/2020-09/40.

Institutional Review Board Statement: Not applicable.

Informed Consent Statement: Not applicable.

Data Availability Statement: Not applicable.

Acknowledgments: The authors are thankful to Branko Švonja for providing plant material, field photography, and initiating this investigation.

Conflicts of Interest: The authors declare no conflict of interest.

References

1. Lesage-Meessen, L.; Bou, M.; Sigoillot, J.C.; Faulds, C.B.; Lomascolo, A. Essential oils and distilled straws of lavender and lavandin: A review of current use and potential application in white biotechnology. *Appl. Microbiol. Biotechnol.* **2015**, *99*, 3375–3385. [CrossRef]
2. Szekely-Varga, Z.; Hitter, T.; Cantor, M. The healing power and the uses in landscape design of lavender (*Lavandula angustifolia* L.). *Hop Med. Plants* **2017**, *25*, 47–55.
3. Pistelli, L.; Najar, B.; Giovanelli, S.; Lorenzini, L.; Tavarini, S.; Angelini, L.G. Agronomic and phytochemical evaluation of lavandin and lavender cultivars cultivated in the Tyrrhenian area of Tuscany (Italy). *Ind. Crops Prod.* **2017**, *109*, 37–44. [CrossRef]
4. Kivrak, S. Essential oil composition and antioxidant activities of eight cultivars of lavender and lavandin from western Anatolia. *Ind. Crops Prod.* **2018**, *117*, 88–96. [CrossRef]
5. Florina, G.E.; Camelia, M.; Teodor, G.V. Lavender-perfume and business in Romania. *Agric. Res. Technol.* **2019**, *19*, 140–144. [CrossRef]
6. Detar, E.; Zambori-Nemeth, E.; Gosztola, B.; Harmath, A.; Ladanyi, M.; Pluhar, Z. Ontogenesis and harvest time are crucial for high quality lavender—Role of the flower development in essential oil properties. *Ind. Crops Prod.* **2021**, *163*, 113334. [CrossRef]
7. Pecanha, D.A.; Freitas, M.S.M.; Vieira, M.E.; Cunha, J.M.; de Jesus, A.C. Phosphorus fertilization affects growth, essential oil yield and quality of true lavender in Brazil. *Ind. Crops Prod.* **2021**, *170*, 113803. [CrossRef]
8. Stanković, I.; Vrandečić, K.; Ćosić, J.; Milojević, K.; Bulajić, A.; Krstić, B. The spreading of Alfalfa mosaic virus in lavandin in Croatia. *Pestic. Phytomed.* **2014**, *29*, 115–122. [CrossRef]
9. Blažeković, B.; Yang, W.; Wang, Y.; Li, C.; Kindl, M.; Pepeljnjak, S.; Vladimir-Knežević, S. Chemical composition, antimicrobial and antioxidant activities of essential oils of *Lavandula* × *intermedia* 'Budrovka' and *L. angustifolia* cultivated in Croatia. *Ind. Crops Prod.* **2018**, *123*, 173–182. [CrossRef]
10. Detar, E.; Zamborine Nemeth, E.; Gosztola, B.; Demjan, I.; Pluhar, Z. Effect of variety and growth year on the essential oil properties of lavender (*Lavandula angustifolia* Mill.) and lavandin (*Lavandula* × *intermedia* Emeric ex Liosel.). *Biochem. Syst. Ecol.* **2020**, *90*, 104020. [CrossRef]
11. Karapandzova, M.; Cvetkovikj, I.; Stefkov, G.; Stoimenov, V.; Crvenov, M.; Kulevanova, S. The influence of duration of the distillation of fresh and dried flowers on the essential oil composition of lavandin cultivated in Republic Macedonia. *Maced. Pharm. Bull.* **2012**, *58*, 31–38. [CrossRef]
12. Jianu, C.; Pop, G.; Gruia, A.T.; Horhat, F.G. Chemical composition and antimicrobial activity of essential oils of lavender (*Lavandula angustifolia*) and lavandin (*Lavandula* × *intermedia*) grown in Western Romania. *Int. J. Agric. Biol.* **2013**, *15*, 772–776.
13. Prusinowska, R.; Śmigielsk, K.B. Composition, biological properties and therapeutic effects of lavender (*Lavandula angustifolia* L.). A review. *Herba Pol.* **2014**, *60*, 56–66. [CrossRef]
14. Bajalan, I.; Pirbalouti, A.G. Variation in chemical composition of essential oil of populations of *Lavandula* × *intermedia* collected from Western Iran. *Ind. Crops Prod.* **2015**, *69*, 344–347. [CrossRef]
15. Pokajewicz, K.; Bialon, M.; Svydenko, L.; Fedin, R.; Hudz, N. Chemical composition of the essential oil of the new cultivars of *Lavandula angustifolia* Mill. bread in Ukraine. *Molecules* **2021**, *26*, 5681. [CrossRef] [PubMed]
16. Marincas, O.; Feher, I. A new cost-effective approach for lavender essential oils quality assessment. *Ind. Crops Prod.* **2018**, *125*, 241–247. [CrossRef]
17. Giray, F.H. An analysis of world lavender oil markets and lessons for Turkey. *J. Essent. Oil-Bear. Plants* **2018**, *21*, 1612–1623. [CrossRef]
18. Ozel, A. Determining leaf yield, some plant characters and leaf essential oil components of different cultivars of lavender and lavandin (*Lavandula* spp.) on the Harran plain ecological conditions. *Appl. Ecol. Environ. Res.* **2019**, *17*, 14087–14094. [CrossRef]
19. Lane, A.W.; Mahmoud, S.S. Composition of essential oil from *Lavandula angustifolia* and *L. intermedia* varieties grown in British Columbia, Canada. *Nat. Prod. Commun.* **2008**, *3*, 1361–1366. [CrossRef]
20. Stanojević, L.; Stanković, M.; Cakić, M.; Nikolić, V.; Nikolić, L.; Ilić, D.; Radulović, N. The effect of hydrodistillation techniques on yield, kinetics, composition and antimicrobial activity of essential oils from flowers of *Lavandula officinalis* L. *Hem. Ind.* **2011**, *65*, 455–463. [CrossRef]
21. Liao, Z.; Huang, Q.; Cheng, Q.; Khan, S.; Yu, X. Seasonal variation in chemical composition of essential oils extracted from lavandin flowers in the Yun-Gui plateau of China. *Molecules* **2021**, *26*, 5639. [CrossRef]
22. Bilia, A.R.; Guccione, C.; Isacchi, B.; Righeschi, C.; Firenzuoli, F.; Bergonzi, M.C. Essential oils loaded in nanosystems: A developing strategy for a successful therapeutic approach. *J. Evid.-Based Complement. Altern. Med.* **2014**, *2014*, 651593. [CrossRef] [PubMed]
23. Mariod, A.A. Chapter 13—Effect of essential oils on organoleptic (smell, taste, and texture) properties of food. In *Essential Oils in Food Preservation, Flavor and Safety*; Preedy, V.R., Ed.; Academic Press: Cambridge, MA, USA, 2016; pp. 131–137. [CrossRef]
24. Aćimović, M.; Tešević, V.; Smiljanić, K.; Cvetković, M.; Stanković, J.; Kiprovski, B.; Sikora, V. Hydrolates—By-products of essential oil distillation: Chemical composition, biological activity and potential uses. *Adv. Technol.* **2020**, *9*, 54–70. [CrossRef]
25. Jakubczyk, K.; Tuchowska, A.; Janda-Milczarek, K. Plant hydrolates—Antioxidant properties, chemical composition and potential applications. *Biomed. Pharmacother.* **2021**, *142*, 112033. [CrossRef]
26. Pino-Otin, M.R.; Navarro, J.; Val, J.; Roig, F.; Mainar, A.M.; Ballestero, D. Spanish *Satureja montana* L. hydrolate: Ecotoxicological study in soil and water non-target organisms. *Ind. Crops Prod.* **2022**, *178*, 114553. [CrossRef]

27. Ferraz, C.A.; Sousa, A.C.; Caramelo, D.; Delgado, F.; de Oliveira, A.P.; Pastorinho, R. Chemical profile and eco-safety evaluation of essential oils and hydrolates from *Cistus ladanifer*, *Helichrysum italicum*, *Ocimum basilicum* and *Thymbra capitata*. *Ind. Crops Prod.* **2022**, *175*, 114232. [CrossRef]
28. Prusinowska, R.; Smigielski, K.B. Hydrosols from lavender (*Lavandula angustifoia*)—Determination of the chemical composition using Dispersive Liquid-Liquid Microextraction (DLLME). *J. Essent. Oil-Bear. Plants.* **2015**, *18*, 519–528. [CrossRef]
29. Garzoli, S.; Petralito, S.; Ovidi, E.; Turchetti, G.; Masi, V.L.; Tiezzi, A.; Trilli, J.; Cesa, S.; Casadei, M.A.; Giacomello, P.; et al. *Lavandula × intermedia* essential oil and hydrolate: Evaluation of chemical composition and antibacterial activity before and after formulation in nanoemulsion. *Ind. Crops Prod.* **2022**, *145*, 1126068. [CrossRef]
30. Politi, M.; Menghini, L.; Conti, B.; Bedini, S.; Farina, P.; Cioni, P.L.; Braca, A.; De Leo, M. Reconsidering hydrosols as main products of aromatic plants manufactory: The lavandin (*Lavandula × intermedia*) case study in Tuscany. *Molecules* **2020**, *25*, 2225. [CrossRef] [PubMed]
31. Garzoli, S.; Masci, V.L.; Franceschi, S.; Tiezzi, A.; Giacomello, P.; Ovidi, E. Headspace/GC-MS analysis and investigation of antibacterial, antioxidant and cytotoxic activity of essential oils and hydrolates from *Rosmarinus officinalis* L. and *Lavandula angustifolia* Miller. *Foods* **2021**, *10*, 1768. [CrossRef] [PubMed]
32. Vujičić, M.D.; Vasiljević, D.A.; Markovi, S.B.; Hose, T.A.; Lukić, T.; Hadžić, O.; Janićević, S. Preliminary geosite assessment model (GAM) and its application on Fruška Gora Mountain, potential geotourism destination of Serbia. *Acta Geogr. Slov.* **2011**, *51*, 361–376. [CrossRef]
33. Renaud, E.; Charles, D. Essential oil quality and composition from 10 cultivars of organically grown lavender and lavandin. *J. Essent. Oil Res.* **2001**, *13*, 269–273. [CrossRef]
34. Giannoulis, K.D.; Evangelopoulos, V.; Gougoilias, N.; Wogiatzi, E. Lavender organic cultivation yield and essential oil can be improved by using bio-stimulants. *Acta Agric. Scand. Sect. B* **2020**, *70*, 648–656. [CrossRef]
35. Seidler-Lozykowska, K.; Mordalski, R.; Kucharski, W.; Kedzia, B.; Bocianowski, J. Yielding and quality of lavender flowers (*Lavandula angustifolia* Mill.) from organic cultivation. *Acta Sci. Pol. Hortorum Cultus* **2014**, *13*, 173–183.
36. Salehi, B.; Mnayer, D.; Özçelik, B.; Altin, G.; Kasapoğlu, K.N.; Daskaya-Dikmen, C.; Sharifi-Rad, M.; Selamoglu, Z.; Acharya, K.; Sen, S.; et al. Plants of the genus *Lavandula*: From farm to pharmacy. *Nat. Prod. Commun.* **2018**, *13*, 1385–1402. [CrossRef]
37. Blažeković, B.; Stanic, G.; Pepeljnjak, S.; Vladimir-Knezevic, S. In vitro antibacterial and antifungal activity of *Lavandula × intermedia* Emeric ex Loisel. 'Budrovka'. *Molecules* **2011**, *16*, 4241–4253. [CrossRef]
38. Vrandečić, K.; Ćosić, J.; Jurković, D.; Ilić, J. Bolesti lavande u istočnom delu Hrvatske. *Glas. Biljn. Zaštite* **2013**, *5*, 391–396. (In Croatian)
39. Aćimović, M.; Pezo, L.; Zeremski, T.; Lončar, B.; Marjanović Jeromela, A.; Stanković Jeremić, J.; Cvetković, M.; Sikora, V.; Ignjatov, M. Weather conditions influence on hyssop essential oil quality. *Processes* **2021**, *9*, 1152. [CrossRef]
40. Mészáros, M.; Markovlc, S.B.; Mijovkć, D.; Jovanovlć, M. Physical geographic characteristics and geo-heritage of Fruska gora mountain (Vojvodina, Serbia). *Acta Geogr. Szeged.* **2004**, *38*, 148–157.
41. Aćimović, M.; Cvetković, M.; Stanković Jeremić, J.; Pezo, L.; Varga, A.; Čabarkapa, I.; Kiprovski, B. Biological activity and profiling of *Salvia sclarea* essential oil obtained by steam and hydrodistillation extraction methods via chemometrics tools. *Flavour Fragr. J.* **2022**, *37*, 20–32. [CrossRef]
42. Aćimović, M.; Ivanović, S.; Simić, K.; Pezo, L.; Zeremski, T.; Ovuka, J.; Sikora, V. Chemical Characterization of *Marrubium vulgare* Volatiles from Serbia. *Plants* **2021**, *10*, 600. [CrossRef] [PubMed]
43. Fernández-Sestelo, M.; Carrillo, J.M. Environmental effects on yield and composition of essential oil in wild populations of spike lavender (*Lavandula latifolia* Medik.). *Agriculture* **2020**, *10*, 626. [CrossRef]
44. Salata, A.; Buczkowska, H.; Nurzynska-Wierdak, R. Yield, essential oil content, and quality performance of *Lavandula angustifolia* leaves, as affected by supplementary irrigation and drying methods. *Agriculture* **2020**, *10*, 590. [CrossRef]
45. Najar, B.; Demasi, S.; Caser, M.; Gaino, W.; Cioni, P.L.; Pistelli, L.; Scario, V. Cultivation substrate composition influences morphology, volatilome and essential oil of *Lavandula angustifolia* Mill. *Agronomy* **2019**, *9*, 411. [CrossRef]
46. Hassiotis, C.N.; Ntana, F.; Lazari, D.M.; Poulios, S.; Vlachonasios, K.E. Environmental and developmental factors affect essential oil production and quality of *Lavandula angustifolia* during flowering period. *Ind. Crops Prod.* **2014**, *62*, 359–366. [CrossRef]
47. Porter, N.G.; Shaw, M.L.; Hurndell, L.C. Preliminary studies of lavender as an essential oil crop for New Zealand. *N. Z. J. Agric. Res.* **1982**, *25*, 389–394. [CrossRef]
48. Desautels, A.; Biswas, K.; Lane, A.; Boeckelmann, A.; Mahmoud, S.S. Suppression of linalool acetate production in *Lavandula × intermedia*. *Nat. Prod. Commun.* **2009**, *4*, 1533–1536. [CrossRef]
49. Baydar, H.; Kineci, S. Scent composition of essential oil, concrete, absolute and hydrosol from lavandin (*Lavandula × intermedia* Emeric ex Loisel.). *J. Essent. Oil-Bear. Plants.* **2009**, *12*, 131–136. [CrossRef]
50. Kara, N.; Baydar, H. Determination of lavender and lavandin cultivars (*Lavandula* sp.) containing high quality essential oil in Isparta, Turkey. *Turk. J. Field Crop.* **2013**, *18*, 58–65.
51. Karaman, R.; Erbas, S.; Baydar, H.; Kaya, M. Allelopathic effect of lavandin (*Lavandula × intermedia* Emeric ex Loilsel. Var Super A) oil on germination and seedling development of some weed and field crops. *Harran Tarım ve Gıda Bilimleri Derg.* **2014**, *18*, 35–41.
52. Garzoli, S.; Turchetti, G.; Giacomello, P.; Tiezzi, A.; Masci, V.L.; Ovidi, E. Liquid and vapour phase of lavandin (*Lavandula × intermedia*) essential oil: Chemical composition and antimicrobial activity. *Molecules* **2019**, *24*, 2701. [CrossRef] [PubMed]

53. Pljevljakušić, D.; Drinić, Z. GC/MS chemical analysis of lavandin (*Lavandula × intermedia*) hydrolat: Sucessive extraction fraction. *Lek. Sirovine* **2020**, *40*, 33–39. [CrossRef]
54. D'Addabbo, T.; Laquale, S.; Argentieri, M.P.; Bellardi, M.G.; Avato, P. Neatocidal activity of essential oil from lavandin (*Lavandula × intermedia* Emeric ex Loisel.) as related to chemical profile. *Molecules* **2021**, *26*, 6448. [CrossRef] [PubMed]
55. Kunicka-Styczynska, A.; Smigielski, K.; Prusinowska, R.; Rajkowska, K.; Kusmider, B.; Sikora, M. Preservative activity of lavender hydrosols in moisturizing body gels. *Lett. Appl. Microbiol.* **2014**, *60*, 27–32. [CrossRef]
56. Silha, D.; Švarcova, K.; Bajer, T.; Kralovec, K.; Tesarova, E.; Mouckova, M.; Bajerova, P. Chemical composition of natural hydrolates and their antimicrobial activity on Arcobacter-like cells in comparison with other microorganisms. *Molecules* **2020**, *25*, 5654. [CrossRef]
57. Azza, S.; Lyoussi, B.; Miguel, M. Antioxidant activity of some Morrocan hydrosols. *J. Med. Plants Res.* **2011**, *5*, 6688–6696. [CrossRef]
58. Prusinowska, R.; Smigielski, K.; Kunicka-Styczynska, A. Hydrolates from lavender (*Lavandula angustifolia*)—Chemical composition, antimicrobial and antioxidant properties. *PhD Interdiscip. J.* **2013**, *3*, 33–39.

horticulturae

MDPI

Review

The Chemistry, Sensory Properties and Health Benefits of Aroma Compounds of Black Tea Produced by *Camellia sinensis* and *Camellia assamica*

Asma Parveen, Chun-Yin Qin, Feng Zhou, Guoping Lai, Piaopiao Long, Mengting Zhu, Jiaping Ke * and Liang Zhang *

State Key Laboratory of Tea Plant Biology and Utilization, Anhui Agricultural University, Hefei 230036, China; asmaparveen12@126.com (A.P.); qin_chunyin@163.com (C.-Y.Q.); zhoufeng99032@126.com (F.Z.); laiguoping20@126.com (G.L.); piaopiao_long@stu.ahau.edu.cn (P.L.); zhumengting@ahau.edu.cn (M.Z.)

* Correspondence: jiapingwin@163.com (J.K.); zhli2091@sina.com (L.Z.); Tel./Fax: +86-551-85786765 (L.Z.)

Abstract: Black tea is frequently consumed worldwide and is renowned for having a distinctive scent. The volatile chemicals in tea are responsible for its aroma, which is important for sensory quality. The enzymatic and non-enzymatic processes that produce the compounds endowing black tea with its distinctive aroma are complex. Black tea is well known for its robust and complex aroma, which can vary based on the type of tea leaves used and processing technologies used. During the production of black tea, several intricate biological and chemical processes contribute to the aroma's development. Different volatile chemicals are generated during the processing of black tea, which includes withering, rolling, fermentation (enzymatic oxidation), and drying. Various methods have been used to analyze and describe the aroma of black tea. Different methods, such as gas chromatography-mass spectrometry, olfactometry, and solid phase extraction, have been used to assess the fragrance of black tea. These methods evaluate characteristics including fruity, flowery, woody, malty, spicy, and smoky flavors. Phenyl ethyl alcohol, one-octen-3-ol, *trans*-linalool oxide (furanoid), geraniol, and nonanal are major aroma-active compounds in black tea. Sensory analytic techniques are used to assess black tea's flavor and scent qualities. This assessment helps figure out the tea's quality, unique characteristics, and even some unpleasant attributes. Black tea is mostly made from *Camellia sinensis* and *Camellia assamica* tea varieties. These two varieties, members of the *Camellia* genus, differ in their development patterns, leaf sizes, and chemical makeup, impacting how black tea's scent is formed. When evaluating black tea made from *Camellia sinensis* and *assamica*, sensory analysis involved assessing the aroma when the tea was dry and, after brewing, noting any differences from other teas. This review focused on how key aromatic compounds are formed during the tea manufacturing process by *Camellia sinensis* and *Camellia assamica* black tea.

Keywords: black tea; aroma; processing aroma; formation mechanism; fermentation (enzymatic oxidation)

Citation: Parveen, A.; Qin, C.-Y.; Zhou, F.; Lai, G.; Long, P.; Zhu, M.; Ke, J.; Zhang, L. The Chemistry, Sensory Properties and Health Benefits of Aroma Compounds of Black Tea Produced by *Camellia sinensis* and *Camellia assamica*. *Horticulturae* **2023**, *9*, 1253. https://doi.org/10.3390/horticulturae9121253

Academic Editor: Charalampos Proestos

Received: 11 October 2023
Revised: 2 November 2023
Accepted: 20 November 2023
Published: 22 November 2023

1. Introduction

Tea, a widely enjoyed beverage with numerous health advantages, is produced from the leaves and leaf buds of flowering plants categorized within the *Camellia* genus, which belongs to the Theaceae family. Tea plants used for producing various types of tea have generally been categorized into two primary varieties: *Camellia sinensis* var. *sinensis*, which is grown in temperate climates, and *Camellia sinensis* var. *assamica*, which is grown in tropical climates. More and more genetic research has clearly clarified that southwest China is an important origin of tea plants [1,2]. In the UK, Sealy named the tea plant "*Camellia sinensis* (L.) O. Kuntze" and identified two variations: "var. *sinensis*" (small-leaf variant) and "var. *assamica*" (large-leaf variant). The scientific name "*Camellia sinensis* (L.) O. Kuntze" has remained consistent despite subsequent studies [3].

Tea cultivation is rooted in southwest China, particularly Yunnan Province, where the tradition of brewing tea infusion is believed to have originated thousands of years ago [4]. Diverse manufacturing processes yield over 300 distinct tea varieties from *Camellia sinensis* leaves. Briefly, tea can be divided into six categories: green tea, white tea, yellow tea, oolong tea, black tea, and dark tea. Therein, black tea production involves the maceration of fresh tea leaves, resulting in the distinctive pigments of black tea, namely theaflavins (TFs) and thearubigins (TRs). These compounds form through the oxidation and polymerization of catechins during tea manufacturing [5]. Black tea covers over 70% of tea production worldwide (Figure 1) [6]. In addition, black tea is particularly popular in Europe and North America [7]. Worldwide, black tea production will reach 4.17 million tons in 2023, up 2.9% annually [8].

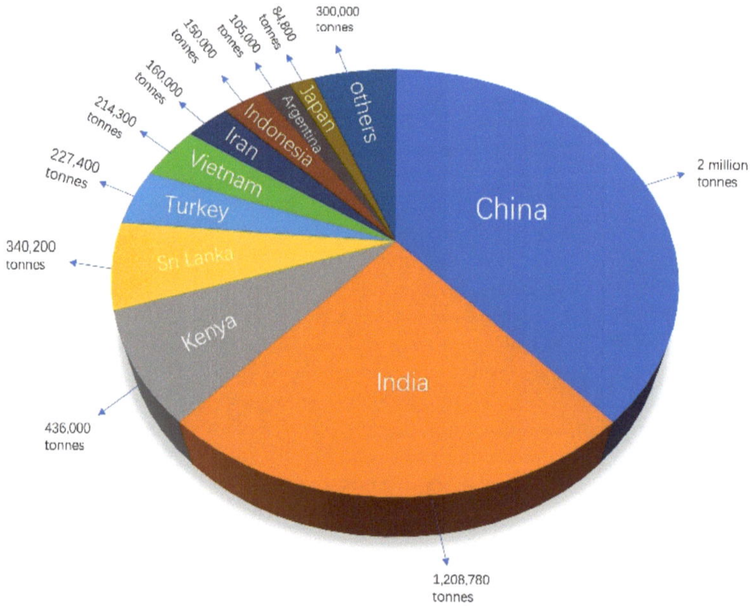

Figure 1. Worldwide Tea Production in 2022.

Tea's aroma is a pivotal factor significantly influencing the character and quality of a tea [9]. Research has meticulously explored fluctuations in the levels of volatile compounds and aroma precursors throughout the fermentation phase of black tea [10,11]. The chemical composition of black tea remains dynamic during tea processing. Black tea's volatile compounds primarily stem from the oxidation, degradation, and hydrolysis of aroma precursors, such as carotenoids, unsaturated fatty acids, and glycosidic aroma compounds [12]. The polyphenol oxidation process during black tea fermentation is not only critical in producing black tea pigments, but also participates in the formation of aroma compounds [13].

The cultivar, growing environment, and processing all result in the variances in volatile chemicals between different types of tea [14–16]. Overall, a complex process involving the interaction of numerous chemicals during tea production brings about the essential fragrance constituents of black tea scent. Black teas have a large number of volatile flavor compounds, although only a small number of them have an odor. Odor Activity Values (OAVs) are used to determine the overall aromatic quality of a substance. Such volatile flavor compounds are considered to be key volatile chemicals, active compounds, or key odorants [17,18].

The idea of tea sensory evaluation and human response are closely associated since sensory tests typically involve the use of the eyes, tongue, and nose. We are able to evaluate the sensory characteristics of product color, appearance, and taste of dry tea and its infusion. Therefore, it becomes appropriate to employ a technique to assess the qualities of the product and consumer acceptability using the senses. Tea aroma, a vital aspect of sensory quality, relies on volatile compounds. Quality attributes for aroma compounds are defined by a taste panel and, on occasion, by analytical methods such as mass spectrometry, gas chromatography, olfactometry (GC-MS/O), and electronic nose, in accordance with established protocols for sensory evaluation. Tea aroma, a vital aspect of sensory quality, relies on volatile compounds. Quality attributes for aroma compounds are defined using methods such as taste panels and analytical techniques like gas chromatography and mass spectrometry (Table 1). The five phases of tea preparation were examined for variations in volatile and non-volatile chemicals using GC–MS based metabolomic analyses [11].

Table 1. Comparative summarization on the different extraction method of tea volatiles.

Extraction Method	Technique	Detector/ Sensor	Application	Reference
Solid phase extraction (SPE)	GC-MS		Used with SPME, and solvent-assisted flavor evaporation (SAFE), respectively, for the analysis of tea body notes, tea infusion, and dry tea aroma from six model manufacturing processes	[19]
	GC	Mass spectrometry (MS), flame ionization (FID) and olfactometry (O)	Used to extract and characterize the aroma compounds in the infusion of Hanzhong black tea	[20]
Solid-phase microextraction (SPME)			Optimized to study the effect of grinding and brewing on tea volatiles of tea (*Camellia sinensis*) and SAFE applied for the analysis of tea volatiles in tea	[21]
	GC-MS		Used to detect volatile compounds in black teas after withering–shaking and shaking–withering processing.	[22]
Stir bar sorptive extraction (SBSE)	GC-O, GC-MS	Gas chromatography-flame photometric detector (GC-FPD)	Used to extracted compound of three famous black teas from around the world—Darjeeling, Keemun, and Ceylon	[23]
Headspace solid-phase microextraction (HS-SPME)	GC-MS		Analyzed the volatile flavor compounds of 31 black tea samples from 7 districts in Guangdong	[24]
	GC×GC-TOF-MS combined with GC-O/GC-TOF-MS		Analyzed the aroma components of 24 samples of the world's four black teas	[25]
	GC-MS		Aroma characteristics and volatile components of Jiangxi Congou black tea.	[26]
	GC-MS		Analyzed volatile flavor compounds in 112 black teas from seven countries by untargeted metabolomics	[27]
	GC-MS, GC-O		Optimized with SAFE method to analyse characteristics of key aroma compounds of three kinds of Chinese representative black tea	[28]
	GC-MS		To extract the volatile compounds when analyzing the volatile profile of black tea	[29]
	Ultra-fast Gas Chromatography (UFGC) analyzer	E-nose detector	Used to identify the variations of Assam black tea (*Camellia Sinensis* (L.) O. Kuntze) with their aroma compounds	[30]
	GC×GC-TOF-MS		Used to identify 158 volatile compounds during the fermentation period of black tea	[13]
	GC-MS, GC-O		Identification of aroma-active components in black teas produced by six Chinese tea cultivars in high-latitude regions	[31]
	GC-MS	E-nose	Main aroma components and aroma differences of black tea produced from two tea cultivars, Fuyun 6 and Jinguanyin	[32]
	GC-MS	Quartz crystal microbalance (QCM)	Used to detect linalool gas in orthodox black tea	[33]

This article details the sensory analysis methods used to assess the black tea scent profile. It explains the various scent constituents and how they impact the flavor of the tea. Additionally, this study highlights the importance of sensory analysis in establishing tea quality. The interactions between the manufacturing processes, including withering, rolling, fermentation, and drying, are also covered in this paper. In addition, this article also discusses the different aroma characteristics and chemical compositions of black tea made from *Camellia sinensis* and *Camellia assamica* (Figure 2), comparing the different types and concentrations of volatile chemicals. This study sheds important light on how different species of tea plants affect the scent that results. We used a thorough strategy with targeted keywords and reviewed multiple databases to achieve a thorough data collection process. The following keywords were utilized, either singly or in combination: processing, fermentation (enzymatic oxidation), black tea, aroma, molecular sensory, quantitative descriptive analysis, aroma extraction dilution analysis, and aroma production mechanism. These keywords were chosen because they were pertinent to the subject of our study and the particular goals we had in mind. The electronic bibliographic databases PubMed, ScienceDirect (Elsevier), Springer, and Nature yielded 466 articles when searched; after removing duplicates, the number of entries was reduced to 389. 108 records remained after the 389 records were sifted using preset inclusion criteria. These databases were chosen because they have a history of holding top-notch research articles on the topics we are interested in. Within these databases, we used advanced search capabilities and search filters to focus our search and find the most pertinent articles. To make sure that our information processing was accurate and consistent, we used standardized data extraction methodologies. We are sure that our strategy has allowed us to gather an extensive and thorough dataset to back up the research results.

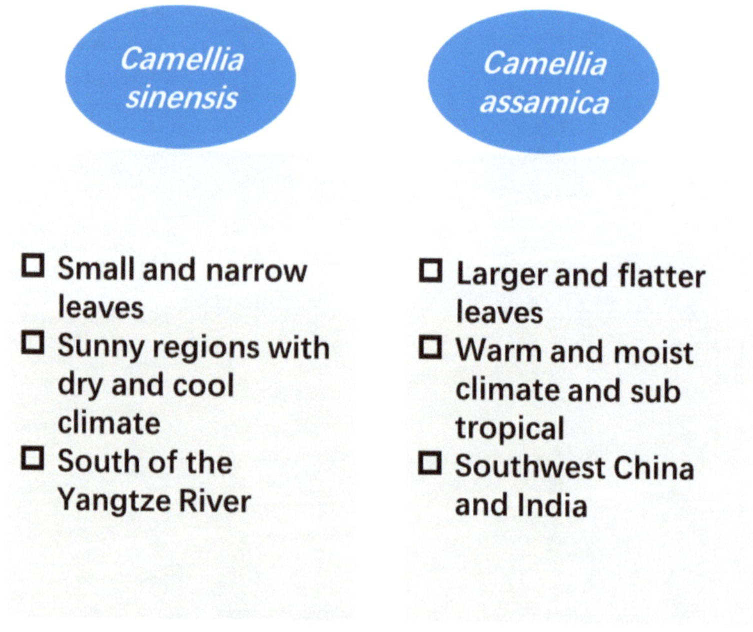

Figure 2. Morphological differences between the two varieties.

2. The Main Aroma Compounds of Black Tea

Extensive research has delved into the chemical composition of tea leaves, with the primary constituents belonging to the polyphenol group accounting for 18 to 36% of the dry weight. Additionally, *C. sinensis* encompasses many chemical constituents, including methylxanthines, amino acids, chlorophyll, carotenoids, lipids, carbohydrates, vitamins, and an impressive roster of over 600 volatile compounds. The chemical composition of black tea comprises both volatile and non-volatile compounds. In tea, volatile organic components (VOCs) exist in exceedingly small quantities, comprising a mere 0.01% of the total dry weight. Despite their minuscule presence, these compounds profoundly influence tea products' flavor due to their remarkably low threshold values and high odor units [34–36]. Some studies revealed volatile flavor compounds with floral and pleasant odor features like geraniol, phenyl ethyl alcohol, β-ionone, and linalool [26].

According to Wang et al. [27], distinctive volatile compounds can distinguish between various black teas. For instance, Indian black teas contain linalool, pentanoic acid, and hexanoic acid, while Chinese black teas feature phenylethyl alcohol. Sri Lankan black teas, on the other hand, are characterized by 1-methyl-naphthalene and β-ionon. Kangra orthodox black tea contains several major volatile constituents, including geraniol, linalool, (Z/E)-linalool oxides, (E)-2-hexenal, phytol, β-ionone, hotrienol, methylpyrazine, and methyl salicylate [36]. A comparative analysis of gas chromatography-olfactometry (GC-O) and odor activity value (OAV) calculations revealed specific odorants that significantly define distinct tea categories. Methyl salicylate in Ceylon tea, (E)-2-octenal in Assam tea (*Camellia assamica*), benzeneacetaldehyde in Keemun tea (*Camellia sinensis*), and linalool and *trans*-linalool oxide (furanoid) in Darjeeling tea (*Camellia sinensis*) emerged as the most definitive odor contributors in their respective tea categories [37].

According to certain researchers' findings, compounds like linalool oxide I, II, and III, E,E-2,4-nonadienal,4,5-dimethyl-3-hydroxy-2,5-dihydrofuran-2-one, 1-octen-3-one, E,Z-2,6-nonadienal, and bis(2-methyl-3-furyl) disulfide displayed higher odor activity values in tea infusions, imparting floral, fatty, caramel, mushroom, cucumber, and cooked beef-like aromas. Overall, floral, mushroom-like, and caramel-like scents predominated in Hanzhong (*Camellia sinensis*) black tea infusions [20]. Su et al. [38] highlighted the positive influence of geraniol, linalool, and methyl salicylate on preserving the floral flavor characteristic of Keemun black tea. Steeping temperature plays a critical role in shaping the aroma of Yingde black tea infusion, as confirmed by a study combining instrumental analysis and sensory evaluations [39]. Xinyang black tea (XYBT) is recognized for its honey sugar-like aroma, a feature developed during fermentation.

Study have found that (E,E,Z)-2,4,6-nonatrienal is the primary odorant component of Darjeeling black tea [37]. The main aroma volatiles found in black tea are geraniol, linalool, (Z/E)-linalool oxides, (E)-2-hexenal, phytol, β-ionone, hotrienol, methylpyrazine and methyl salicylate [36].

Some researchers have also extensively demonstrated that Yunnan Congou black teas of varying grades exhibited similar aroma compositions but differed significantly in content. Superior Yunnan Congou black tea showcased higher levels of linalool, linalool oxide, α-terpineol, and geraniol than their inferior counterparts. Conversely, inferior Yunnan black tea had elevated levels of leaf aldehyde, benzaldehyde, and benzeneacetaldehyde. In conclusion, higher grade Yunnan Congou black tea exhibited superior fragrance quality. In contrast, all grades of Congou black tea have a robust fragrance, strong aroma, and distinctive taste compared to Dianhong black tea. Essential aroma-active compounds responsible for the fundamental Congou black tea aroma included linalool, (E)-furan linalool oxide, (Z)-pyran linalool oxide, methyl salicylate, β-myrcene, and phenylethyl alcohol, as identified through GC-O analysis of representative samples [40,41].

According to Wang et al. [31], phenylethyl alcohol, geraniol, linalool, α-ionone, *cis*-3-hexenyl hexanoate, and methyl salicylate play pivotal roles as major contributors to the floral fragrance of black tea. Notably, they are prominent in Jinxuan, Longjing, and Baihaozao varieties, constituting a substantial proportion of their aroma profiles. Sensory-directed aroma recombination and omission tests further confirmed the importance of phenylacetaldehyde, linalool, geraniol, and 3-ethyl-2,5-dimethylpyrazine in shaping the sensory characteristics of high-grade Dianhong black tea, which is dominated by floral, sweet, and caramel-like odors [42]. Moreover, linalool and benzeneacetaldehyde play significant roles in creating the sweet aroma of Assam black tea. These compounds exhibit significantly high OAVs and contribute to the sweet and floral attributes of the tea's aroma [25]. Black teas with a sweat and floral aroma had much higher alcohol content than those with a sweet aroma alone. Linalool and its oxides, known as linalool oxides, were among the newly found alcohol components and were crucial odorants for black tea, which adds flowery and citrus flavors. Lilac, citrus, and mushrooms are among the scents connected to oct-1-en-3-ol, created by the oxidative breakdown of linoleic acid [39]. According to Schuh and Schieberle [37], black tea contains several aroma-active compounds, including geraniol, phenyl ethyl alcohol, methyl salicylate, phenylacetaldehyde, (*E*)-2-octenal, linalool, 1-octen-3-ol, benzyl alcohol, (*Z*)-jasmone, trans-linalool oxide (furanoid), and trans-β-ionone [25,43]. Linalool oxide has a sweet scent, while geraniol, phenylacetaldehyde, benzaldehyde, methyl salicylate, phenyl ethanol, and hexanal have a flowery fragrance. Additionally, as indicated, the ingredients give black tea its characteristic scent [44]. Forty-six aroma substances were identified in the two cultivars through GC-MS analysis, with nine exhibiting significant differences. These include heptaldehyde, furfural, 2,2,6-trimethylcyclohexanone, 2-pentadec-2-enylfuran, *cis*-3-hexenyl acetate, carveol, methyl salicylate, 5-methyl-2-hexanone, and hexanal in black tea from the Yunnan, Fujian, and Yichang regions [45].

The chemical compounds responsible for flavor and aroma in black tea are as follows: linalool and linalool oxide are responsible for the sweet flavor and citrus/lemon odor. Geraniol attributed rose odor and a floral flavor, and phenyl acetaldehyde is associated with distinctive hyacinth notes and floral flavors. Benzaldehyde brings fruity and almond smells—the fruity and minty smell given off by methyl salicylate, while phenyl ethanol is the source of the fruity and honey smell. Hexanal brings the grassy and fresh smell [46].

The color and taste of black tea are associated with TFs and TRs produced by catechins as they oxidize. They are responsible for giving black tea the dark color and astringent taste. Linalool is responsible for a floral and sweet aroma in black tea. Black tea has undergone complete fermentation (enzymatic oxidation), and its floral aroma contributes significantly to the aroma's complexity. This article examines some of the most significant major fragrance components and their manufacturing technique (Table 2 and Figure 3).

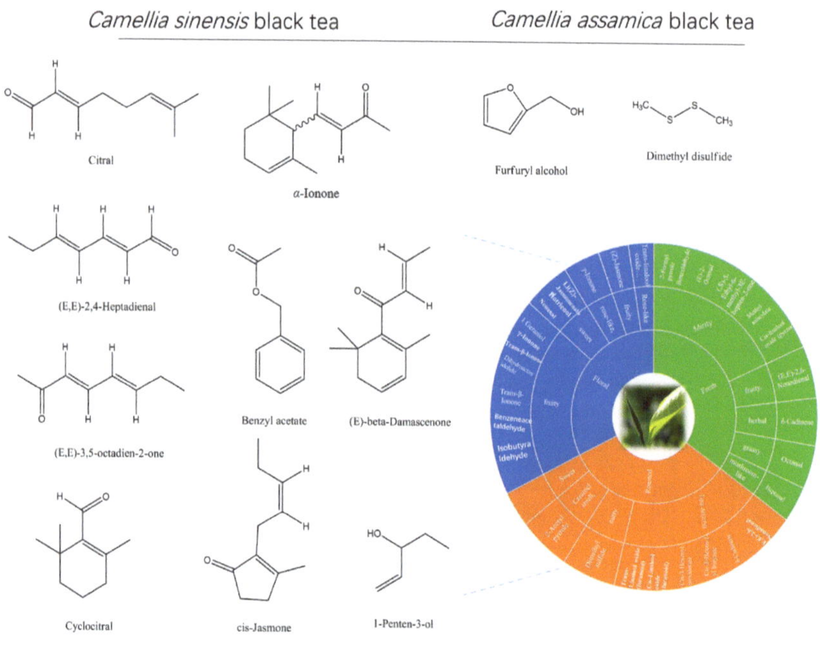

Figure 3. Main typical aroma compounds in black tea of *Camellia sinensis* and *Camellia assamica*.

Table 2. The critical aroma compounds of various types of black tea.

Group	Compounds	Ots (ug/L)	*C.sinensis*	*C.assamica*	Aroma Quality	Identification	References
Aldehydes	(*E*)-2-Hexenal	82	✓	✓	Green apple-like, bitter almond-like, green grass	MS, RI, STD	[20,21,28,47]
	Benzeneacetaldehyde	52	✓	✓	Rose-like floral, honey	MS, RI, STD	[21,25,28]
	Citral	5	✓	×	Fruity, citrus	MS, RI, STD	[24,47,48]
	Benzaldehyde	320	✓	✓	Nutty, bitter almond, oily, green, floral	MS, RI, STD	[21,25,28]
	(*E,E*)-2,4-heptadienal	56	✓	×	Fatty, green, vegetable	MS, RI, STD	[23,28]
	Furfural	9.56	✓	✓	Sweet, woody, almond smell	MS, RI, STD	[23,28]
	Hexanal	2.4	✓	✓	Fresh green grass	MS, RI, STD	[23,28]
	Neral	n.m	-	-	Lemon peel, citrus	MS, RI, STD	[28]
	(*E*)-2-Pentenal	310	✓	×	Fruity	MS, RI, STD	[20,43]
	(*E,E*)-2,4-Nonadienal	0.06	✓	✓	Fatty	MS, RI, STD	[20,25,37,47]
	Nonanal	1	✓	✓	Floral, fresh	MS, RI, STD	[25]
	(*E,Z*)-2,6-Nonadienal	0.0045	✓	✓	Fresh, cucumber-like	MS, RI, STD	[20,25,37,47]
	(*E*)-2-Nonenal	0.39	✓	✓	Fatty, green	MS, RI, STD	[23,25,37]
	(*E*)-2-Octenal	3	✓	✓	Fatty, greece odor	MS, RI, STD	[24,25,48]
	5-Methyl furfural	500	✓	✓	Sweet, maple	MS, RI, STD	[28]
	2-Methylbutanal	1.5	✓	✓	Musty, cocoa	MS, RI, STD	[23,28,48]
	3-Methylbutanal	0.5	✓	✓	Fruity, chocolate	MS, RI, STD	[23,25,28,48]
	2-Methylpropanal	1.9	✓	✓	Fresh aldehyde	MS, RI, STD	[25,48]
	beta-Cyclocitral	3	✓	×	Tropical saffron, herbal	MS, RI, STD	[23,28]
	(1R)-(-)-Myrtenal	n.m	-	-	sweet, minty	MS, RI, STD	[28]
	2-Phenyl-2-Butenal	n.m	-	-	Floral, black tea	MS, RI, STD	[28]
Ketones	6-Methyl-5-hepten-2-one	160	✓	✓	Citrus, green	MS, RI, STD	[25,28,43,47]
	α-Ionone	58	✓	×	Woody, floral, violet incense	MS, RI, STD	[20,24,25,28,47,48]
	β-Ionone	21	✓	✓	Woody, floral, fruity, violet odor	MS, RI, STD	[20,23–25,28,37,43,47,48]
	2-Nonanone	5	✓	✓	Green	MS, RI, STD	[25]
	Coumarin	n.m	-	-	Sweet, hay, bean	MS, RI, STD	[23,28]
	(*E,E*)-3,5-Octadien-2-one	150	✓	×	Fruity, green	MS, RI, STD	[28,48]
	3-methylnonane- 2,4-dione	0.01	✓	✓	Hay-like	MS, RI, STD	[28,37]

Table 2. *Cont.*

Group	Compounds	Ots (ug/L)	*C.sinensis*	*C.assamica*	Aroma Quality	Identification	References
Ketones	1-Octen-3-one	0.5	✓	✓	Mushroom-like	MS, RI, STD	[20,25,37,47]
	β-Damascenone	0.004	✓	×	Floral, fruity	MS, RI, STD	[28,37]
	cis-Jasmone	24	✓	×	Woody, herbal	MS, RI, STD	[25,28,43]
	2-Heptanone	0.14	✓	×	Fruity, cheese	MS, RI, STD	[28,43]
	Isophorone	n.m	-	-	Sweet, woody	MS, RI, STD	[28]
Alcohols	Linalool	19	✓	✓	Citrus, floral	MS, RI, STD	[20,23–25,28,48]
	Geraniol	27	✓	✓	Sweet floral	MS, RI, STD	[20,23–25,28,37,43,47,48]
	Phenylethyl alcohol	772	✓	✓	Rose, floral	MS, RI, STD	[20,24,25,28,37,43,48]
	(*Z*)-3-Hexenol	70	✓	✓	Fresh grass	MS, RI, STD	[25,28,37]
	Benzyl alcohol	11,076	✓	✓	Floral, bitter almond-like	MS, RI, STD	[20,23–25,28,43]
	cis-Linaloloxide	320	✓	✓	Citrus, floral, sweet woody	MS, RI, STD	[24,28]
	Nerol	528	✓	×	Neroli, citrus	MS, RI, STD	[24,28]
	α-Terpineol	404	✓	✓	Lilac, woody, terpene	MS, RI, STD	[23,28,37,43,47]
	1-Hexanol	500	✓	✓	Oily, fruity	MS, RI, STD	[24,28,43]
	(*Z*)-3-Hexen-1-ol	8	✓	×	Green, grass, sweet	MS, RI, STD	[23,25,43]
	(*E*)-2-hexen-1-ol	1900	✓	✓	Green, Leaf, walnut, woody	MS, RI, STD	[23,24,37,43,48]
	trans-Linaloloxide	320	✓	✓	Citrus, floral	MS, RI, STD	[25,28]
	2-Pentenol	400	✓	×	Fruity	MS, RI, STD	[28]
	1-Penten-3-ol	400	✓	×	Green, radish	MS, RI, STD	[20,28]
	2-Ethyl-1-hexanol	300	✓	✓	Citrus, sweet	MS, RI, STD	[28]
	Furfuryl alcohol	4.5	×	✓	Sweet, caramel	MS, RI, STD	[28]
	1-Octen-3-ol	45	✓	✓	Mushroom, earthy	MS, RI, STD	[20,23,25,28,43]
	Hotrienol	110	✓	✓	Lavender	MS, RI, STD	[25,28]

Table 2. *Cont.*

Group	Compounds	Ots (ug/L)	*C.sinensis*	*C.assamica*	Aroma Quality	Identification	References
Acids	Benzoic acid	n.m	-	-	Faint, balsam	MS, RI, STD	[28]
	Geranic acid	n.m	-	-	Green, woody	MS, RI, STD	[28]
	(*E*)-2-Hexenoic acid	1900	✓	✓	Fruity, sweet	MS, RI, STD	[28]
	(*E*)-3-Hexenoic acid	n.m	-	-	Green, woody	MS, RI, STD	[28]
	Hexanoic acid	1000	✓	✓	Fatty, cheesy, sweet odor	MS, RI, STD	[24,28]
	Butanoic acid	1000	✓	×	Sharp acetic	MS, RI, STD	[28]
Esters	Methyl salicylate	75	✓	✓	Winter green-like	MS, RI, STD	[23,25,28,48]
	Dihydroactinidiolide	0.0021	✓	×	Musk, rose-like, fruit, woody	MS, RI, STD	[23,25,28,43,48]
	Methyl hexanoate	10	✓	×	Fruity, pineapple	MS, RI, STD	[28]
	Benzyl acetate	30	✓	×	Floral, fruity	MS, RI, STD	[28]
	γ-Butyrolactone	50	✓	✓	Creamy, oily	MS, RI, STD	[28]
Hydrocarbons	β-Ocimene	48	✓	✓	Citrus, tropical	MS, RI, STD	[28]
	β-Myrcene	1.2	✓	✓	Pepper woody, sweet citrus, balsamic aroma	MS, RI, STD	[23,25,28]
	D-Limonene	200	✓	✓	Citrus, orange, light flowers	MS, RI, STD	[23,28,48]
	Styrene	50	✓	✓	Sweet floral, balsam	MS, RI, STD	[28,48]
	Longifolene	n.m	-	-	Sweet, woody	MS, RI, STD	[28]
	Naphthalene	n.m	-	-	Balmy	MS, RI, STD	[28]
Sulfide	Dimethyl sufide	0.84	✓	✓	Corn	MS, RI, STD	[28]
	Dimethyl disulfide	1.1	×	✓	Sulfurous, vegetable	MS, RI, STD	[25,28]
Pyrazines	2-Methylpyrazine	60	✓	✓	Popcorn, nutty	MS, RI, STD	[25,28]
	2,6-Dimethyl pyrazine	6	✓	×	Roasted, coffee, roasted nut, roast beef	MS, RI, STD	[25,28]
	2-Ethyl-pyrazine	4	✓	✓	Nutty, musty	MS, RI, STD	[28]
Others	1-Ethyl-1H-pyrrole-2-carbaldehyde	n.m	-	-	Burnt, roasted, smoky	MS, RI, STD	[28]
	2-Formyl-1H-pyrrole	n.m	-	-	Musty, beefy	MS, RI, STD	[28]
	2-Acetyl pyrrole	n.m	-	-	Musty, sweet	MS, RI, STD	[28]
	2-Ethylfuran	100	✓	✓	Bread, sweet	MS, RI, STD	[28]
	2-Pentylfuran	4.8	✓	✓	Fruity, green	MS, RI, STD	[28]

Ots = Odor thresholds in water. n.m: not mentioned. × indicated this aroma compound was not detected. ✓ indicated this aroma was detected.

Black tea produced by *Camellia sinensis* and *Camellia assamica* contains a variety of fragrance components, including:

- ➢ The floral-smelling molecule linalool is typically found in black tea, among other forms of tea. It improves the overall flavor and aroma of the tea.
- ➢ Another floral component found in black tea is geraniol. It has a rose-like scent and enhances the tea's fragrant profile.
- ➢ Black tea contains a chemical called *cis*-jasmone, which has a fruity scent. It contributes to the tea's characteristic aroma.
- ➢ Black tea typically contains the chemical methyl salicylate, which smells like wintergreen. It imparts a pleasant, minty undertone to the entire smell.
- ➢ Black tea contains the chemical *trans*-2-hexenal, which has a grassy, green scent. It adds to the tea's energizing, earthy aroma.
- ➢ Black tea includes a component known as furfural, which has a distinctive caramel aroma. It imparts a sweet and roasted note to the entire scent.
- ➢ Black tea is given a sweet and floral aroma with benzyl alcohol.
- ➢ Black tea contains benzyl acetate, which contributes to black tea's fruity and floral aroma.

2.1. The Main Processing Technologies of Black Tea

Black tea scent is created by a complicated process that includes numerous phases and chemical reactions. Tea leaves undergo a series of steps during processing, including withering, rolling, fermentation, and drying (Figure 4).

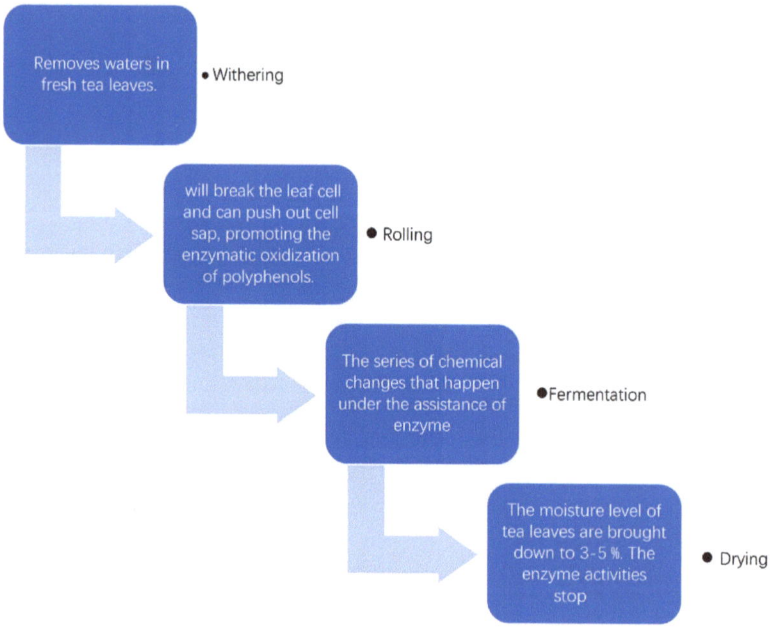

Figure 4. Tea Processing stages.

2.1.1. Withering

Fresh tea leaves (*Camellia sinensis*) are first withered by being exposed to air, having their moisture content decreased from 76 to 60–62%. Enzymes (phenol oxidase) in the tea leaves break down materials, including proteins and polyphenols, during withering and producing volatile compounds. High-molecular-weight phenol oxidase forms with catechol oxidation activity are typically generated during this step [49]. The withering process in black tea manufacturing plays a pivotal role in controlling the development of terpenoids, leading to an enhanced flavor quality in brewed tea. This process significantly influences black tea's aroma and overall sensory characteristics [50]. The withering procedure encourages the softness of the tea leaves, and postharvest enzymatic reactions start to take place, which accelerates the chemical changes [51].

2.1.2. Rolling

Rolling is the next process after withering the leaves to liberate more volatile chemicals. Tea leaves (*Camellia sinensis*) that have withered are ruptured during the rolling process, which also damages the tea cells and releases oxidase, which initiates oxidation. The oxidation of the green tea leaves causes them to become bright copper in color while rolling. This technique determines the black tea's curling. The cell walls of the leaves are destroyed during this process, allowing the enzymes and other chemicals to mingle and interact more easily. Fresh tea leaves will have their surfaces curled and disrupted by a light rolling motion and most TFs are produced during rolling [52]. Black tea's aroma is largely produced by the hydrolysis of glycosidically bound fragrance precursors because the structure of the leaf cells is changed, and the substances are completely mixed during the rolling process [53]. A deep rolling may promote the release of endogenous enzymes and tea extract, causing the tea polyphenols and polyphenol oxidase (PPO) to mix in black tea [52].

2.1.3. Fermentation (Enzymatic Oxidation)

Fermentation plays an integral role in the black tea (*Camellia sinensis*) manufacturing process, exerting significant influence over the ultimate quality of the finished tea product. Unfermented tea leaves undergo a remarkable transformation during this critical phase, shifting from their initial green hue to a rich, coppery brown. Simultaneously, the grassy aroma evolves into a more delicate floral scent. This transformation hinges on a complex series of biochemical reactions, with the precise timing of fermentation cessation being paramount. Proven criteria are required to conclude the fermentation at the right moment, as both under-fermentation and over-fermentation can result in a decline in the quality of the final tea product [54]. Essential oils and amino acids are some examples of volatile and non-volatile substances, respectively [55].

Fermentation is the most crucial of the various stages of black tea manufacturing. This step delves into the process of protein degradation and the metabolic pathways of proteinaceous amino acids during enzymatic reactions that occur throughout fermentation, elucidating their impact on tea quality [56,57]. Enzymatic oxidation, sometimes referred to as fermentation, is the stage of tea processing that happens when the tea leaves' cell walls disintegrate and their interiors become airborne. Oxygen exposure triggers a chemical interaction with the polyphenol-containing organic enzymes found inside the leaves. Among these enzymes, polyphenol oxidase is the most significant. Depending on the situation, the fermentation time is a crucial parameter in the production process of black tea (*Camellia assamica*). Some researchers focused on compounds like TFs and TRs and found that the ideal fermentation period for black tea changes depending on the type of fresh leaves utilized [44,58]. Enzymatic processes in tea leaves accelerate during fermentation, reducing the amount of bitter and astringent chemicals [47]. An electronic nose equipped with quartz crystal microbalance sensors can be used to track the ideal period for black tea fermentation since the fermentation process alters the fragrant volatiles in the tea [56]. Although fermentation is the term used to describe the part of the tea processing where the tea is left to brown and cure, this is actually a misnomer. When biological agents like yeast or bacteria break down organic material, taste, alcohol, and carbon dioxide are produced. This is known as true fermentation.

2.1.4. Drying

The last step is the drying process in black tea (*Camellia sinensis*), which ends the fermentation process and concentrates the volatile components that give the tea its aroma. The leaves are moved via "tiered dryers" on metal conveyor belts at the peak of the fermentation process. The tea is dried in heated air at 80–90 °C for about 20 min, which causes the cell fluid to adhere to the leaves and gives the tea its deep brown to black color. The leaves have a final humidity of between 5 and 6%.

Different drying techniques can result in the formation of various volatiles. Drying plays a crucial role in tea production as it can greatly affect the aroma of the tea due to the volatile chemicals created, altered, or lost by the tea molecules [59]. Some researchers investigated the effects of different drying techniques, such as conventional hot-air drying, microwave drying, far-infrared drying, halogen-lamp drying and halogen lamp–microwave combination drying processes in black tea. The study's findings suggested that black tea that was microwave-treated exhibited the most favorable outcomes. In addition to flavonoids, essential oils and amino acids play pivotal roles in defining tea's characteristic taste and aroma. Notably, the content of essential oils increases during the withering, rolling, and fermentation phases, diminishing during the drying step. However, this decline is compensated for by the Maillard reaction, an interaction between amino acids and sugars during drying that contributes positively to the tea's flavor and color [60]. Hot-air drying is the most widely utilized drying technique for black tea. Following oxidation, rolling, and withering, hot air is used to dry the tea leaves in a regulated setting. In order to obtain the desired taste and fragrance profiles, this method is frequently used in the manufacture of black tea since it provides for fine control over temperature and humidity. Hot-air drying has gained popularity because it is more effective and can maintain consistent tea quality, even if traditional air drying is still utilized in some areas. Traditional drying techniques using hot air provide benefits such as ease of use and low cost, but have drawbacks such as extended drying times and significant nutritional loss. Drying takes place for about 10–15 min at 130–150 °C, followed by cooling outside of the drying apparatus for an hour, and finally drying for about 10 min at 80–90 °C.

2.2. *The Formation Pathway of Critical Aroma Compounds*

Many volatile substances, such as aldehydes, ketones, hydrocarbons, and terpenoids, contribute to black tea's scent. These aromatic chemicals are created during tea preparation through a complicated series of chemical interactions (Figure 5). The volatile compounds that give finished teas their scents can be present in freshly picked leaves and those produced during production from precursors [12,61,62].

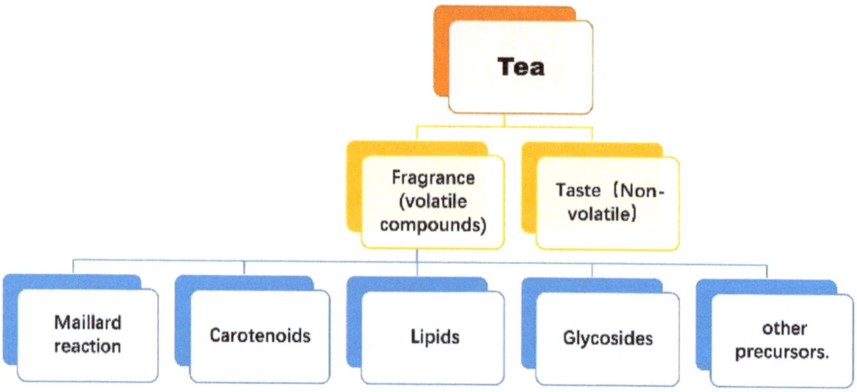

Figure 5. The formation pathway of critical aroma compounds.

Volatile terpenoids (VTs), volatile fatty acid derivatives (FADVs), volatile amino acid derivatives (AADVs), and carotenoid-derived volatiles (CDVs) are the four main groups of tea volatiles based on their sources of synthesis [9,20]. When tea is processed, hydrolases like β-primeverosidase liberate volatiles typically attached to glycosidic chains in the fresh tea leaves [63]. Notably, specific compounds known as glycosidically bound volatiles (GBVs) play a significant role in the scents. According to Schwab et al. [61], these molecules consist of a sugar molecule attached to a volatile molecule. The degradation of carotenoids significantly increases the concentration of volatile flavor compounds (VFC), like β-ionone, α-ionone, β-damascone, and α-damascone, during the black tea leaf fermentation process [64].

Catechins and other flavonoids found in new tea leaves undergo oxidative degradation, which is the first step in synthesizing fragrance compounds in black tea. Tea leaves are stretched out and exposed to air when withering, which causes some of the polyphenols to oxidize partially. The polyphenol oxidase (PPO) enzyme catalyzes this oxidation process. The catechins are consequently transformed into intricate secondary metabolites like TFs and TRs. During metabolic processes, young isolated leaves undergo hydrolysis of proteins, fatty acids, glycosides, and other polysaccharides [65,66]. These compounds operate as the substance for the ultimate development of the tea's flavors.

The taste and aroma of black tea are greatly affected by the presence of TFs and TRs. They are oxidative condensation products of catechins, diverse and highly polymerized molecules. TRs are dark brown to black pigments, whereas TFs are yellow to orange-red pigments with an astringent taste [10,67]. The tea leaves are dried and rolled in the final stage of processing, which results in the creation of scent molecules. Intriguingly, the Maillard reaction (a non-enzymatic browning reaction involving amino acids and reducing sugars) occurs during this step and produces furfural, a furan-like substance. This compound is responsible for the sweet and caramel-like fragrance that is a defining feature of black tea. Furthermore, β-cyclocitral imparts a refreshing fruity aroma to tea [22].

The fermentation process of black tea can produce many aroma components. In order to understand the creation of scent in black tea, a comprehensive analysis must be conducted to identify the key volatile molecules (FADVs, AADVs, VTs, or CDVs) involved in the fermentation process. This process is responsible for black tea's sweet, floral, and fruity fragrances. For instance, by raising the quantity of appropriate precursors in fresh leaves by cultivation techniques, it could be used to enhance black tea's scent quality [13].

2.3. The Variation in Critical Aroma Compounds during Processing

The quality of black tea can be enhanced by improving the processing methods. During the processing of black tea leaves (*Camellia sinensis* L.), intricate chemical changes take place [68]. The use of partial least-squares discriminant analysis and odor activity value analysis determined the five key components, including 3-carene, geraniol, β-myrcene, τ-cadinol, and β-ionone. In addition, a sensory investigation revealed that withering–shaking resulted in a black tea (*Camellia sinensis*) with a fruitier flavor while withering–withering produced a more floral taste [69].

During the rolling and fermentation of black tea (*Camellia assamica*), linalool, methyl salicylate, and C_6-aldehydes produce a rose-like flavor, whereas unsaturated fatty acids decrease. During the withering and fermentation of black tea (*Camellia assamica*), *trans*-2-hexenal increases while *cis*-3-hexenal decreases [67,70]. Fruity and floral fragrances are significantly influenced by the volatile derivatives of phenylpropanoid/benzenoid [71]. According to Wang et al. [72], the fermentation process alters the chemical composition of tea, resulting in changes to its profile and volatile compounds. The sum of the five volatile chemicals (*E*)-benzaldehyde, indole, methyl-5-hapten-2-one, methyl salicylate, and 2-hexenal, were demonstrated to be capable of distinguishing between fermented and unfermented teas. According to Wu et al. [73], certain flavor components had dramatically different contents. Variously fermented teas suggested that certain active ingredients may distinguish different fermentation levels.

During withering, the breakdown of proteins may result in higher levels of amino acids. Proteins are broken down by an enzyme called peptidase [74]. Glycoside-bound volatiles in tea leaves (*Camellia sinensis*) are converted into free volatiles by β-primeverosidases, β-glucosidases, and glycosyl transferases, as extensively researched and functionally characterized [75,76]. It is essential to follow certain rules to fully appreciate the various sensory qualities of black tea, such as its color, aroma, taste, and health benefits. It is crucial to carefully monitor the fermentation process [44].

2.3.1. The Effects of Manufacturing Parameters on Critical Aroma Compounds

According to the research, the quality of tea leaves changed after processing due to considerable changes in the composition and concentration of the chemicals in tea leaves [7]. Black tea processing involves several chemical reactions that significantly alter the tea's flavor and aroma profile. Some of the essential fragrance molecules and how they change as black tea (*Camellia sinensis*) is processed are as follows. Linalool: A terpene alcohol called linalool is responsible for black tea's flowery and zesty scent. During the withering process, the concentration of linalool, a flavonoid glycoside compound, rises but falls during later processing phases. As the withering process progresses, the concentration of linalool, a flavonoid glycoside compound, increases. However, during the later processing phases, it decreases [77]. Geraniol: A terpene alcohol with a rose-like scent in black tea. During withering and fermentation, geranyl pyrophosphate oxidizes to produce geraniol. Its concentration increases in the first stages of fermentation [78]. Benzaldehyde/Phenylacetaldehyde has an almond-like, sweet scent in various nuts and fruits. This compound contributes to black tea's nutty and fruity notes in its aroma [79].

The volatility of black tea may be affected by the temperature of the fermenting process. The relative volatile content might be more volatile at higher fermentation temperatures, including alcohols and alkenes at 31 °C, and esters and alkanes at 28 °C [80]. The process of withering can significantly impact fragrance components, for instance, increasing the concentrations of nerol, hexyl alcohol, and (*E*)-2-hexenoic acid, as well as increasing salicylic acid, benzyl alcohol, (*Z*)-2-pentenol, (*E*)-2-hexenal, and benzaldehyde acid and other aroma compound precursors [63]. According to Vuong et al. [81], amino acids contribute significantly to black tea's nutritional value and overall quality. Lipoprotein synthesis is also crucial for aroma compound production [82]. Lipids degrade, which results in a drop in unsaturated fatty acid content to produce fragrance chemicals through oxidative cleavage during withering [83]. According to Soheilifard et al. [84], sensory evaluations revealed that withering the tea leaves for 16 h resulted in superior sensory quality attributes in the final tea product. This rolling action facilitated the release of sap and essential oils from the tea leaves, ultimately enhancing the tea's flavor. According to Zheng et al. [85], fermentation increased the amounts of methyl salicylate and linalool, with its distinctive scent. The amount of (*E*)-2-hexenal chemicals was reduced by drying [68]. During the rolling and fermentation processes, there was a concurrent increase in *trans*-2-hexenal, linalool, and methyl salicylate. Regrettably, most of these volatile compounds were lost during the firing [65].

The quality of black tea greatly depends on the fermentation stage during processing. The suitable temperature of 28 °C is crucial for black tea fermentation. If the fermentation is carried out at high temperatures and for a longer duration such as 2–4 h or 55–110 min, it results in black tea with high levels of TF and intense color. However, when fermentation takes place at high temperatures, it leads to higher TR and total color levels, with lower values of TF, sensory rating, and brightness. Research showed that during the drying, the final step in tea production, the tea leaves were exposed to high temperatures over an extended period. As a result, more volatile chemicals were lost during the drying process, but the overall content of volatile compounds increased [86]. The main fragrance components in tea leaves converged and gathered throughout the drying process, creating its distinctive aroma [87]. The tea's aroma was greatly influenced by processing, and a thorough examination of how the tea aroma changed throughout processing substantially impacted how the tea was processed. The tea aroma quality formation depends on the tea leaves' volatile compounds.

2.3.2. The Effects of Season, Horticultural Practice, and Growing Area on the Aroma Compounds

Liu et al. [88] investigated how varying seasons exhibited distinct volatile chemicals and aroma precursors of Yingde black tea (*Camellia assamica*). The types and quantities of volatile compounds typically exhibited a declining trend followed by an increasing trend when the harvesting season changed, confirming that the development of tea scent is a dynamic process. Moreover, using large amounts of nitrogen fertilizer raised the amount of unsaturated fatty acids and decreased the aroma of black tea. To make high-quality tea requires the provision of the necessary amounts of macronutrients, nitrogen, phosphorus, and potassium along with other minor nutrients, since the quality of tea is strongly correlated with caffeine, catechins, amino acids, and volatile compounds in harvestable tea shoots (two or three succulent leaves along with a bud) [89]. Some researchers effectively determined the distinct scent compounds linked to distinct tea origins, offering insightful information about the aroma qualities of black teas from several geographical areas [90].

2.4. *The Health Benefits of Critical Aroma Compounds of Black Tea*

2.4.1. Neuro-Regulation Effects

Some evidence indicates the beneficial effects of black tea (*Camellia sinensis*) consumption on mental alertness, planning abilities, and work-related focus [91].

The aging process represents a significant risk factor for neurodegenerative conditions such as Alzheimer's disease (AD), Parkinson's disease (PD), and amyotrophic lateral sclerosis (ALS). Remarkably, black tea, a caffeine beverage, exhibits an inverse association with Parkinson's disease risk. This investigation was carried out within the context of the Singapore Chinese Health Study, scrutinizing various factors about the occurrence of Parkinson's disease [92]. A research initiative, spearheaded by Associate Professor Louis Tan, a Senior Consultant at the Department of Neurology within the National Neuroscience Institute (NNI), in collaboration with other neurologists from NNI, found that black tea may harbor neuroprotective agents that contribute to its beneficial effects against Parkinson's disease [93]. Regular consumption of black tea may lower the risk of developing Parkinson's disease. The concept of combining iron chelation with antioxidant therapy emerges as a promising approach for neuroprotection. Tea flavonoids (black tea) have gained recognition for their potential to chelate divalent metals, exhibit antioxidant properties, and demonstrate anti-inflammatory effects. They possess the capacity to traverse the blood–brain barrier, guarding against neuronal death in diverse cellular and animal models of neurological diseases [93–95].

2.4.2. Anti-Microbial Effects

Tea's (*Camellia sinensis*) antibacterial and antioxidant properties were attributed to the main mechanism by which fermentation enhances antioxidant activity which is the production of more free and absorbable phenolic compounds and flavonoids as a consequence of tea processing. Additionally, fermentation causes the structural disintegration of plant cell walls, which releases or synthesizes a variety of antioxidant chemicals. After fermentation, the caffeine content is slightly reduced, improving the tea's flavor. Additionally, the range of aromatic compounds expands after fermentation, giving the tea a fresh aroma [96]. These teas (*Camellia sinensis*) have anti-inflammatory and weight-loss properties. They also contain L-theanine, which can regulate brain activity and human attention [97]. Black tea's caffeine might also improve mental clarity.

Maintaining a healthy gut can help prevent various illnesses and health issues. Drinking black tea (*Camellia assamica*) might increase the number and type of beneficial intestinal microorganisms. The tea polyphenols may serve as probiotics and food for the beneficial bacteria in the gut. Additionally, these polyphenols may stop the development of other dangerous bacteria in the gut. In addition to possibly lowering the incidence of colorectal and esophageal/stomach malignancies, black tea may help heal stomach ulcers [98]. Most bacteria have the potential to cause illnesses and even death. Black tea's phytonutrients and antioxidants may have antibacterial characteristics [99]. Black tea (*Camellia sinensis*) has antioxidant and antibacterial qualities that shield against staphylococcus infections [100,101]. Black tea's fluoride also prevents tooth caries [102]. Consequently, drinking black tea (*Camellia sinensis*) may also benefit oral health.

According to Del Rio et al. [103], higher dietary phenolic intake has been linked to lower risks of various chronic diseases. Tea's antioxidant and antibacterial properties are improved after fermentation. Microbially fermented tea also possesses anti-inflammatory and weight-loss properties. Microbially fermented tea has positive medicinal and physiological benefits when consumed over time. Fermented tea's anti-microbial qualities have a promising future in food preservation [20]. Amino acids affect the nutritional value and quality of black tea. Fermentation is the process that is most important in the creation of black tea.

Numerous studies indicate that tea can enhance psychological and neurological functions. Recent research reveals that tea catechins can enter the brain and protect neurons from cell death caused by neurological diseases. This is due to their antioxidant, anti-inflammatory, and divalent metal-chelating properties, as demonstrated in animal and cellular models [104]. As the polyphenols in black tea are said to prevent the formation of dangerous germs, drinking black tea also performs an antibacterial function [105]. Another study that showed a notable reduction in the growth of various dangerous bacteria offers further evidence of the antibacterial capability of black tea components [106].

2.4.3. Digestive Tract Protective Effects

The polyphenols in black tea (*Camellia sinensis*) prevent obesity by limiting fat and complex sugar digestion and absorption. Tea aroma components, notably TFs, exhibit strong antioxidant action. In the digestive system, they can aid in scavenging dangerous free radicals that can cause cell damage and raise the possibility of developing digestive problems like gastric ulcers and colorectal cancer [107].

Certain tea fragrance components, such as catechins and theanine (*Camellia sinensis*), have been discovered to have antibacterial activities. In the digestive tract, they can stop the formation of dangerous bacteria like *Helicobacter pylori*, which is linked to gastric ulcers and gastritis. Tea aroma components can support the maintenance of a healthy balance of gut flora and safeguard the digestive tract by preventing the overgrowth of these bacteria [108,109]. Theanine is crucial for the T-cell-mediated immune system's efficient operation since it helps keep these cells activated. A biologically active substance called theine has been shown to activate the central nervous and cardiovascular systems, relax blood vessels, enhance digestion, increase circulation, and stop platelet aggregation [110].

For gastroprotective effects it has been demonstrated that tea (*Camellia sinensis*) aroma chemicals, such as tannins, have gastroprotective properties. They may be able to fortify the mucosal lining of the digestive tract, acting as a defense against harm from irritants and stomach acids. This step can help prevent gastroesophageal reflux disease (GERD) and stomach ulcers [111]. By encouraging the release of gastric juices, regulating intestinal flora, which lowers the risk of gastrointestinal disorders, and activating peristaltic motions, organic acids aid in digestion [110].

3. Conclusions

Tea has been made from tea (*Camellia sinensis*) leaves for over 50 centuries. Tea plant varieties mainly belong to two major groups: *Camellia sinensis* var. *sinensis* (CSS; Chinese type) and *Camellia sinensis* var. *assamica* (CSA; Assam type). Both types of tea have distinctive fragrance profiles that add to the special qualities of black tea. The interplay of numerous metabolic pathways is a complex process that results in the synthesis of fragrance molecules in black tea. When making black tea, the process of fermentation is essential because it causes the leaves to oxidize. This process causes some chemical molecules, like catechins, to break down while creating other ones, including TFs and TRs. These substances help give black tea its distinctive flavor and aroma.

The significant fragrance components present in various kinds of black tea belong to *Camellia sinensis var. sinensis* and *Camellia sinensis var. assamica*, covered in this article. Overall, several procedures and chemical interactions are needed to produce the scent of black tea. In conclusion, the Maillard reaction, fermentation, oxidative degradation of catechins, oxidative condensation of TFs, and TRs all contribute to creating important scent components in black tea. The essential elements of the black tea scent are currently being processed. According to sensory analysis, the scent of black tea from *Camellia assamica* is comparable to that of the more well-known *Camellia sinensis*.

➢ Notwithstanding, there are a few subtle variances in the aromatic patterns of both types of tea. For instance, black tea from *Camellia assamica* often has a stronger, maltier aroma than *Camellia sinensis*. Furthermore, according to the cultivar and specific growing conditions, *Camellia assamica* teas have a slightly fruity or floral aroma. Conclusively, the aroma sensory analysis of black tea from *Camellia sinensis* and *Camellia assamica* involves the analysis of volatile compounds formed during tea processing.

Numerous fragrance components in black tea have been shown to provide a range of health advantages. The aroma of black tea can activate the nervous system, increasing energy and improving mental clarity. The aromatic elements in black tea may benefit the digestive tract. They can aid in maintaining a healthy gut environment, promote healthy digestion, reduce digestive problems, including bloating and indigestion, and increase immunity. Black tea's aromatic constituents may help to relax the body and mind. They can aid in lowering tension and stress and fostering relaxation, all of which can benefit mental health. While black tea may have many health advantages, it is crucial to remember that individual responses may differ, and excessive drinking should be avoided because it may have negative consequences, including a faster heartbeat, sleeplessness, or digestive problems. All in all, researchers, tea experts, and enthusiasts interested in black tea's sensory evaluation and aroma development will find this review useful.

Author Contributions: Conceptualization, L.Z.; methodology, A.P.; software, A.P.; validation, A.P., J.K. and L.Z.; formal analysis, A.P.; investigation, A.P.; resources, L.Z.; data curation, A.P.; writing—original draft preparation, A.P.; writing—review and editing, A.P., C.-Y.Q., F.Z., G.L., P.L. and M.Z.; visualization, A.P. and L.Z.; supervision, L.Z.; project administration, L.Z.; funding acquisition, L.Z. All authors have read and agreed to the published version of the manuscript.

Funding: This work was supported by the Anhui Key Research and Development Plan (202104b11020001).

Institutional Review Board Statement: Not applicable.

Informed Consent Statement: Not applicable.

Data Availability Statement: All datasets presented in this study are included in the article.

Conflicts of Interest: The authors declare no conflict of interest.

References

1. Wei, C.; Yang, H.; Wang, S.; Zhao, J.; Liu, C.; Gao, L.; Xia, E.; Lu, Y.; Tai, Y.; She, G.; et al. Draft genome sequence of *Camellia sinensis var. sinensis* provides insights into the evolution of the tea genome and tea quality. *Proc. Natl. Acad. Sci. USA* **2018**, *115*, E4151–E4158. [CrossRef]
2. Zhang, Z.B.; Xiong, T.; Chen, J.H.; Ye, F.; Cao, J.J.; Chen, Y.R.; Zhao, Z.W.; Luo, T. Understanding the Origin and Evolution of Tea (*Camellia sinensis* [L.]): Genomic Advances in Tea. *J. Mol. Evol.* **2023**, *91*, 156–168. [CrossRef]
3. Meng, X.-H.; Li, N.; Zhu, H.-T.; Wang, D.; Yang, C.-R.; Zhang, Y.-J. Plant Resources, Chemical Constituents, and Bioactivities of Tea Plants from the Genus Camellia Section Thea. *J. Agric. Food Chem.* **2019**, *67*, 5318–5349. [CrossRef]
4. Hayat, K.; Iqbal, H.; Malik, U.; Bilal, U.; Mushtaq, S. Tea and Its Consumption: Benefits and Risks. *Crit. Rev. Food Sci. Nutr.* **2015**, *55*, 939–954. [CrossRef]

5. Zhao, D.; Shah, N.P. Antiradical and tea polyphenol-stabilizing ability of functional fermented soymilk-tea beverage. *Food Chem.* **2014**, *158*, 262–269. [CrossRef] [PubMed]
6. Abudureheman, B.; Yu, X.; Fang, D.; Zhang, H. Enzymatic Oxidation of Tea Catechins and Its Mechanism. *Molecules* **2022**, *27*, 942. [CrossRef] [PubMed]
7. Zhang, L.; Ho, C.T.; Zhou, J.; Santos, J.S.; Armstrong, L.; Granato, D. Chemistry and Biological Activities of Processed *Camellia sinensis* Teas: A Comprehensive Review. *Compr. Rev. Food Sci. Food Saf.* **2019**, *18*, 1474–1495. [CrossRef]
8. FAOSTAT-Agriculture. Statistical Databases. Available online: https://www.fao.org/3/cc0238en/cc0238en.pdf (accessed on 1 October 2023).
9. Yang, Z.; Baldermann, S.; Watanabe, N. Recent studies of the volatile compounds in tea. *Food Res. Int.* **2013**, *53*, 585–599. [CrossRef]
10. Sanderson, G.W.; Grahamm, H.N. Formation of black tea aroma. *J. Agric. Food Chem.* **1973**, *21*, 576–585. [CrossRef]
11. Wu, H.; Huang, W.; Chen, Z.; Chen, Z.; Shi, J.; Kong, Q.; Sun, S.; Jiang, X.; Chen, D.; Yan, S. GC-MS-based metabolomic study reveals dynamic changes of chemical compositions during black tea processing. *Food Res. Int.* **2019**, *120*, 330–338. [CrossRef] [PubMed]
12. Ho, C.-T.; Zheng, X.; Li, S. Tea aroma formation. *Food Sci. Hum. Wellness* **2015**, *4*, 9–27. [CrossRef]
13. Chen, Q.; Zhu, Y.; Liu, Y.; Liu, Y.; Dong, C.; Lin, Z.; Teng, J. Black tea aroma formation during the fermentation period. *Food Chem.* **2022**, *374*, 131640. [CrossRef]
14. Owuor, P.O.; Obanda, M.; Nyirenda, H.E.; Mandala, W.L. Influence of region of production on clonal black tea chemical characteristics. *Food Chem.* **2008**, *108*, 263–271. [CrossRef]
15. Owuor, P.O.; Obanda, M. The changes in black tea quality due to variations of plucking standard and fermentation time. *Food Chem.* **1998**, *61*, 435–441. [CrossRef]
16. Ravichandran, R.; Parthiban, R. The impact of processing techniques on tea volatiles. *Food Chem.* **1998**, *62*, 347–353. [CrossRef]
17. Zhu, J.; Wang, J.; Yuan, H.; Ouyang, W.; Li, J.; Hua, J.; Jiang, Y. Effects of Fermentation Temperature and Time on the Color Attributes and Tea Pigments of Yunnan Congou Black Tea. *Foods* **2022**, *11*, 1845. [CrossRef]
18. Wang, X.; Zeng, L.; Liao, Y.; Zhou, Y.; Xu, X.; Dong, F.; Yang, Z. An alternative pathway for the formation of aromatic aroma compounds derived from l-phenylalanine via phenylpyruvic acid in tea (*Camellia sinensis* (L.) O. Kuntze) leaves. *Food Chem.* **2019**, *270*, 17–24. [CrossRef]
19. Feng, Z.; Li, Y.; Li, M.; Wang, Y.; Zhang, L.; Wan, X.; Yang, X. Tea aroma formation from six model manufacturing processes. *Food Chem.* **2019**, *285*, 347–354. [CrossRef] [PubMed]
20. Chen, X.; Chen, D.; Jiang, H.; Sun, H.; Zhang, C.; Zhao, H.; Li, X.; Yan, F.; Chen, C.; Xu, Z. Aroma characterization of Hanzhong black tea (*Camellia sinensis*) using solid phase extraction coupled with gas chromatography-mass spectrometry and olfactometry and sensory analysis. *Food Chem.* **2019**, *274*, 130–136. [CrossRef] [PubMed]
21. Lau, H.; Liu, S.Q.; Xu, Y.Q.; Lassabliere, B.; Sun, J.; Yu, B. Characterising volatiles in tea (*Camellia sinensis*). Part I: Comparison of headspace-solid phase microextraction and solvent assisted flavour evaporation. *LWT* **2018**, *94*, 178–189. [CrossRef]
22. Shakoor, A.; Zhang, C.; Xie, J.; Yang, X. Maillard reaction chemistry in formation of critical intermediates and flavour compounds and their antioxidant properties. *Food Chem.* **2022**, *393*, 133416. [CrossRef] [PubMed]
23. Xiao, Z.; Cao, X.; Zhu, J.; Chen, F.; Niu, Y. Characterization of the key aroma compounds in three world-famous black teas. *Eur. Food Res. Technol.* **2022**, *248*, 2237–2252. [CrossRef]
24. Wang, Q.; Qin, D.; Jiang, X.; Fang, K.; Li, B.; Wang, Q.; Pan, C.; Ni, E.; Li, H.; Chen, D.; et al. Characterization of the Aroma Profiles of Guangdong Black Teas Using Non-Targeted Metabolomics. *Foods* **2023**, *12*, 1560. [CrossRef] [PubMed]
25. Kang, S.; Yan, H.; Zhu, Y.; Liu, X.; Lv, H.P.; Zhang, Y.; Dai, W.D.; Guo, L.; Tan, J.F.; Peng, Q.H.; et al. Identification and quantification of key odorants in the world's four most famous black teas. *Food Res. Int.* **2019**, *121*, 73–83. [CrossRef]
26. Yue, C.; Yang, P.; Qin, D.; Cai, H.; Wang, Z.; Li, C.; Wu, H. Identification of volatile components and analysis of aroma characteristics of Jiangxi Congou black tea. *Int. J. Food Prop.* **2020**, *23*, 2160–2173. [CrossRef]
27. Wang, Q.; Qin, D.; Huang, G.; Jiang, X.; Fang, K.; Wang, Q.; Ni, E.; Li, B.; Pan, C.; Li, H.; et al. Identification and characterization of the key volatile flavor compounds in black teas from distinct regions worldwide. *J. Food Sci.* **2022**, *87*, 3433–3446. [CrossRef] [PubMed]
28. Niu, Y.; Ma, Y.; Xiao, Z.; Zhu, J.; Xiong, W.; Chen, F. Characterization of the Key Aroma Compounds of Three Kinds of Chinese Representative Black Tea and Elucidation of the Perceptual Interactions of Methyl Salicylate and Floral Odorants. *Molecules* **2022**, *27*, 1631. [CrossRef] [PubMed]
29. Ma, L.; Qiao, Y.; Du, L.; Li, Y.; Huang, S.; Liu, F.; Xiao, D. Evaluation and Optimization of a Superior Extraction Method for the Characterization of the Volatile Profile of Black Tea by HS-SPME/GC-MS. *Food Anal. Methods* **2017**, *10*, 2481–2489. [CrossRef]
30. Kumar, S.; Panchariya, P.C.; Prasad, P.B.; Sharma, A. In Non-Destructive Classification of Assam Black Tea Using Ultra-fast Gas Chromatography (UFGC) Coupled with Soft Independent Modeling of Class Analogy (SIMCA). *Sens. Transducers* **2015**, *186*, 104–111.
31. Wang, B.; Chen, H.; Qu, F.; Song, Y.; Di, T.; Wang, P.; Zhang, X. Identification of aroma-active components in black teas produced by six Chinese tea cultivars in high-latitude region by GC–MS and GC–O analysis. *Eur. Food Res. Technol.* **2022**, *248*, 647–657. [CrossRef]
32. Yan, T.; Lin, J.; Zhu, J.; Ye, N.; Huang, J.; Wang, P.; Jin, S.; Zheng, D.; Yang, J. Aroma analysis of Fuyun 6 and Jinguanyin black tea in the Fu'an area based on E-nose and GC–MS. *Eur. Food Res. Technol.* **2022**, *248*, 947–961. [CrossRef]

33. Sharma, P.; Ghosh, A.; Tudu, B.; Bhuyan, L.P.; Tamuly, P.; Bhattacharyya, N.; Bandyopadhyay, R.; Chatterjee, A. Detection of linalool in black tea using a quartz crystal microbalance sensor. *Sens. Actuators B Chem.* **2014**, *190*, 318–325. [CrossRef]
34. Muthumani, T.; Senthil Kumar, R.S. Influence of fermentation time on the development of compounds responsible for quality in black tea. *Food Chem.* **2007**, *101*, 98–102. [CrossRef]
35. Chaturvedula, V.S.P.; Prakash, I. The aroma, taste, color and bioactive constituents of tea. *J. Med. Plants Res.* **2011**, *5*, 2110–2124.
36. Joshi, R.; Gulati, A. Fractionation and identification of minor and aroma-active constituents in Kangra orthodox black tea. *Food Chem.* **2015**, *167*, 290–298. [CrossRef] [PubMed]
37. Schuh, C.; Schieberle, P. Characterization of the key aroma compounds in the beverage prepared from Darjeeling black tea: Quantitative differences between tea leaves and infusion. *J. Agric. Food Chem.* **2006**, *54*, 916–924. [CrossRef]
38. Su, D.; He, J.; Zhou, Y.-Z.; Li, Y.-L.; Zhou, H.-J. Aroma effects of key volatile compounds in Keemun black tea at different grades: HS-SPME-GC-MS, sensory evaluation, and chemometrics. *Food Chem.* **2021**, *373*, 131587. [CrossRef]
39. Liu, H.; Xu, Y.; Wen, J.; An, K.; Wu, J.; Yu, Y.; Zou, B.; Guo, M. A comparative study of aromatic characterization of Yingde Black Tea infusions in different steeping temperatures. *LWT* **2021**, *143*, 110860. [CrossRef]
40. Zhang, J.S. Comparative analysis of aroma compositions in different-grades Yunnan Congou black tea. *J. South. Agric.* **2012**, *43*, 489–492.
41. Mao, S.; Lu, C.; Li, M.; Ye, Y.; Wei, X.; Tong, H. Identification of key aromatic compounds in Congou black tea by partial least-square regression with variable importance of projection scores and gas chromatography-mass spectrometry/gas chromatography-olfactometry. *J. Sci. Food Agric.* **2018**, *98*, 5278–5286. [CrossRef]
42. Ma, L.; Gao, M.; Zhang, L.; Qiao, Y.; Li, J.; Du, L.; Zhang, H.; Wang, H. Characterization of the key aroma-active compounds in high-grade Dianhong tea using GC-MS and GC-O combined with sensory-directed flavor analysis. *Food Chem.* **2022**, *378*, 132058. [CrossRef] [PubMed]
43. Xiao, Z.; Wang, H.; Niu, Y.; Liu, Q.; Zhu, J.; Chen, H.; Ma, N. Characterization of aroma compositions in different Chinese congou black teas using GC-MS and GC-O combined with partial least squares regression. *Flavour Fragr. J.* **2017**, *32*, 265–276. [CrossRef]
44. Jolvis Pou, K.R. Fermentation: The Key Step in the Processing of Black Tea. *J. Biosyst. Eng.* **2016**, *41*, 85–92. [CrossRef]
45. Yun, J.; Cui, C.; Zhang, S.; Zhu, J.; Peng, C.; Cai, H.; Yang, X.; Hou, R. Use of headspace GC/MS combined with chemometric analysis to identify the geographic origins of black tea. *Food Chem.* **2021**, *360*, 130033. [CrossRef]
46. Zhang, L.; Xu, Y.; Liu, Z. Tea: From Historical Documents to Modern Technology. *Molecules* **2023**, *28*, 2992. [CrossRef]
47. Chen, Y.; Zeng, L.; Liao, Y.; Li, J.; Zhou, B.; Yang, Z.; Tang, J. Enzymatic Reaction-Related Protein Degradation and Proteinaceous Amino Acid Metabolism during the Black Tea (*Camellia sinensis*) Manufacturing Process. *Foods* **2020**, *9*, 66. [CrossRef]
48. Yao, H.; Su, H.; Ma, J.; Zheng, J.; He, W.; Wu, C.; Hou, Z.; Zhao, R.; Zhou, Q. Widely targeted volatileomics analysis reveals the typical aroma formation of Xinyang black tea during fermentation. *Food Res. Int.* **2023**, *164*, 112387. [CrossRef]
49. Omiadze, N.T.; Mchedlishvili, N.I.; Rodrigez-Lopez, J.N.; Abutidze, M.O.; Sadunishvili, T.; Pruidze, N.G. Biochemical processes at the stage of withering during black tea production. *Appl. Biochem. Microbiol.* **2014**, *50*, 394–397. [CrossRef]
50. Mahanta, P.K.; Baruah, S. Relationship between process of withering and aroma characteristics of black tea. *J. Sci. Food Agric.* **1989**, *46*, 461–468. [CrossRef]
51. Liang, Y.; Lu, J.; Zhang, L.; Wu, S.; Wu, Y. Estimation of black tea quality by analysis of chemical composition and colour difference of tea infusions. *Food Chem.* **2003**, *80*, 283–290. [CrossRef]
52. Lee, L.-S.; Kim, Y.-C.; Park, J.-D.; Kim, Y.-B.; Kim, S.-H. Changes in major polyphenolic compounds of tea (*Camellia sinensis*) leaves during the production of black tea. *Food Sci. Biotechnol.* **2016**, *25*, 1523–1527. [CrossRef] [PubMed]
53. Wang, D.; Kurasawa, E.; Yamaguchi, Y.; Kubota, K.; Kobayashi, A. Analysis of glycosidically bound aroma precursors in tea leaves. 2. Changes in glycoside contents and glycosidase activities in tea leaves during the black tea manufacturing process. *J. Agric. Food Chem.* **2001**, *49*, 1900–1903. [CrossRef] [PubMed]
54. Bhattacharyya, N.; Seth, S.; Tudu, B.; Tamuly, P.; Jana, A.; Ghosh, D.; Bandyopadhyay, R.; Bhuyan, M. Monitoring of black tea fermentation process using electronic nose. *J. Food Eng.* **2007**, *80*, 1146–1156. [CrossRef]
55. Co, H.; Sanderson, G.W. Biochemistry of tea fermentation: Conversion of amino acids to black tea aroma constituents. *J. Food Sci.* **1970**, *35*, 160–164. [CrossRef]
56. Sharma, P.; Ghosh, A.; Tudu, B.; Sabhapondit, S.; Baruah, B.D.; Tamuly, P.; Bhattacharyya, N.; Bandyopadhyay, R. Monitoring the fermentation process of black tea using QCM sensor based electronic nose. *Sens. Actuators B Chem.* **2015**, *219*, 146–157. [CrossRef]
57. Tozlu, B.; Okumuş, H. A new approach to automation of black tea fermentation process with electronic nose. *Automatika* **2018**, *59*, 374–382. [CrossRef]
58. Samanta, T.; Cheeni, V.; Das, S.; Roy, A.; Ghosh, B.C.; Mitra, A. Assessing biochemical changes during standardization of fermentation time and temperature for manufacturing quality black tea. *J. Food Sci. Technol.* **2015**, *52*, 2387–2393. [CrossRef]
59. Polat, A.; Şat, İ.G.; Ilgaz, Ş. Comparison of black tea volatiles depending on the grades and different drying temperatures. *J. Food Process. Preserv.* **2018**, *42*, e13653. [CrossRef]
60. Can Ağca, A.; Vural, N.; Şarer, E. Determination of volatile compounds in green tea and black tea from Turkey by using HS-SPME and GC-MS. *İstanbul J. Pharm.* **2020**, *50*, 111–115. [CrossRef]
61. Schwab, W.; Davidovich-Rikanati, R.; Lewinsohn, E. Biosynthesis of plant-derived flavor compounds. *Plant J. Cell Mol. Biol.* **2008**, *54*, 712–732. [CrossRef]

62. Dudareva, N.; Klempien, A.; Muhlemann, J.K.; Kaplan, I. Biosynthesis, function and metabolic engineering of plant volatile organic compounds. *New Phytol.* **2013**, *198*, 16–32. [CrossRef]
63. Wang, D.; Yoshimura, T.; Kubota, K.; Kobayashi, A. Analysis of glycosidically bound aroma precursors in tea leaves. 1. Qualitative and quantitative analyses of glycosides with aglycons as aroma compounds. *J. Agric. Food Chem.* **2000**, *48*, 5411–5418. [CrossRef] [PubMed]
64. Ravichandran, R. Carotenoid composition, distribution and degradation to flavour volatiles during black tea manufacture and the effect of carotenoid supplementation on tea quality and aroma. *Food Chem.* **2002**, *78*, 23–28. [CrossRef]
65. Tomlins, K.I.; Mashingaidze, A. Influence of withering, including leaf handling, on the manufacturing and quality of black teas: A review. *Food Chem.* **1997**, *60*, 573–580. [CrossRef]
66. Wang, H.; Provan, G.J.; Helliwell, K. Tea flavonoids: Their functions, utilization and analysis. *Trends Food Sci. Technol.* **2000**, *11*, 152–160. [CrossRef]
67. Selvendran, R.R.; Reynolds, J.; Galliard, T. Production of volatiles by degradation of lipids during manufacture of black tea. *Phytochemistry* **1978**, *17*, 233–236. [CrossRef]
68. Deka, H.; Sarmah, P.P.; Devi, A.; Tamuly, P.; Karak, T. Changes in Major Catechins, Caffeine, and Antioxidant Activity During CTC Processing of Black Tea from North East India RSC Advances. 2021, pp. 11457–11467. Available online: http://europepmc.org/abstract/MED/35423631 (accessed on 1 October 2023).
69. Lin, J.; Tu, Z.; Zhu, H.; Chen, L.; Wang, Y.; Yang, Y.; Lv, H.; Zhu, Y.; Yu, L.; Ye, Y. Effects of Shaking and Withering Processes on the Aroma Qualities of Black Tea. *Horticulturae* **2022**, *8*, 549. [CrossRef]
70. Ravichandran, R.; Parthiban, R. Lipid occurrence, distribution and degradation to flavour volatiles during tea processing. *Food Chem.* **2000**, *68*, 7–13. [CrossRef]
71. Mühlemann, J.; Klempien, A.; Dudareva, N. Floral volatiles: From biosynthesis to function. *Plant Cell Environ.* **2014**, *37*, 1936–1949. [CrossRef] [PubMed]
72. Wang, L.F.; Lee, J.Y.; Chung, J.O.; Baik, J.H.; So, S.; Park, S.K. Discrimination of teas with different degrees of fermentation by SPME-GC analysis of the characteristic volatile flavour compounds. *Food Chem.* **2008**, *109*, 196–206. [CrossRef] [PubMed]
73. Wu, Q.; Dong, Q.; Sun, W.-J.; Huang, Y.; Wang, Q.; Zhou, W. Discrimination of Chinese teas with different fermentation degrees by stepwise linear discriminant analysis (S-LDA) of the chemical compounds. *J. Agric. Food Chem.* **2014**, *62*, 9336–9344. [CrossRef]
74. Choudhury, M.N.D.; Bajaj, K.L. Biochemical changes during withering of tea shoots. *Two Bud* **1980**, *27*, 13–16.
75. Ohgami, S.; Ono, E.; Horikawa, M.; Murata, J.; Totsuka, K.; Toyonaga, H.; Ohba, Y.; Dohra, H.; Asai, T.; Matsui, K.; et al. Volatile Glycosylation in Tea Plants: Sequential Glycosylations for the Biosynthesis of Aroma beta-Primeverosides Are Catalyzed by Two *Camellia sinensis* Glycosyltransferases. *Plant Physiol.* **2015**, *168*, 464–477. [CrossRef] [PubMed]
76. Zhou, Y.; Zeng, L.; Gui, J.; Liao, Y.; Li, J.; Tang, J.; Meng, Q.; Dong, F.; Yang, Z. Functional characterizations of β-glucosidases involved in aroma compound formation in tea (*Camellia sinensis*). *Food Res. Int.* **2017**, *96*, 206–214. [CrossRef]
77. Zhou, J.; Fang, T.; Li, W.; Jiang, Z.; Zhou, T.; Zhang, L.; Yu, Y. Widely targeted metabolomics using UPLC-QTRAP-MS/MS reveals chemical changes during the processing of black tea from the cultivar *Camellia sinensis* (L.) O. Kuntze cv. Huangjinya. *Food Res. Int.* **2022**, *162*, 112169. [CrossRef]
78. Gohain, B.; Borchetia, S.; Bhorali, P.; Agarwal, N.; Bhuyan, L.P.; Rahman, A.; Sakata, K.; Mizutani, M.; Shimizu, B.-I.; Gurusubramaniam, G.; et al. Understanding Darjeeling tea flavour on a molecular basis. *Plant Mol. Biol.* **2012**, *78*, 577–597. [CrossRef] [PubMed]
79. Takakura, Y.; Ono, T.; Danjo, K.; Nozaki, H. Efficient enzymatic production of benzaldehyde from l-phenylalanine with a mutant form of 4-hydroxymandelate synthase. *Biosci. Biotechnol. Biochem.* **2022**, *86*, 1718–1725. [CrossRef] [PubMed]
80. Qu, F.; Zeng, W.; Tong, X.; Feng, W.; Chen, Y.; Ni, D. The new insight into the influence of fermentation temperature on quality and bioactivities of black tea. *LWT* **2020**, *117*, 108646. [CrossRef]
81. Vuong, Q.V.; Bowyer, M.C.; Roach, P.D. L-Theanine: Properties, synthesis and isolation from tea. *J. Sci. Food Agric.* **2011**, *91*, 1931–1939. [CrossRef]
82. Mahanta, P.K.; Tamuli, P.; Bhuyan, L.P. Changes of fatty acid contents, lipoxygenase activities, and volatiles during black tea manufacture. *J. Agric. Food Chem.* **1993**, *41*, 1677–1683. [CrossRef]
83. Wright, A.J.; Fishwick, M.J. Lipid degradation during manufacture of black tea. *Phytochemistry* **1979**, *18*, 1511–1513. [CrossRef]
84. Soheilifard, F.; Ghassemzadeh, H.; Salvatian, S. Impact of withering time duration on some biochemical properties and sensory quality attributes of black tea. *Biol. Forum. Int. J.* **2015**, *7*, 1045–1049.
85. Zheng, X.Q.; Li, Q.S.; Xiang, L.P.; Liang, Y.R. Recent Advances in Volatiles of Teas. *Molecules* **2016**, *21*, 338. [CrossRef]
86. Teshome, K. Effect of tea processing methods on biochemical composition sensory quality of black tea (*Camellia sinensis* (L), O. Kuntze): A review. *J. Hortic. For.* **2019**, *11*, 84–95.
87. Qu, F.; Zhu, X.; Ai, Z.; Ai, Y.; Qiu, F.; Ni, D. Effect of different drying methods on the sensory quality and chemical components of black tea. *LWT* **2019**, *99*, 112–118. [CrossRef]
88. Liu, H.; Xu, Y.; Wu, J.; Wen, J.; Yu, Y.; An, K.; Zou, B. GC-IMS and olfactometry analysis on the tea aroma of Yingde black teas harvested in different seasons. *Food Res. Int.* **2021**, *150*, 110784. [CrossRef]
89. Piyasena, K.G.N.P.; Hettiarachchi, L.S.K. Comparison of tea quality parameters of conventionally and organically grown tea, and effects of fertilizer on tea quality: A mini-review. *Food Chem. Adv.* **2023**, *3*, 100399. [CrossRef]

90. Huang, J.; Yan, T.; Yang, J.; Xu, H. Aroma Components Analysis and Origin Differentiation of Black Tea Based on ATD-GC-MS and E-Nose. *Horticulturae* **2023**, *9*, 885. [CrossRef]
91. De Bruin, E.A.; Rowson, M.J.; Van Buren, L.; Rycroft, J.A.; Owen, G.N. Black tea improves attention and self-reported alertness. *Appetite* **2011**, *56*, 235–240. [CrossRef] [PubMed]
92. Mandel, S.A.; Amit, T.; Weinreb, O.; Youdim, M.B. Understanding the broad-spectrum neuroprotective action profile of green tea polyphenols in aging and neurodegenerative diseases. *J. Alzheimers Dis.* **2011**, *25*, 187–208. [CrossRef]
93. Tan, L.C.S.; Koh, W.-P.; Yuan, J.-M.; Wang, R.; Au, W.L.; Tan, J.J.H.; Tan, E.K.; Yu, M.C. Differential effects of black versus green tea on risk of Parkinson's disease in the Singapore Chinese Health Study. *Am. J. Epidemiol.* **2007**, *167*, 553–560. [CrossRef] [PubMed]
94. Tan, E.; Tan, C.; Fook-Chong, S.M.C.; Lum, S.Y.; Chai, A.; Chung, H.; Shen, H.; Zhao, Y.; Teoh, M.L.; Yih, Y.; et al. Dose-dependent protective effect of coffee, tea, and smoking in Parkinson's disease: A study in ethnic Chinese. *J. Neurol. Sci.* **2003**, *216*, 163–167. [CrossRef] [PubMed]
95. Checkoway, H.; Powers, K.; Smith-Weller, T.; Franklin, G.M.; Longstreth, W.T., Jr.; Swanson, P.D. Parkinson's disease risks associated with cigarette smoking, alcohol consumption, and caffeine intake. *Am. J. Epidemiol.* **2002**, *155*, 732–738. [CrossRef]
96. Hu, T.; Shi, S.; Ma, Q. Modulation effects of microorganisms on tea in fermentation. *Front. Nutr.* **2022**, *9*, 931790. [CrossRef] [PubMed]
97. Nobre, A.C.; Rao, A.; Owen, G.N. L-theanine, a natural constituent in tea, and its effect on mental state. *Asia Pac. J. Clin. Nutr.* **2008**, *17* (Suppl. S1), 167–168.
98. Banerjee, S.; Chatterjee, J. Efficient extraction strategies of tea (*Camellia sinensis*) biomolecules. *J. Food Sci. Technol.* **2015**, *52*, 3158–3168. [CrossRef]
99. Chen, Z.; Wang, Y.; Guo, C.; Li, H. The Antibacterial Activity of Natural Product Black Tea Theaflavins on Streptococcus Mutans. *IOP Conf. Ser. Mater. Sci. Eng.* **2019**, *612*, 022003. [CrossRef]
100. Mughal, T.; Tahir, A.; Qureshi, S.; Nazir, T.; Rasheed, M. Antibacterial activity of black tea against streptococcus mutans and its synergism with antibiotics. *J. Appl. Pharm.* **2010**, *2*, 60–67.
101. Patil, M.; Patil, K.T.; Ngabire, D.; Seo, Y.B.; Kim, G.D. Phytochemical, antioxidant and antibacterial activity of black tea (*Camellia sinensis*). *Int. J. Pharmacogn. Phytochem. Res.* **2016**, *8*, 341–346.
102. Sarkar, S.; Sett, P.; Chowdhury, T.; Ganguly, D.K. Effect of black tea on teeth. *J. Indian Soc. Pedod. Prev. Dent.* **2000**, *18*, 139–140.
103. Del Rio, D.; Rodriguez-Mateos, A.; Spencer, J.P.E.; Tognolini, M.; Borges, G.; Crozier, A. Dietary (Poly)phenolics in Human Health: Structures, Bioavailability, and Evidence of Protective Effects Against Chronic Diseases. *Antioxid. Redox Signal.* **2013**, *18*, 1818–1892. [CrossRef] [PubMed]
104. Khan, N.; Mukhtar, H. Tea polyphenols for health promotion. *Life Sci.* **2007**, *81*, 519–533. [CrossRef]
105. Hervert-Hernández, D.; Goñi, I. Dietary Polyphenols and Human Gut Microbiota: A Review. *Food Rev. Int.* **2011**, *27*, 154–169. [CrossRef]
106. Chan, E.C.; Tie, P.; Soh, E.; Law, Y. Antioxidant and antibacterial properties of green, black, and herbal teas of Camellia sinensis. *Pharmacogn. Res.* **2011**, *3*, 266. [CrossRef]
107. Pan, M.-H.; Tung, Y.-C.; Yang, G.; Li, S.; Ho, C.-T. Molecular mechanisms of the anti-obesity effect of bioactive compounds in tea and coffee. *Food Funct.* **2016**, *7*, 4481–4491. [CrossRef]
108. Yee, Y.K.; Koo, M.; Szeto, M.-l. Chinese tea consumption and lower risk of Helicobacter infection. *J. Gastroenterol. Hepatol.* **2002**, *17*, 552–555. [CrossRef] [PubMed]
109. Heinrich, H.; Goetze, O.; Menne, D.; Iten, P.X.; Fruehauf, H.; Vavricka, S.R.; Schwizer, W.; Fried, M.; Fox, M. Effect on gastric function and symptoms of drinking wine, black tea, or schnapps with a Swiss cheese fondue: Randomized controlled crossover trial. *BMJ* **2010**, *341*, c6731. [CrossRef]
110. Mitra, S.; Khandelwal, S. Health Benefits of Tea. In *Exploring the Nutrition and Health Benefits of Functional Foods*; IGI Global: Hershey, PA, USA, 2017; pp. 99–116.
111. Adhikary, B.; Yadav, S.K.; Roy, K.; Bandyopadhyay, S.K.; Chattopadhyay, S. Black Tea and Theaflavins Assist Healing of Indomethacin-Induced Gastric Ulceration in Mice by Antioxidative Action. *Evid. Based Complement. Altern. Med.* **2011**, *2011*, 546560. [CrossRef] [PubMed]

MDPI AG
Grosspeteranlage 5
4052 Basel
Switzerland
Tel.: +41 61 683 77 34

Horticulturae Editorial Office
E-mail: horticulturae@mdpi.com
www.mdpi.com/journal/horticulturae

www.ingramcontent.com/pod-product-compliance
Lightning Source LLC
Chambersburg PA
CBHW041402210926
51850CB00013B/3152
* 9 7 8 3 7 2 5 8 3 6 3 9 0 *